NONLINEAR SYSTEMS

VOLUME I, DYNAMICS and CONTROL

Ronald R. Mohler

Oregon State University

PRENTICE HALL, Englewood Cliffs, New Jersey 07632

Library of Congress Cataloging-in-Publication Data

Mohler, Ronald R.
Nonlinear systems / Ronald R. Mohler.
p. cm.
Includes bibliographical references.
Contents: v. 1. Dynamics and control -- v. 2. Applications to bilinear control.
ISBN 0-13-623489-5 (v. 1). -- ISBN 0-13-623521-2 (v. 2)
1. Control theory. 2. Nonlinear theories. I. Title.
QA402.3.M634 1991
003'.75--dc20 89-26633
CIP

Editorial/production supervision and
interior design: PATRICE FRACCIO
Cover design: EDSAL ENTERPRISES
Manufacturing buyer: LORI BULWIN

Printed in the United States of America
10 9 8 7 6 5 4 3 2 1

ISBN 0-13-623489-5

Prentice-Hall International (UK) Limited, *London*
Prentice-Hall of Australia Pty. Limited, *Sydney*
Prentice-Hall Canada Inc., *Toronto*
Prentice-Hall Hispanoamericana, S.A., *Mexico*
Prentice-Hall of India Private Limited, *New Delhi*
Prentice-Hall of Japan, Inc., *Tokyo*
Simon & Schuster Asia Pte. Ltd., *Singapore*
Editora Prentice-Hall do Brasil, Ltda., *Rio de Janeiro*

To Pamela, Susan, Anita, Jennifer, Lisa

CONTENTS

PREFACE

This book on dynamics and control is the first volume of a set on nonlinear systems. The second volume is on applications to bilinear control and includes a variety of multidisciplinary processes. The objective of this two-volume sequence on deterministic systems is to integrate some of the recent developments of nonlinear control theory with the by-now-classical nonlinear system theory. Although a certain mathematical rigor is essential to the development, the emphasis is to applications in engineering as well as other fields, such as biomedicine. The advent of economical microprocessors makes the subject timely with respect to practical engineering design and implementation. For example, high-performance adaptive nonlinear systems can now be realized in many cases without degradation in reliability. I originally planned to meet this goal in a single volume, but as the text developed I realized that having two volumes would be more reasonable. The first volume is intended to fill the needs of an applications-oriented introductory course on nonlinear systems or control. The second volume builds on the first book and is a bit more advanced than the first. I have found the material in the second volume quite useful for a special topics course in system theory, methodology, and design.

The book is written primarily for the graduate student and practicing engineer who have a basic knowledge of linear system theory. It should form a base and reference for design and for continued research in nonlinear dynamics and control. The presentation has a multidisciplinary orientation that should appeal to most engineers and applied mathematicians with an interest in system analysis. According to my experience, Vol. 1 may be covered in a three-hour semester course or a four-hour quarter course. Much of the material has been taught in recent years at the Oregon State University. More completely, however, the book reflects my past teaching and practice at the Naval Postgraduate School, the University of Oklahoma, the University of

New Mexico, the Los Alamos National Laboratory, and the Hughes Aircraft Company, along with visiting positions at the University of California at Los Angeles, the University of Rome, and the University of London.

After the introduction in Chapter 1, certain basic principles of mathematical modeling are studied in Chapter 2 as a base for nonlinear systems and control. Applications include rigid-body dynamics, robotics, nuclear fission, heat-transfer, and electric-power systems. Then, linearization, bilinearization, and Liapunov's first method of stability for equilibrium points are presented in Chapter 3. The theory is continued in Chapter 4 for state trajectories, with a summary of appropriate time-variant linear system theory, including Floquet theory and limit-cycle stability. Nonlinear state-plane methodologies, including conservative systems and limit-cycle analyses by index theory and the Bendixson and Poincaré–Bendixson theorems, are studied in Chapter 5.

Time-domain and frequency-domain analysis and design are studied in Chapters 6 and 7, respectively. The presentation starts with the celebrated Liapunov direct, or second, method for stability analysis and proceeds to develop a basis for control and regulator design. Closed-loop stability is presented by means of traditionally convenient frequency-domain analysis in the form of the Popov line and circle criteria and by describing functions. Generally, the former gives a somewhat conservative estimate by stabilizing parameter values, while the latter can be too optimistic. Examples include the stability analysis of an electrical, power-generating system under the influence of a fault. Also, singular-perturbation theory is summarized in Chapter 6.

Stability, controllability, and other structural properties are studied in more detail in Vol. 2. Here the emphasis is given to bilinear systems—an appealing class of systems from both a mathematical and a practical perspective. It is shown that nonlinear control can be utilized to enhance system controllability. Observer design is presented and a successful application is made to a field-controlled dc motor with an inertial load. Subsequent topics in Vol. 2 include nonlinear-system characterization and solution (or its approximation) by Volterra series and Walsh functions, together with optimal and adaptive control and case-study designs.

I am aware that the scope is quite wide. Indeed, such is the intent, in order to show the multidisciplinary role of nonlinear dynamics and control, whether it be an electric machine or a malignant tumor. In Vol. 1, particular nonlinear phenomena of interest may be selected from Chapter 2 with others omitted. Then the essence of a traditional nonlinear systems course is presented in Chapters 3 through 7.

It is impossible to acknowledge the contributions of all the many colleagues, former professors, and students whom have made this undertaking possible. However, I am most grateful to the late Art Brickman for providing the motivation to study control theory as an undergraduate student. I am grateful to A. V. Balakrishnan, G. I. Marchuk, and A. Ruberti for their professional collaboration and advice. Although virtually all of my former students have contributed indirectly, several especially noteworthy topics, which are presented here, were developed by R. E. Rink, W. J. Kolodziej, C. S. Hsu, V. R. Karanam, R. Baheti, R. Li, W. D. Smith, and J. Saugen. F. J. Bugnon contributed exercises, read the entire manuscript, and noted numerous corrections. I would like to thank A. Pacut for his suggestions and T. Kwon for his general assistance.

Oregon State University, Australian National University, and the International Institute for Applied Systems Analysis provided the excellent environment and support conducive to the completion of the manuscript. I am most grateful that my research on related topics is sponsored by the National Science Foundation and the Office of Naval Research. Typing and processing of the text was completed under sometimes trying circumstances thanks to the excellent contributions of Kay Hearder, April Melton, Laurie Dockendorf, and Edith Gruber. Thanks also for the patience and understanding of my wife, Nancy.

Ron Mohler

CHAPTER 1

INTRODUCTION to NONLINEAR SYSTEM MODELING and CONTROL

This book is intended to integrate some of the promising, more recent developments in nonlinear control and adaptive processes or variable-structure systems with, by now, traditional nonlinear system theory. While certain mathematical rigor is essential to these topics, the goal here is not to overburden the reader with mathematical details to the extent that it becomes impossible to bridge the gap with practical applications. A multidisciplinary approach is presented in order to demonstrate the diverse application of nonlinear system theory. Indeed, such application is the very motivation for this study. At the same time, it is hoped that a cross-fertilization of ideas will evolve from the variety of applications. For example, engineers may improve the effectiveness of their designs by a better understanding of nature. Similarly, disease treatment and prosthesis (design of artificial organs and limbs) are potentially fruitful areas of biomedical application.

In some cases the very nature of the process that is to be controlled or studied dictates nonlinear system analysis. In other cases, the desired system performance may only be realized by a nonlinear design synthesis. In general, control and model complexity depends on the specifications of the overall system design or on questions that the model is expected to address in an analysis and simulation. For example, whereas classical linear constant-coefficient control theory may be quite adequate for the design of an altitude- and Mach-hold mode of a high-performance aircraft (by perturbation theory, which is studied in Chapter 3), it may be quite inadequate for minimum-fuel or minimum-time trajectories between different flight conditions within an admissible Mach–altitude envelope.

Linear system theory has led to useful methodologies for the control engineer. At the same time, its limitations must be understood. For example, the reader may recall the simple rank test of a controllability matrix to determine linear system controllability, which is reviewed here in Chapter 3. Unfortunately, such controllability may require infinite control variation. In fact, as shown in Vol. 2, a linear constant-coefficient system, with homogeneous part asymptotically stable, cannot be completely controllable in the presence of control magnitude constraints. By appropriate synthesis with a simple multiplier (control times state), however, such systems can be made completely controllable. Of course, in practice, one is usually interested in controllability of some region of state space which may or may not be attainable by a linear system design. In the case of the human immune system, which controls the level of replicating alien intruders as studied in Vol. 2, nonlinear control is essential to maintain the required region of controllability corresponding to healthy states. Indeed, it would appear that human life could not exist in this alien world if its control processes were synthesized according to classical linear control theory. The immune control system, as well as presenting a challenge to the control systems analyst, may provide a clue to more effective control synthesis, such as in pursuit-evasion scenarios of engineering significance.

There is an old axiom, "you don't get something for nothing," which seems to be particularly true in engineering design. In this case, an added complexity in the synthesis of a nonlinear control policy may result from the analysis. However, the advent of cheap, efficient microprocessors makes the role of nonlinear control especetially timely. In some cases, the nonlinear policy is generated directly from measured natural nonlinear system variables with little complexity in the synthesis.

1.1 SYSTEM OVERVIEW

System representations or models may be classified into numerous categories according to mathematical structure and physical realizability. A typical classification is summarized in Table 1.1. For example, systems may be distributed, and are commonly represented by partial differential equations, or lumped, commonly represented by a finite number of ordinary differential equations. They may be stochastic (random) or deterministic, linear or nonlinear, discrete or continuous, autonomous or nonautonomous. Others may involve time delay. Physical systems usually are assumed to be causal in response and, furthermore, cannot diverge to infinity in finite time (roughly, finite escape time) or must have finite energy. Some of these classes are defined and discussed in more detail in Section 1.2 and in Chapters 2 and 3, but the reader is referred to [Z1] or [P1] or most any other good book on basic mathematical system theory for a more complete formulation.

Although linear system theory and the ensuing control design has been well established for several decades, with numerous new contributions appearing monthly, nonlinear systems, in general, do not have a convenient uniform theory. On the other hand, the field is ripe for continued research and development of design methodologies. The treatise presented here can be only a step in that direction. Unfortunately, many of the classical notions that have evolved from our familiarity with linear constant-parameter systems are not valid for nonlinear systems. Of course, by its very

TABLE 1.1 CERTAIN SYSTEM CLASSES: GENERAL SYSTEMS WITH APPROPRIATE MATHEMATICAL OPERATORS ACCORDING TO CATEGORIES UNDER DIFFERENTIAL SYSTEMS

- Differential systems
  - Distributed systems
    - With partial differential equations
  - Stochastic systems
    - With random processes in differential equations
  - Deterministic nonlinear systems
    - With nonlinear differential equations (ordinary for lumped case)
  - Finite-state systems
    - With a finite number of ordinary state differential equations
  - Autonomous systems
    - Unforced differential equations with constant coefficients
  - Linear systems
    - Linear differential equations
  - Linear time-invariant systems
    - Linear differential equations with constant coefficients, may be forced or free (i.e., nonautonomous or autonomous)
- Discrete systems
  - With appropriate difference equations according to categories above.

definition, superposition generally is not valid for nonlinear systems. Intuitively, we like to assume that the system response to various individual stimuli is comprised of a summation of their appropriate individual responses. For nonlinear systems, such reasoning generally will lead to grossly erroneous conclusions. In biomedicine, for example, a good example would be the multiplication of inputs that affect crucial parameters in the immune system whereby cancer may arise by an appropriate combination of genetics, age, diet, contaminants, and so on.

Even our concepts of stability from linear system theory may not always be convenient for nonlinear system design. For example, it will be shown in Vol. 2 that a judicious or "controlled" use of "instability" in a variable-structure system can be useful for improved controllability and performance. Except for cases with imaginary eigenvalues, bounded-input, bounded-output (BIBO) stability corresponds to free (unforced or zero input) system stability for linear time-invariant (constant-parameter) systems. In the case of nonlinear systems, however, stability in general is strongly dependent on the magnitude of the initial condition as well as the magnitude of any input. A nonlinear free system may be stable and the forced system unstable. Nonlinear systems, generally, have multiple equilibrium conditions, which result in multiple equilibrium points for state-space realizations. Some equilibrium states may be stable, whereas others may be unstable.

With the exception of very specific input frequencies for the harmonic oscillator (imaginary eigenvalue case), a linear system with a sinusoidal input will exhibit a sinusoidal component of the same frequency. For the asymptotically stable linear time-invariant system, appropriate computations on this basis yield a system frequency response which completely characterizes the system dynamics. (Indeed, for such systems, any periodic input will result eventually in the same periodicity of the output as that shown by the Fourier series.) Obviously, such frequency response methods generally are not valid for nonlinear systems. In fact, periodic inputs to nonlinear systems may yield lower harmonics, higher harmonics, or nonperiodic outputs. There may exist jump phenomena in the sinusoidal frequency response of certain nonlinear systems such that the discontinuity in amplitude and phase shift occurs at different frequencies, depending on whether the response is taken with increasing or decreasing frequencies. The analytical model for such a system exhibits nonunique solutions over the frequency band between these points of discontinuity.

In general, nonlinear mathematical models do not have unique solutions. Such systems would seem to have limited practical application, but have been utilized for approximate description over certain ranges of the variables. In other cases, such models result from an unrealistic low-order approximation. Systems with bifurcation characteristics, and "singularly perturbed" systems, depict sudden structural changes and have been applied with increasing popularity. Finite escape time may occur for a nonlinear model, but would again have limited practical application. While linear unforced time-invariant (autonomous) systems may exhibit sustained oscillations or a continuum of closed-solution paths in state space, nonlinear autonomous systems may exhibit isolated closed-solution paths called limit cycles. They have interesting stability problems and will be studied in more detail in Chapters 4 and 5.

1.2 SYSTEM DEFINITIONS AND FORMULATIONS

Now, let us mathematically formulate those classes of systems that will be studied throughout this book. Despite their limitations, we consider linear systems first since many concepts for nonlinear systems will be related to popular linear-system properties and performance. Generally, we will be concerned with differential, finite-order causal systems. For the linear nth-order scalar case, the system may be represented by the following ordinary differential equation:

$$a_n(t)\,\frac{d_n y(t)}{dt^n} + a_{n-1}(t)\,\frac{d^{n-1}y(t)}{dt^{n-1}} + \cdots + a_0(t)y(t)$$
$$= b_q(t)\,\frac{d^q u(t)}{dt^q} + \cdots + b_0(t)u(t), \tag{1.1}$$

where $y(t)$, and $u(t)$ are the scalar output and the scalar input or control, respectively, and $q \le n$. In case the coefficients, $a_0(\cdot), \ldots, a_n(\cdot), b_0(\cdot), \ldots, b_q(\cdot)$, are constant, the system is said to be a linear time-invariant system (LTIS), and Laplace transforms may be applied in the traditional manner. Usually, it will be assumed that these coefficients are at least continuous in time, and that $u(t)$ is at least piecewise continuous, thereby assuming a unique solution with n initial conditions (see Appendix A and [C1]).

For the LTIS with impulse response $h(t)$, the output solution to (1.1) is given by

$$y(t) = \int_{-\infty}^{t} h(t - \tau)u(\tau)\, d\tau, \tag{1.2}$$

or with a given initial condition solution $y_0(t - t_0)$

$$y(t) = y_0(t - t_0) + \int_{t_0}^{t} h(t - \tau)u(\tau)\, d\tau. \tag{1.3}$$

Both (1.2) and (1.3) are readily derived from a superposition of weighted impulse responses. The corresponding transfer function is given by the Laplace transform of the impulse response $h(t)$, or $y(s)/u(s)$ with $y(t_0) = 0$. Consequently, the transfer function is a rational function written as a ratio of polynomials

$$G(s) = \mathcal{L}^{-1}[h(t)] = \frac{P(s)}{Q(s)}, \tag{1.4}$$

where $P(\cdot)$, $Q(\cdot)$ are of order q, and n, respectively, with $q \leq n$. Furthermore, it is sometimes convenient for analysis and design to factor $P(s)$ and $Q(s)$ by first-order terms $(\tau_i\, s + 1)$ and second-order terms $(s^2/\omega_{ni}^2 + 2\zeta_i/\omega_{ni}s + 1)$ which locate the corresponding real and complex poles and zeros in terms of time constants τ_i, damping coefficients ζ_i, and natural frequencies ω_{ni}, respectively.

It is interesting to note here that even certain distributed systems may be solved from (1.3), but the transfer function will not be a ratio of polynomials as given by (1.4). Typically, it will consist of transcendental functions of s. In a simplified heat exchanger, such a transfer function may relate temperature variations at a given location to heat variations applied at another point or surface boundary. Similarly, it could relate voltages and/or currents at crucial points in an idealized transmission line. Examples of such problems are studied in [H1].

For linear time-variant systems (LTVS) [i.e., (1.1) for the general case], the impulse response depends on two variables: when the impulse is applied and when its response is measured (rather than merely the time difference as for LTIS). Consequently, (1.3) for LTVS with impulse response, $h(t, \sigma)$ becomes

$$y(t) = y_0(t, t_0) + \int_{t_0}^{t} h(t, \sigma)u(\sigma)\, d\sigma.$$

Again, the Laplace transformation has limited utility since it must be defined over a space of two complex variables. However, future research may very well prove such higher-order transforms useful in computer-aided design (CAD).

Obviously, distributed systems would need an infinite number of dependent variables (i.e., an infinite order or an infinite number of states) to characterize its dynamic behavior in general. Here, however, we are interested primarily in finite-order or "finite-state" (actually, a finite number of state variables) systems. In general, complex systems may have numerous inputs or controls $u_j(t)$, $j = 1, \ldots, m$, and outputs $y_i(t)$, $i = 1, \ldots, p$. A matrix of impulse responses, with (i,j)th element $h_{ij}(t)$, relates the ith output to the jth input, as shown by Fig. 1.1. Then, in vector-matrix notation,

$$\mathbf{y}(t) = \mathbf{y}_0(t - t_0) + \int_{t_0}^{t} \mathbf{H}(t-\tau)\mathbf{u}(\tau)\, d\tau \tag{1.5}$$

or

$$\mathbf{y}(s) = \mathbf{H}(s)\mathbf{u}(s), \tag{1.6}$$

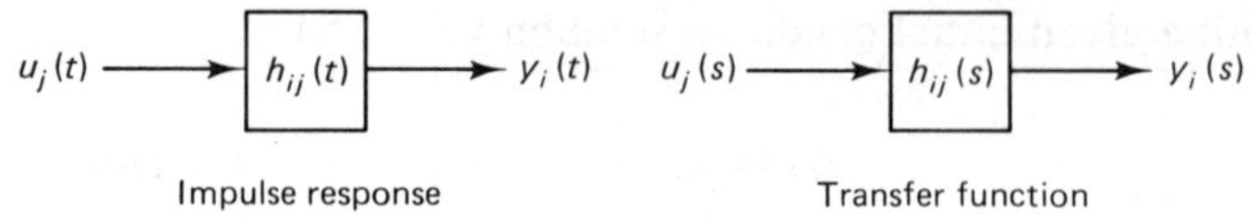

Figure 1.1 Scalar input–output diagrams.

where $\mathbf{y}(t)$ is a p-dimensional output vector, $\mathbf{y} \in \mathbb{R}^p$, $\mathbf{u}(t)$ is an m-dimensional input vector, $\mathbf{u} \in \mathbb{R}^m$, and $\mathbf{H}(\cdot)$ is an appropriate real matrix mapping from $\mathbb{R}^m$ to $\mathbb{R}^p$. Such a mapping is shown by the input–output diagram in Fig. 1.2.

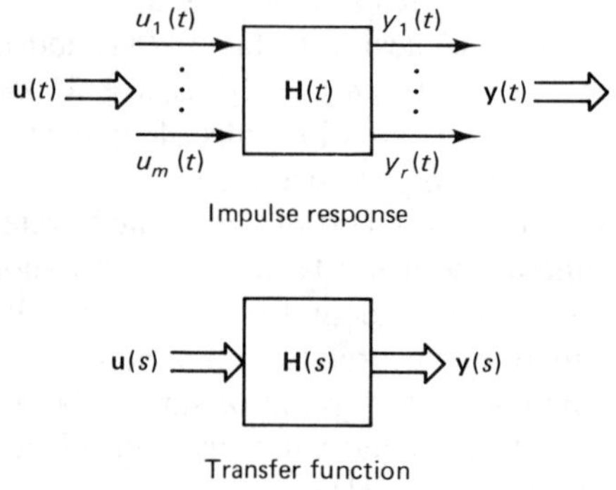

Figure 1.2 Vector–matrix input–output diagrams.

In most cases such multiple-input, multiple-output systems (MIMO) have a state-space representation such that

$$\frac{d\mathbf{x}(t)}{dt} = \mathbf{A}\mathbf{x}(t) + \mathbf{B}\mathbf{u}(t), \tag{1.7}$$

$$\mathbf{y}(t) = \mathbf{C}\mathbf{x}(t) + \mathbf{D}\mathbf{u}(t), \tag{1.8}$$

where $\mathbf{x}$ is an n-dimensional state vector, $\mathbf{x} \in \mathbb{R}^n$, $\mathbf{A} \in \mathbb{R}^{n\times n}$, $\mathbf{B} \in \mathbb{R}^{n\times m}$, $\mathbf{C} \in \mathbb{R}^{p\times m}$, $\mathbf{D} \in \mathbb{R}^{p\times m}$. The corresponding state diagram is shown in Fig. 1.3. Again, Laplace transforms may conveniently be applied to each variable to obtain $\mathbf{x}(s)$ and $\mathbf{y}(s)$ from $\mathbf{u}(s)$. Solutions to (1.7) and (1.8) for the time-invariant and time-variant case will be studied in Chapters 3 and 4, respectively. For LTVS, $\mathbf{A}$, $\mathbf{B}$, $\mathbf{C}$, and $\mathbf{D}$ are appropriate matrix functions of time. Such LTVS may result from perturbation models on time-variant trajectories, just as LTIS may evolve from fixed-point perturbation models as discussed in Chapter 3.

Generally, there are an infinite number of state-space realizations for given input–output vector pairs. Of course, there must be a set of n independent variables that span the n-dimensional real space, $\mathbb{R}^n$. In practice, a convenient set of state variables usually evolves in a natural manner. Of course, mathematical convenience sometimes suggests a linear transformation to establish a simpler linear structure. Such transformations are reviewed in Chapter 3.

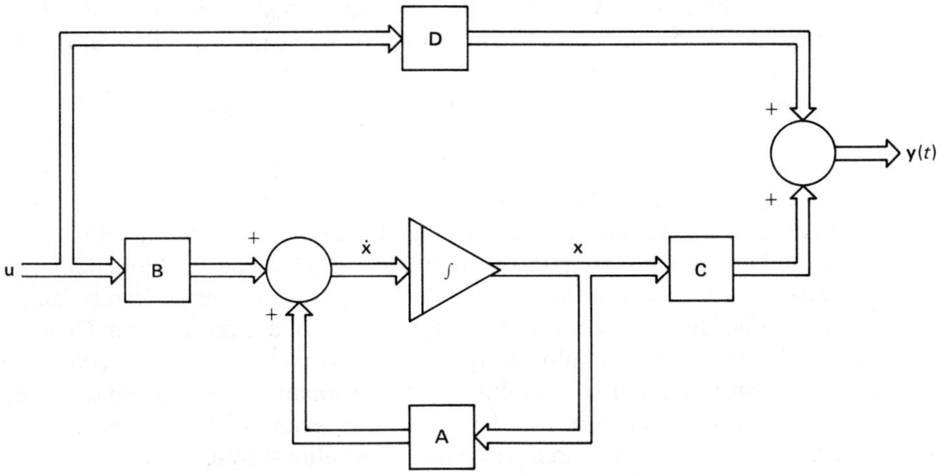

Figure 1.3 Linear system state diagram.

For a simple single-input, single-output system (SISO) without input differentiation, let (1.1) take the form

$$\frac{d^n y(t)}{dt^n} = -a_{n-1}\frac{d^{n-1}y(t)}{dt^{n-1}} - \cdots - a_0 y(t) + bu(t).$$

Then the state $\mathbf{x}(t)$ may conveniently be defined by components

$$x_1(t) = y(t), \quad x_2(t) = \frac{dy}{dt} = \frac{dx_1}{dt}, \quad \cdots, \quad x_n = \frac{d^{n-1}y}{dt} = \frac{dx_{n-1}}{dt},$$

so that

$$\frac{dx_n(t)}{dt} = -a_{n-1}x_n - \cdots - a_0 x_1(t) + bu(t). \tag{1.9}$$

In the vector–matrix formulation, the state equations are given by (1.7) with

$$\mathbf{A} = \begin{bmatrix} 0 & 1 & 0 & \cdot & \cdot & \cdot & 0 \\ 0 & 0 & 1 & 0 & \cdot & \cdot & 0 \\ \cdot & & & & & & \\ \cdot & & & & & & \\ \cdot & & & & & & \\ 0 & 0 & \cdot & \cdot & \cdot & 0 & 1 \\ -a_0 & -a_1 & \cdot & \cdot & \cdot & \cdot & -a_{n-1} \end{bmatrix}$$

$$\mathbf{B} = b = \begin{bmatrix} 0 \\ \cdot \\ \cdot \\ \cdot \\ 0 \\ b \end{bmatrix}.$$

Here the output is simply given by the state x_1, that is, by (1.8) with $\mathbf{C} = \mathbf{c}$, a zero matrix except for the nth element, which is unity, and $\mathbf{D} = \mathbf{d} = 0$, the zero matrix. Such state variables are sometimes called *normal variables*, *phase variables*, or *canonical phase variables* by different authors. We shall use the latter term. A set of state variables that is sometimes mathematically convenient is the canonical normal set, sometimes called *normal variables* or *canonical variables*. These are formulated by means of the eigenvector matrix transformation to a diagonal form that is particularly useful for real eigenvalues. The reader probably is familiar with such techniques from a basic linear system course or a linear algebra course. More detailed examples, including a second transformation which in polar coordinates can be quite useful for autonomous linear systems (i.e., unforced LTIS), are presented in Chapter 3.

In some cases, nonlinear systems (NLS) may be analyzed conveniently by an appropriate selection of coordinates, transformation, or state-space representation. Unfortunately, such selection for NLS is generally difficult. There is considerable research activity in this area particularly for bilinear systems; see, for example, [B1] and [W1] on the utility of Lie algebra in this respect. Convenient applications have included dynamics that are limited to the surface of a sphere.

Bilinear systems (BLS) comprise perhaps the simplest class of NLS. These are systems that are linear in control and linear in state but not jointly linear in state and control. In other words, they involve products of state and control. Hence it is readily seen that superposition does not apply. Time-invariant bilinear systems, for our purposes, have a state form as follows:

$$\frac{d\mathbf{x}(t)}{dt} = \mathbf{A}\mathbf{x}(t) + \mathbf{B}\mathbf{u}(t) + \sum_{k=1}^{m} \mathbf{N}_k u_k(t)\mathbf{x}(t), \tag{1.10}$$

where $\mathbf{N}_k$, $k = 1, \cdots, m$ are $n \times n$ matrices (i.e., $\mathbf{N}_k \in \mathbb{R}^{n\times n}$). Sometimes, the last term in (1.10) is written simply as $\mathbf{N}(\mathbf{u}, \mathbf{x})$ or $\mathbf{Nu} \circ \mathbf{x}$. The corresponding BLS state diagram is shown in Fig. 1.4. The output may be generated by $\mathbf{y}(t) = \mathbf{Cx}(t)$, or more generally, by (1.8) with $\mathbf{D}$ nonzero.

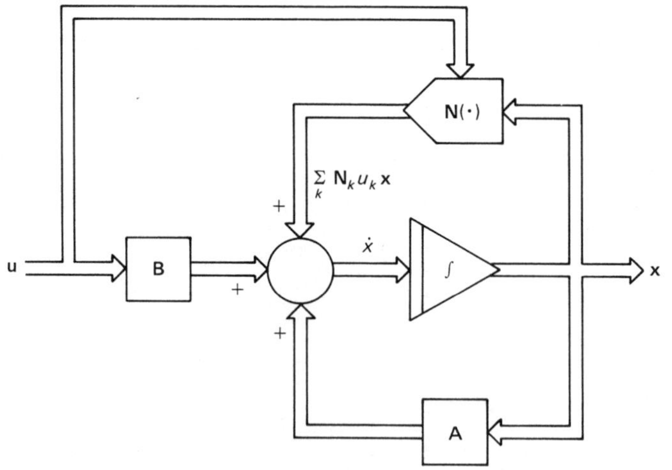

Figure 1.4 Bilinear system state diagram.

Bilinear systems (BLS) are investigated in more detail in Chapter 2 and in Vol. 2. References [M1] and [M2] present still more basic analysis and applications of BLS. In Chapter 2 we present physical BLS models, followed by BLS approximations (beyond linear approximations) for nonlinear systems in Chapter 3. In subsequent chapters we show that BLS can be useful to the design engineer to synthesize systems with controllability and performance characteristics that are not attainable by linear system techniques with certain physical constraints. In this respect, BLS typify adaptive or variable-structure systems as well as nonlinear systems in general.

In some practical systems (e.g., nuclear fission reactors and immunology as studied in Chapter 2 and Vol. 2) bilinear systems are coupled together so that the total system is more highly nonlinear. Such system coupling may be linear or nonlinear, depending on the particular process under investigation. For example, the output of one subsystem, say reactor power, may become an input to another subsystem, such as heat transfer. In the nuclear power reactor, which is studied briefly in Chapter 2, appropriate temperature from the heat exchange becomes a feedback variable via temperature reactivity (or neutron multiplication) to the nuclear fission process, which generates power. Here such systems are called *coupled bilinear systems*.

An interesting class of coupled bilinear systems results if the nonlinearly appearing states are measured. This case may be interpreted as a BLS with measurable nonlinear feedback terms generating the closed-loop controls. In other words, with the scalar control $u(\cdot)$ a function of measured output $\mathbf{y}(t)$, (1.10) takes the following form:

$$\frac{d\mathbf{x}(t)}{dt} = [\mathbf{A} + \mathbf{N}(\mathbf{y}(t))]\mathbf{x}(t) + \mathbf{g}(\mathbf{y}(t)), \tag{1.11}$$

and the output $\mathbf{y}(t)$ is composed of only those components of $\mathbf{x}(t)$ that are measured. Since for a given set of measurements $\mathbf{y}(t)$, the system is linear (i.e., conditionally linear), (1.11) is called a *conditionally linear system*. Equation (1.11) can also be interpreted as a model that is linear in nonmeasurable states but nonlinear in the measurable states. It seems to be a reasonable way to approximate a nonlinear process with little known information about structure since the nonlinearities are based on measured information. Such measurements may detect a deviation from linearity and make an appropriate model change in structure. Conditionally linear systems will be studied further in subsequent chapters. Finite-state estimators (nonlinear filters) have been derived for stochastic representations of such systems. The interested reader is referred to [K1] and [M3] for more details. Similar to linear systems and bilinear systems, deterministic state observers may be generated for unmeasured states of the conditionally linear system for purposes of feedback controller design.

A class of nonlinear systems that is amenable to frequency-domain analysis is one which consists of a linear system with appropriately constrained nondynamic nonlinear element in the feedback regulator. Such a nonlinear scalar regulator is shown by Fig. 1.5. In some cases the stability of such systems is analyzed by describing functions or by Popov's method as studied in Chapter 7. The latter is based on the existence of a Liapunov function, which is developed for linear and nonlinear systems in Chapter 6.

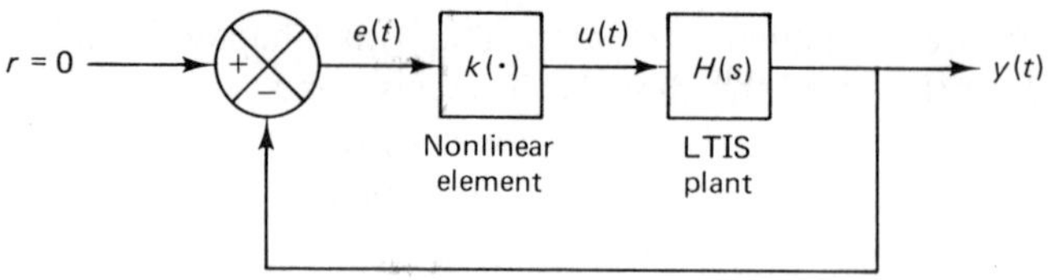

Figure 1.5 Simple nonlinear regulator.

More generally, a finite-state differential system is defined by the nonlinear vector differential equation:

$$\frac{d\mathbf{x}(t)}{dt} = \mathbf{f}(\mathbf{x}(t), \mathbf{u}(t)), \tag{1.12}$$

where $\mathbf{f}(\cdot)$ is a real nonlinear mapping from $\mathbb{R}^n \times \mathbb{R}^m$ to $\mathbb{R}^n$. Similarly, the output may be given by

$$\mathbf{y}(t) = \mathbf{g}(\mathbf{x}(t), \mathbf{u}(t)), \tag{1.13}$$

where again $\mathbf{x} \in \mathbb{R}^n$, $\mathbf{u} \in \mathbb{R}^m$, and $\mathbf{y} \in \mathbb{R}^p$. Time-variant nonlinear systems of this class are given by (1.12) and (1.13) with $\mathbf{f}(t, \mathbf{x}(t), \mathbf{u}(t))$ and $\mathbf{g}(t, \mathbf{x}(t), \mathbf{u}(t))$.

The solution of nonlinear differential equations is quite difficult in general and usually impossible. Such nice results as the convolution integral relationships (1.2) and (1.3) for linear systems generally are not possible. On the other hand, the Volterra series (see Vol. 2) does present a generalization of this result for the zero-state response of nonlinear systems. But again, the failure of superposition for nonlinear systems does not allow a simple addition of zero-state and zero-input responses for the general solution. Indeed, a portion of the response will depend on both. It will be seen in Vol. 2, however, that the BLS Volterra kernels may conveniently be synthesized from the first-order kernel or exponential state-transition matrix associated with the linear-system portion of the BLS. Then higher-order transforms of the kernels characterize the system dynamics in the frequency domain. Such kernels are convenient to understand the structure of certain nonlinear systems, and variable-structure or adaptive systems are studied in Vol. 2. To better appreciate adaptive control, however, elements of optimal control theory are first summarized in Vol. 2.

Discrete systems, that is, primarily difference-equation representations of the continuous systems above, will be studied briefly in Chapter 4. In general, the nonlinear discrete system of interest here may take the following state form:

$$\mathbf{x}(k + 1) = \mathbf{f}(k; \mathbf{x}(k), \mathbf{u}(k)), \tag{1.14}$$

$$\mathbf{y}(k) = \mathbf{g}(k; \mathbf{x}(k), \mathbf{u}(k)), \tag{1.15}$$

where $k = 1, 2, 3, \ldots$ refers to the kth sampling time, kT. Here T is the sampling period, which is assumed constant. In the linear time-invariant discrete case, (1.14) and (1.15) become

$$\mathbf{x}(k + 1) = \mathbf{A}\mathbf{x}(k) + \mathbf{B}\mathbf{u}(k), \tag{1.16}$$

$$\mathbf{y}(k) = \mathbf{C}\mathbf{x}(k) + \mathbf{D}\mathbf{u}(k), \tag{1.17}$$

where **A**, **B**, **C**, and **D** are appropriate constant matrices, but generally they are not identical with a direct discrete formulation of (1.7) and (1.8). In Chapter 4 it is shown that stability of the unforced version of (1.16) is quite similar to that of periodically time-variant linear unforced systems. Discrete systems are particularly important for computer implementation of adaptive policies, but only a brief introduction can be given here. As noted above, unfortunately, we can only expect to analyze bits and pieces of general nonlinear systems due to the limited theory at our disposal. On the other hand, the possibilities are almost unlimited, suggesting a potentially fruitful area for future research and computer-aided design.

1.3 FUTURE RESEARCH AND DESIGN

In this era of rapidly growing technology, the potential for ultrahigh-performance system design is almost unlimited. Fast, relatively cheap distributed computing systems with a large memory are evolving so that, indeed, on-line nonlinear-system identification and adaptive control will be realized. Such smart (intelligent) systems will be required to detect system faults, failures, or malfunctions, and then to self-correct, adaptively control, or in extreme situations, shut the process down. A good example is presented by a future power plant for a manned space station. Present designs envision nuclear, solar-supplemented plants for electric power generation, heating, and so on. Certainly, the success and even survival of manned space stations may well depend on the efficiency and the reliability of such power plants. Due to vehicle payload limitations, redundancy of components should be minimized—yet performance and reliability are crucial. As should become more apparent in reading the present book, nonlinear systems analysis and design are essential to this project. High-performance mathematical policies frequently require that the system operate on the verge of failure (e.g., bang-bang time-optimal control with thermal state constraints). While the occurrence of malfunctions must be minimized (and some degree of mathematical optimality sacrificed), ultimately a few malfunctions will occur, in which case, rapid detection and self-correction must be made. The integration of nonlinear analysis, on-line system identification, fault detection, robotic self-correction, and adaptive control is essential. Systems analysts, designers, and certainly control engineers for such projects should have some basic knowledge of all these topics. Unfortunately, it is not possible to describe all such areas adequately in one concise book. Consequently, such nonlinear statistical analysis as is required of such systems for detection, estimation, and tracking (among other topics) will not be studied here. Rather, deterministic nonlinear system analysis, including stability studies and control design, is emphasized.

The example above is but one of many future applications that is relevant to this work. Hypersonic aircraft for earth transportation, manned planetary space missions, automation and robotics, residential security and livability in hostile contaminated environments, mass rapid ground-transportation systems, socioeconomic planning and decision making, and systematic health care are but a few additional examples. Certain aspects of some of these problems will be studied throughout this book. For example, the immune system, nuclear fission, heat transfer, and rigid-body dynamic models, such as are relevant to disease diagnosis and treatment, to nuclear and solar power plant design, to residential heating, to aircraft, and to robotic control, are modeled in Chapter 2 and Vol. 2. Two excellent sources on progress and future directions in control theory and nonlinear system theory are presented in [L1], [V1], and [V2]. These should be useful sources of information for beginning researchers of nonlinear systems and control theory.

A typical control design formulation will utilize appropriate theoretical developments introduced in the following procedure.

1. Formulate the mission, design specification, and constraints of the system or process.
2. Create a mathematical model together with numerical constraints and parameter values available for this model.
3. Collect experimental data (possibly creating new experiments) for unknown system parameters and constraints. In some cases, part of this process must be performed on-line for adaptive control reactions to a priori unknown disturbances.
4. Derive a simple model for analytical purposes. This may first involve a nonlinear low-order approximation to, say, a complex high-order process (possibly a distributed stochastic system), which may be followed by a sequence of approximating perturbation linearized models.
5. Perform an optimal control analysis of the simple model with its corresponding physical constraints represented mathematically. This procedure may start with the linear approximation for the more complicated nonlinear, but still simplified, low-order model.
6. Create a computer simulation of the complicated high-order system controlled by the optimal controller that was derived for the low-order nonlinear system.
7. Formulate an adaptive system that may be a suboptimal compromise from robustness, optimization, and reliability considerations. The robustness consideration may arise from a variety of possible system disturbances.
8. Synthesize the control system with available components to approximate the above-derived mathematical synthesis as closely as possible.
9. Verify experimentally the system as derived in steps 1 to 8.
10. Repeat the steps above to further tune the nonlinear controller design.

Indeed, the biggest improvement in performance may very well come from the feedback of information from this control-design procedure to the plant design process. The performance available is usually most constrained by the plant design (e.g., state constraints) rather than the precise control implementation. For example, performance may be very sensitive with respect to a particular state constraint such as thermal stress or dynamic pump characteristics in a jet engine. Of course, such design procedures will vary according to the application at hand. In some chemical process control scenarios, little a priori information may be available on particular process dynamics. On-line identification may have to proceed sequentially with particular adaptive control adjustments.

In any event, the reader should be aware of the overwhelming significance of understanding the process to be controlled. This is true whether on-line adaptive control or fixed-structure control is utilized. Consequently, Chapter 2 follows with a discussion of a multidisciplinary survey of certain nonlinear models. Although nonlinear systems are emphasized here, it must be cautioned that we are somewhat limited in our knowledge of these systems, and the repertoire of available linear methods should not be ignored. Sometimes a linear control synthesis may result as a useful approximation of an optimal nonlinear controller.

CHAPTER 2

MATHEMATICAL MODEL FOUNDATIONS

A good understanding of the process that is being analyzed, designed, or controlled is of the utmost importance. It is not possible to present a complete treatise of system dynamics in the space available here. Rather, a few basic principles are presented, and then mathematical models are formulated for a multidisciplinary variety of naturally nonlinear processes in mechanics, aerodynamics, heat transfer, nuclear power generation, and physiology. Others, including electric machines, are studied through examples later. Modeling, dynamic analysis, and control are all active areas of research for these processes. As noted in Chapter 1, a model always depends on its use. For example, a detailed aircraft aeroelastic stress model will be different from an aircraft trajectory-control model, and the latter will probably be different from an aircraft pitch-hold control design. In any case, however, model formulation is one of the first steps—probably the most important step—in the analysis and design.

2.1 DYNAMIC MODEL PRINCIPLES

The *generalized conservation principle* seems to be the most common approach in multidisciplinary modeling. *Conservation of energy* is perhaps the most popular approach, and its application to conservative systems is the simplest. *Conservative systems* are those for which the total energy is constant. That is, kinetic energy T, plus potential energy V, is a constant E:

$$T(t) + V(t) = E. \tag{2.1}$$

For example, in Fig. 2.1, an initial displacement of the mass from equilibrium corresponds to some E, with zero initial velocity (or kinetic energy). Then the mass

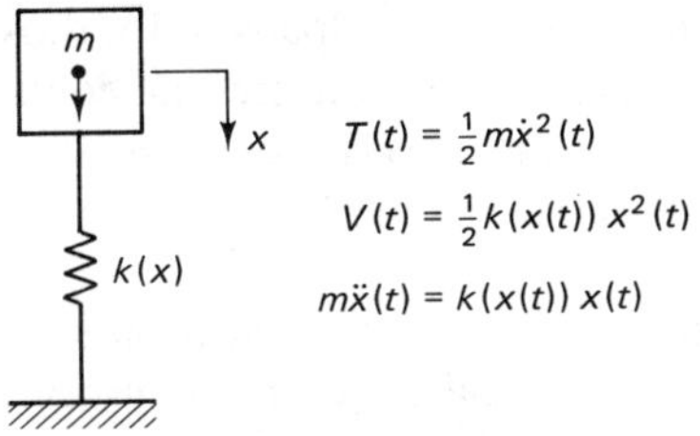

Figure 2.1 Conservative second-order mechanical system.

will oscillate between points of maximum $T = E$ (or minimum $V = 0$) and minimum $T = 0$ (or maximum $V = E$). This problem is discussed in more detail below.

More generally, however, some of the energy may be tranformed into work or into other forms of energy, or from work and/or from other forms of energy. In other words,

$$E_i(t) = T_i(t) + V_i(t) + \sum_{j \neq i} W_{ij}(t) - \sum_{k \neq i} W_{ki}(t) + \sum_{j \neq i} E_{ij}(t) - \sum_{k \neq i} E_{ki}(t), \tag{2.2}$$

where $E_i(\cdot)$ is the internal energy applied to the ith process or element; $T_i(\cdot)$ and $V_i(\cdot)$ the kinetic and potential energies, respectively, of the ith process or element; $W_{ij}(\cdot)$ and $W_{ki}(\cdot)$ the energy losses from the ith process or gains to the ith process, due to work upon the jth process or by the kth process, respectively; and $E_{ij}(\cdot)$ and $E_{ki}(\cdot)$ the energy transfers (and possible transformations) from the jth to the ith process and from the ith to the kth process, respectively. Here i, j, and k may take on integer values corresponding to the appropriate processes or elements. For example, in a nuclear power reactor, heat released from fission is transferred to a fluid, which results in an increase in its energy. This energy, for the most part, may be imparted onto a turbine which, in turn, drives a generator, and the process goes on and on. Again, of course, an appropriate energy balance may be made for the appropriate process or element according to the problem of interest. The nuclear reactor example is considered in more detail below.

Even Newton's second law in application really consists of a *conservation of forces or moments and torques*, referred to as *d'Alembert's principle*. For example, in Fig. 2.1, if a force $F(t)$ is applied to the constant mass m, it is balanced by the acceleration force md^2x/dt^2 and the restraining spring force $k(x)x$, or

$$F(t) = m\,\frac{d^2x(t)}{dt^2} + k(x(t))x(t), \tag{2.3}$$

where $k(\cdot)$ may be a nonlinear spring stiffness and $x(t)$ is the mass displacement, which may be taken conveniently with respect to the equilibrium position (i.e., with $d^2x/dt^2 = dx/dt = 0$). Similarly, a summation of moments about a shaft axis in a polar coordinate system may result in

$$T(t) = J\frac{d^2\theta}{dt^2} + k(\theta)\theta, \tag{2.4}$$

where $T(\cdot)$ is the applied shaft torque and θ is an angle of motion with its argument (t) dropped for brevity; J may be the polar moment of inertia of the shaft, flywheel, and so on, and $k(\theta)$ is the shaft "stiffness." The latter term is usually unimportant, but can be important for cases with large torques or loads or for those with a somewhat elastic shaft. Equations (2.3) and (2.4) are linear differential equations (i.e., linear systems) if the stiffness or spring characteristic, $k(\cdot)$, is not a function of x or θ. Of course, if $k(\cdot)$ is a function of t and not of x or θ, (2.3) and (2.4) would be linear time-variant systems, LTVS, as noted in Chapter 1.

Similar to a conservation of forces, with Newton's second law we can interpret the node or loop equations of electrical circuits as conservation of current, $i = dq/dt$ in Fig. 2.2, or voltages that formulate Kirchhoff's laws. Again, certain network elements may be functions of dependent variables, current or voltage, and thereby yield a nonlinear system. For example, a simple conservative nonlinear network is shown in Fig. 2.2 with the capacitance given as a function of charge. Other typical nonlinear network elements may arise from transformer hysteresis, tunnel diode characteristics, or solid-state-device dead zone and limiting, which will be studied in Chapter 7.

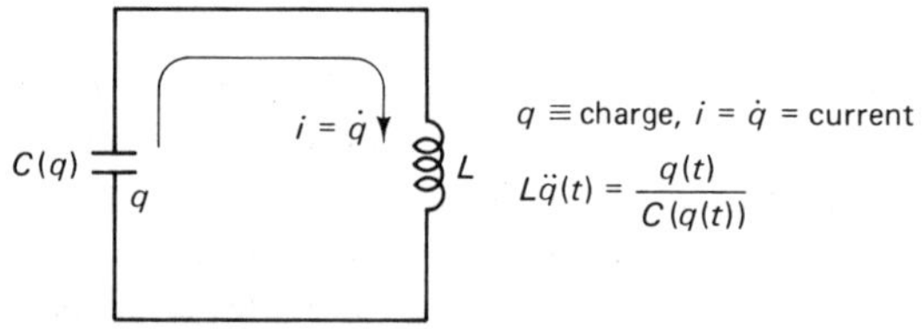

Figure 2.2 Conservative second-order electrical system.

In diffusion processes, a conservation of appropriate fluxes may be important in a model formulation. For fluid-flow dynamics or other *mass-transport processes,* mass flow rates would be conserved at a junction of lines in a fashion analogous to currents in an electrical circuit. Similarly, fluid storage compartments in the line introduce capacitance with pressure drop across the element analogous to voltage drop in the electrical network. Change in capacitance as a function of pressure and therefore mass flow rate by the perfect gas equation, such as in a pneumatic bellows, can lead to a nonlinear system analogous to Fig. 2.2. Such a model is presented in Chapter 3.

Again similar to current conservation equations at nodes, one can derive *population conservation models* for biological species, for biological cells, for molecules, for neutrons, or for other particles. For example, a simple biospecies population model may have the following form:

$$\frac{dp_i}{dt} = bp_i - dp_i + \sum_{j \neq i} m_{ij}(p_j - p_i) + \sum_{k \neq i} m_{ki}(p_i - p_k), \tag{2.5}$$

where $p_i(p_j)$ is the population in the ith(jth) region or compartment; b and d are the birth- and death-rate coefficients, respectively; and $m_{ij}(m_{ki})$ are the immigration (emigration) coefficients, respectively. Here it is assumed that migration rates vary directly with intercompartmental population differences. Although (2.5) is linear in its simplest form, in reality, all the coefficients depend on population $p_i(t)$, making the system nonlinear. If these parameters become demographic control variables, (2.5) becomes *a bilinear control system* (1.10). Such control may take the form of resource allocation according to species and regions. Further, complications occur for *predator–prey models* whereby one population feeds on another, thereby affecting the model parameters (e.g., death rate of prey). Such models are discussed in Vol. 2 with regard to cells and molecules in the immune system. Sometimes it is convenient to formulate a dynamical model from its Lagrangian formulation, which uses generalized coordinates. Only a summary of this elegant, compact method is presented here. Further details are presented in almost any of the many excellent texts available on classical mechanics (see, e.g., [G1] and [G2]).

Again, consider the simple conservative mechanical system given in Fig. 2.1. Let the momentum p and spring force f_s be given as functions of velocity v and position x, respectively, so that

$$p = mv = m\frac{dx}{dt} \tag{2.6}$$

and

$$f_s = k(x)x. \tag{2.7}$$

The kinetic energy of the mass m is given by

$$T = \int p(v)\,dv = \tfrac{1}{2}mv^2 \tag{2.8}$$

and its potential energy by

$$V = \int f_s(x)\,dx. \tag{2.9}$$

Now the Lagrangian is defined as

$$L(v, x) = T(v) - V(x). \tag{2.10}$$

Then it can be shown from the calculus of variations (using first variations) that minimization of $L(v, x)$ over the time duration of interest T,

$$J = \int_0^T L(v(t), x(t))\,dt, \tag{2.11}$$

requires that

$$\frac{d}{dt}\left(\frac{\partial L}{\partial v}\right) - \frac{\partial L}{\partial x} = 0 \tag{2.12}$$

for a conservative system. Such motion, minimizing the time integral of the difference between kinetic and potential energies, is sometimes referred to as the path of "least resistance" in the conversion between kinetic and potential energies.

For Fig. 2.1, equation (2.12) yields

$$\frac{d}{dt}(mv) + k(x)x = 0, \tag{2.13}$$

or for m constant,

$$m\frac{d^2x}{dt^2} + k(x)x = 0. \tag{2.14}$$

If a forcing function $F(t)$ is added to Fig. 2.1, externally pushing the mass down and a dissipation force f_d is included, (2.14) becomes from a force balance

$$m\frac{d^2x}{dt^2} + k(x)x + f_d = F(t). \tag{2.15}$$

For a traditional mechanical damper (dashpot) in parallel with the spring, $f_d = s\,dx/dt$, where s is normally assumed constant. To include such generalizations in the Lagrangian formulation, one can think of the dissipation term as part of a generalized kinetic energy, and the external force as part of generalized potential energy. Another interpretation follows from (2.2) with work by the system (dissipation of energy) and work on the system (adding energy). Then it follows that (2.12) becomes

$$\frac{d}{dt}\left(\frac{\partial L}{\partial v}\right) - \frac{\partial L}{\partial x} = F(t) - s\dot{x}. \tag{2.16}$$

Traditionally, the Lagrangian formulation is derived for conservative systems in terms of independent momentum variables p_i (or velocity v_i) and position variables q_i such that (2.12) becomes

$$\frac{d}{dt}\left(\frac{\partial L}{\partial p_i}\right) - \frac{\partial L}{\partial q_i} = 0. \tag{2.17}$$

If $i = 1, \ldots, N$, in vector form, (2.17) is written as

$$\frac{d}{dt}\left(\frac{\partial L}{\partial \mathbf{p}}\right) - \frac{\partial L}{\partial \mathbf{q}} = \mathbf{0}, \tag{2.18}$$

where $\mathbf{p} \in \mathbb{R}^N$ and $\mathbf{q} \in \mathbb{R}^N$ are appropriate momentum and position vectors and $\mathbf{0}$ is the null vector. The Lagrangian formulation will be used in the next section, where a generalized vector force or torque $\mathbf{F}(t)$ is added to (2.18) as in (2.16). Consequently,

$$\frac{d}{dt}\left(\frac{\partial L}{\partial \mathbf{p}}\right) - \frac{\partial L}{\partial \mathbf{q}} = \mathbf{F}(t). \tag{2.19}$$

2.2 MECHANICS

Here individual element dynamics are studied first. Then the elements are put together to form simple linkages as for a robot manipulator. Finally, fundamentals of aerodynamics control are presented.

2.2.1 Particle Dynamics

The motion of a point particle relative to an inertial coordinate system is the simplest element and represents a fundamental starting point for mechanical systems and their electrical analogs. According to *Newton's first law,* a particle with uniform velocity will persist along a straight trajectory with constant velocity if no forces are acting on the particle. Adding forces to the particle, *Newton's second law* (as discussed above) states that the time rate of change of the linear momentum of the particle is defined as the product of its mass and its velocity. (Of course, at velocities approaching the speed of light, relativity appropriately alters this law.) Finally, *Newton's third law* deals with the interaction of two particles and states that the force due to the first particle on the second one is equal to, but in the opposite direction from, the force due to the second particle on the first. That is, for every action there is an equal reaction—again the force-conservation principle.

In the usual three-dimensional inertial space the *linear momentum* is given by

$$\mathbf{p} = m\mathbf{v}, \tag{2.20}$$

where m is the single particle mass with velocity $\mathbf{v}$. Under the influence of a single vector force $\mathbf{F}$ only, the particle moves according to

$$\mathbf{F} = \frac{d(m\mathbf{v})}{dt}, \tag{2.21}$$

which with constant mass m and $\mathbf{v} = \dot{\mathbf{x}}$, the time rate of change of position relative to some inertial reference, becomes

$$\mathbf{F} = m\ddot{\mathbf{x}}. \tag{2.22}$$

The kinetic energy is given by

$$T = \tfrac{1}{2} m|\mathbf{v}|^2 = \tfrac{1}{2} m(\dot{\mathbf{x}} \cdot \dot{\mathbf{x}}). \tag{2.23}$$

Work done by the force on the particle in a motion $d\mathbf{x}$ generally is defined by $(\mathbf{F} \cdot d\mathbf{x})$ or work per unit time by $(\mathbf{F} \cdot \dot{\mathbf{x}})$. Then the time integral of work from (2.22) with $d\mathbf{x} = \dot{\mathbf{x}}\, dt$ is equal to the change in kinetic energy, or

$$\int_{t_1}^{t_2} (m\ddot{\mathbf{x}} \cdot \dot{\mathbf{x}})dt = \tfrac{1}{2} m|\mathbf{v}|^2 \Big|_{t_1}^{t_2} = T_2 - T_1. \tag{2.24}$$

Then if the system is conservative with total constant energy $E = T(t) + V(t)$, it is seen from (2.24) and (2.23) that

$$\mathbf{F} = -\nabla V(\mathbf{x}) = -\begin{bmatrix} \dfrac{\partial V}{\partial x_1} \\ \dfrac{\partial V}{\partial x_2} \end{bmatrix}, \tag{2.25}$$

where ∇ is the vector-gradient operator at $\mathbf{x}$. That is, the field of force, being derived from the field potential, is conservative. Note that for such systems the integral given by (2.24) is equal to $T_2 - T_1 = V_1 - V_2$, where $V_1 = V(t_1)$, $V_2 = V(t_2)$, and consequently this integral depends only on the positions of the particle at times $t = t_1$ and $t = t_2$, not on the path traversed between these times. This is a requirement of the conservative system.

The system presented by Fig. 2.1 or its electrical analog in Fig. 2.2 is a good example with motion defined by (2.14). Here the constant energy is given by

$$E = \tfrac{1}{2} m\dot{x}^2 + V(x) \tag{2.26}$$

or

$$\dot{x} = \pm\left[\frac{2}{m}(E - V(x))\right]^{1/2}. \tag{2.27}$$

For a linear spring with constant stiffness k, from (2.9),

$$V(x) = \tfrac{1}{2}kx^2. \tag{2.28}$$

The solution to (2.14) with $x(0) = 0$, $\dot{x}(0) = \left(\frac{2E}{m}\right)^{1/2}$, in the linear time-invariant case, is readily obtained to be

$$x(t) = \left(\frac{2E}{k}\right)^{1/2} \sin\left(\frac{k}{m}\right)^{1/2} t. \tag{2.29}$$

From (2.25), the following can be shown, by choosing the origin at the center of the field so that the force $\mathbf{F}$ acts along a line between the origin and the particle position:

$$\mathbf{F} = f(r)\mathbf{x}, \tag{2.30}$$

where r is the radial distance from the origin to the particle. In other words, (2.30) follows for so-called conservative central-force fields. For nonconservative central-force fields,

$$\mathbf{F} = f(\mathbf{x})\mathbf{x}. \tag{2.31}$$

$f(\cdot)$ is an appropriate scalar function in both cases. From (2.30) and (2.25), it follows that

$$V(\mathbf{x}) = V(r). \tag{2.32}$$

For the oscillator example above,

$$V = \tfrac{1}{2}kr^2,$$

and for a gravitational or Coulomb central-force field,

$$V = -\frac{k}{r}. \tag{2.33}$$

It can be shown from Newton's second law (2.21), that particle acceleration in a central-force field (2.31) will be in the direction of the force, and hence particle motion will remain in the plane through the origin which also contains the velocity vector. Hence the particle motion will stay within this orbital plane.

Angular momementum with respect to the origin is defined by the cross product

$$\mathbf{M} = \mathbf{x} \times \mathbf{p}. \tag{2.34}$$

Then for a central force field it follows from (2.31) and (2.22) that

$$\frac{d\mathbf{M}}{dt} = 0. \tag{2.35}$$

Again, (2.34) and (2.35) yield a constant of the central-force field motion, and verify that $\mathbf{x}(t)$ remains in the plane orthogonal to the constant angular momentum vector $\mathbf{M}$. Furthermore, by selecting one axis along $\mathbf{M}$ and introducing polar coordinates (r, θ), Kepler's second law may be derived. That is,

$$M = mr^2\dot{\theta}, \tag{2.36}$$

where M is the absolute magnitude of the angular momentum which is constant here. Sometimes (2.36) is stated that the radius vector describes equal areas in equal times. The kinetic energy in polar coordinates for the particle of mass m in the central force-field becomes, from (2.8),

$$T = \tfrac{1}{2}m\dot{r}^2 + \tfrac{1}{2}mr^2\dot{\theta}^2, \tag{2.37}$$

which is a summation of radial and transverse motion components. Next, an example is given for an earth satellite trajectory calculation.

Example 2.1

Consider Fig. 2.3, where a satellite of mass m moves in the earth's gravitational field. Since the earth's mass $m_e \gg m$, the satellite has negligible effect on the earth's motion. The most convenient formulation consists of polar coordinates in a central-force field. Obviously, the motion will be constrained to its orbital plane. The kinetic energy is given by (2.37) and the potential energy by (2.33), or

$$V(r) = \frac{-\gamma m m_e}{r}, \tag{2.38}$$

where γ is the universal gravitational constant. Then the Lagrangian is given by

$$L = T - V = \frac{1}{2}m(\dot{r}^2 + r^2\dot{\theta}^2) + \frac{\gamma m m_e}{r}, \tag{2.39}$$

and the Euler–Lagrange equations here become

$$\frac{d}{dt}\left(\frac{\partial L}{\partial \dot{r}}\right) - \frac{\partial L}{\partial r} = 0, \tag{2.40}$$

$$\frac{d}{dt}\left(\frac{\partial L}{\partial \dot{\theta}}\right) - \frac{\partial L}{\partial \theta} = 0. \tag{2.41}$$

Consequently, it follows that the radial momentum yields

$$\frac{d}{dt}(m\dot{r}) - mr\dot{\theta}^2 + \frac{\gamma m m_c}{r^2} = 0, \tag{2.42}$$

and the angular momentum yields from (2.36)

$$\frac{d}{dt}(mr^2\dot{\theta}) = 0. \tag{2.43}$$

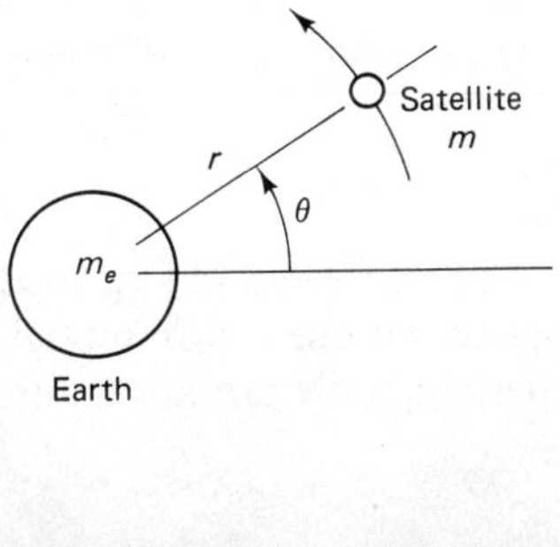

Figure 2.3 Earth satellite trajectory.

Equation (2.43) is a statement of the conservation of satellite angular momentum (2.36). The total energy is given by the constant

$$E = T + V = m\left[\frac{1}{2}\left(\dot{r}^2 + r^2\dot{\theta}^2\right) - \frac{\gamma m_e}{r}\right]. \tag{2.44}$$

Obviously, (2.42) and (2.43) form a nonlinear system, and its control will be studied by perturbation analysis in Chapter 3. Examine trajectory characteristics further for initial conditions r_0, θ_0, $\dot{r}_0$, and $\dot{\theta}_0$. The constant angular momentum magnitude (2.36) is given by

$$M = mr_0^2\,\dot{\theta}_0 = mr^2\,\dot{\theta},$$

so that substitution into (2.44) yields

$$\frac{1}{2}\left[\left(\frac{dr}{d\theta}\right)^2\frac{M^2}{r^4} + \frac{M^2}{mr^2}\right] - \frac{\gamma m m_e}{r} = E. \tag{2.45}$$

Now (2.45) can be solved by separation of variables, which with some algebra takes the following form:

$$r = \frac{K}{1 - \epsilon\cos(\theta - \theta_0 + \phi)},$$

where

$$K = \frac{M^2}{\gamma m_e m^2},$$

$$\epsilon = 1 + \frac{2EK}{\gamma m m_e},$$

$$\phi = \cos^{-1}\left(1 - \frac{K}{\epsilon}\right),$$

$$E = \frac{1}{2}m(\dot{r}_0^2 + r_0^2\,\dot{\theta}_0^2) - \frac{\gamma m m_e}{r_0}.$$

Obviously, the trajectory type depends on the initial energy and angular momentum, but in any case the motion is described by a conic section in polar form: that is, a circle (with $\epsilon = 0$), ellipse ($0 < \epsilon < 1$), parabola ($\epsilon = 1$), or hyperbola ($\epsilon > 1$), depending on total energy E and angular momentum M, both of which remain constant for this conservative system. The borderline case, $\epsilon = 1$, corresponds to $E = 0$ or the degenerate case $M = 0$. The former case ($E = 0$) yields the "satellite" escape velocity

$$v_e = \sqrt{\frac{2\gamma m_e}{r_0}}.$$

The zero angular-momentum degenerate case corresponds to a straight-line trajectory. Obviously, (2.42) and (2.43) could be derived, respectively, from the conservation of forces in the radial direction (Newton's second law) and the conservation of angular momentum (Kepler's second law). This method, however, is a bit more awkward, showing the convenience of Lagrange's equation and generalized coordinates.

2.2.2 Rigid-Body Dynamics

Only particle dynamics have been considered up to this point. Even so, the derived models were found to comprise nonlinear differential equations in general. Next we consider systems of point particles that are not collinear and for which the distance between any two particles is rigidly fixed—hence the name *rigid bodies*. Rigid bodies may have six degrees of freedom. This includes translational motion along the three coordinates and rotational motion about each of the three axes. A mathematical model can be derived by force and moment conservations or by Lagrange's equation with the kinetic energy of a rigid body expressed in terms of its mass and translational velocities, and its moments of inertia and rotational velocities referred to its principal axis. First, however, the kinematic (geometrical) relationships of a maneuvering rigid body will be examined. It will be seen that the resulting coordinate transformation to a fixed reference is very nonlinear.

In the analysis of particle dynamics above, motion was studied with respect to an inertial coordinate system with appropriately defined vectors. Any system moving with a uniform velocity relative to an inertial coordinate system can be considered a new inertial coordinate system. In general, translation of coordinates is a simple linear mathematical operation compared to coordinate rotations, which involve "direction-cosine relationships."

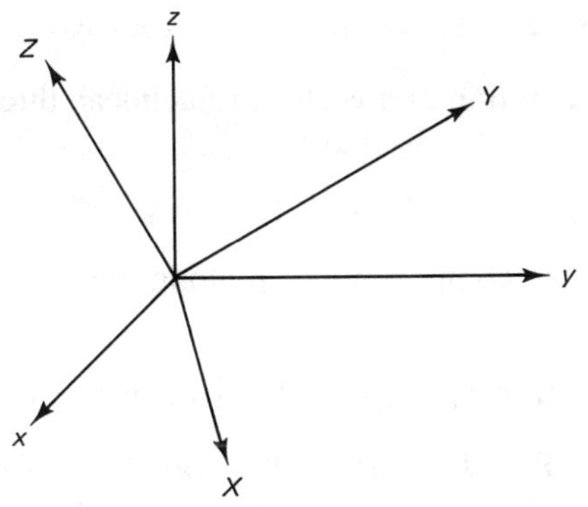

Figure 2.4 Rotation of body axes (XYZ) relative to reference (xyz).

For convenience, assume that an orthogonal X, Y, Z coordinate system is defined with its origin at the center of mass (i.e., center of mass coordinates as shown in Fig. 2.4). This triplet (X, Y, Z) corresponds to the translational degrees of freedom. The reference coordinates x, y, z are fixed in space, and three angles ϕ, θ, and ψ characterize the orientation of body axes (X, Y, Z) relative to reference (x, y, z). Since it is convenient to perform multiple rotations by matrix operations, let us find a 3×3 matrix $\mathbf{R}$ that will transform the coordinates of a point $\mathbf{p}_{XYZ} = [p_X, p_Y, p_Z]^T$ into its representation in the original reference coordinates, $\mathbf{p}_{xyz} = [p_x, p_y, p_z]^T$. That is, find the rotated-coordinate representation of the point $\mathbf{p}$. A simple rotation about the x axis is shown in Fig. 2.5. Here it is desired to find the rotation matrix $\mathbf{R}_{x,\phi}$ so that

$$\mathbf{p}_{xyz} = \mathbf{R}_{x,\phi}\, \mathbf{p}_{XYZ}. \tag{2.46}$$

(See [F1] and [F2] for more details.) From the simple projections onto y, z coordinates it is seen that

$$\mathbf{R}_{x,\phi} = \begin{bmatrix} 1 & 0 & 0 \\ 0 & \cos\phi & -\sin\phi \\ 0 & \sin\phi & \cos\phi \end{bmatrix}. \tag{2.47}$$

Now let us leave this single rotation to consider a more general case.

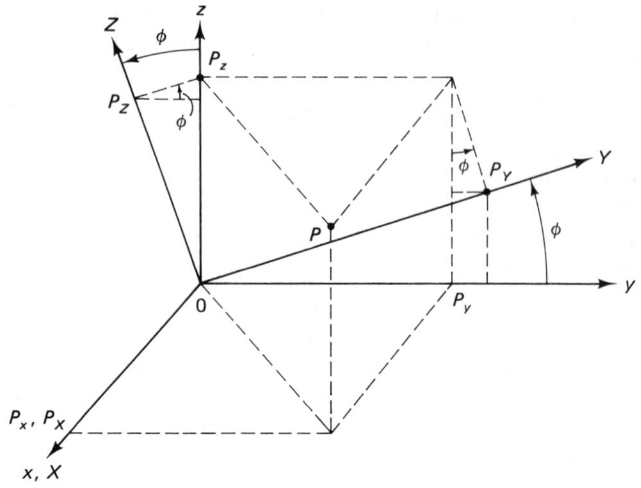

Figure 2.5 Rotation by angle ϕ about x axis.

From the traditional definition of a vector in Euclidean three-dimensional space with the unit basis vectors $\mathbf{i}_X$, $\mathbf{j}_Y$, $\mathbf{k}_Z$, we have

$$\mathbf{p}_{XYZ} = p_X\mathbf{i}_X + p_Y\mathbf{j}_Y + p_Z\mathbf{k}_Z, \tag{2.48}$$

where p_X, p_Y, p_Z represent the components of $\mathbf{p}$ along the X, Y, and Z axes, respectively. Consequently,

$$\begin{aligned} p_x &= \mathbf{i}_x \cdot \mathbf{p} = \mathbf{i}_x \cdot \mathbf{i}_X p_X + \mathbf{i}_x \cdot \mathbf{j}_Y p_Y + \mathbf{i}_x \cdot \mathbf{k}_Z p_Z, \\ p_y &= \mathbf{j}_y \cdot \mathbf{p} = \mathbf{j}_y \cdot \mathbf{i}_X p_X + \mathbf{j}_y \cdot \mathbf{j}_Y p_Y + \mathbf{j}_y \cdot \mathbf{k}_Z p_Z, \\ p_z &= \mathbf{k}_z \cdot \mathbf{p} = \mathbf{k}_z \cdot \mathbf{i}_X p_X + \mathbf{k}_z \cdot \mathbf{j}_Y p_Y + \mathbf{k}_z \cdot \mathbf{k}_Z p_Z \end{aligned}$$

or

$$\mathbf{P}_{xyz} = \begin{bmatrix} p_x \\ p_y \\ p_z \end{bmatrix} = \begin{bmatrix} \mathbf{i}_x \cdot \mathbf{i}_X & \mathbf{i}_x \cdot \mathbf{j}_Y & \mathbf{i}_x \cdot \mathbf{k}_Z \\ \mathbf{j}_y \cdot \mathbf{i}_X & \mathbf{j}_y \cdot \mathbf{j}_Y & \mathbf{j}_y \cdot \mathbf{k}_Z \\ \mathbf{k}_z \cdot \mathbf{i}_X & \mathbf{k}_z \cdot \mathbf{j}_Y & \mathbf{k}_z \cdot \mathbf{k}_Z \end{bmatrix} \mathbf{p}_{XYZ}, \tag{2.49}$$

where $\mathbf{p}_{XYZ}^{\mathrm{T}} = [p_X \quad p_Y \quad p_Z]$. In simple vector–notation,

$$\mathbf{p}_{xyz} = \mathbf{R}\mathbf{p}_{XYZ} \tag{2.50}$$

or

$$\mathbf{P}_{XYZ} = \mathbf{R}^{\mathrm{T}}\mathbf{p}_{xyz}, \tag{2.51}$$

where $\mathbf{R}^{\mathrm{T}} = \mathbf{R}^{-1}$. That is, the transpose of $\mathbf{R}$ is equal to the inverse of $\mathbf{R}$ since the transformation is one of rotation (orthonormal) only. For the single rotation by angle ϕ about the x-axis above,

$$\mathbf{R}_{x,\phi} = \mathbf{R}(\mathbf{i}_x = \mathbf{i}_X), \tag{2.52}$$

which agrees with (2.47).

Now a composite rotation ϕ about the x axis is followed by a rotation ψ about the z axis and then a rotation θ about the y axis. Consequently, the resultant rotation matrix for this precise order is given by

$$\mathbf{R} = \mathbf{R}_{y,\theta}\mathbf{R}_{z,\psi}\mathbf{R}_{x,\phi}, \tag{2.53}$$

where $\mathbf{R}_{x,\phi}$ is given by (2.47),

$$\mathbf{R}_{z,\psi} = \begin{bmatrix} \cos\psi & -\sin\psi & 0 \\ \sin\psi & \cos\psi & 0 \\ 0 & 0 & 1 \end{bmatrix}, \tag{2.54}$$

and

$$\mathbf{R}_{y,\theta} = \begin{bmatrix} \cos\theta & 0 & \sin\theta \\ 0 & 1 & 0 \\ -\sin\theta & 0 & \cos\theta \end{bmatrix}. \tag{2.55}$$

Consequently, with $C \equiv \cos(\cdot)$ and $S \equiv \sin(\cdot)$, this composite transformation is represented by the following rotational matrix:

$$\mathbf{R}_{\theta\psi\phi} = \mathbf{R}_{y,\theta}\mathbf{R}_{z,\psi}\mathbf{R}_{x,\phi} \tag{2.56}$$

or

$$\mathbf{R}_{\theta\psi\phi} = \begin{bmatrix} C\theta C\psi & S\theta S\phi - C\theta S\psi C\phi & C\theta S\psi S\phi + S\theta C\phi \\ S\psi & C\psi C\theta & -C\psi S\phi \\ -S\theta C\psi & S\theta S\psi C\phi + C\theta S\phi & C\theta C\phi - S\theta S\psi S\phi \end{bmatrix}. \tag{2.57}$$

It is most important to realize that the composite rotation matrix depends on the order of rotation. For example, with the order θ, ψ, ϕ about the y, z, and x axes, respectively,

$$\mathbf{R}_{\phi\psi\theta} = \mathbf{R}_{x,\phi}\mathbf{R}_{z,\psi}\mathbf{R}_{y,\theta} \tag{2.58}$$

or

$$\mathbf{R}_{\phi\psi\theta} = \begin{bmatrix} C\psi C\psi & S\psi & C\psi S\theta \\ C\phi S\psi C\theta + S\phi S\theta & C\phi C\psi & C\phi S\psi S\theta - S\phi C\theta \\ S\phi S\psi C\theta - C\phi S\theta & S\phi C\psi & S\phi S\psi S\theta + C\phi C\theta \end{bmatrix}. \tag{2.59}$$

Here the angles pitch, yaw, and roll of an aircraft, missile, or other body correspond to θ, ψ, and ϕ, respectively. In aerodynamic control, however, it is the traditional convention to have the z axis pointed down with positive yaw ψ appropriately designated with the right-hand rule as for θ and ϕ about the y and x axes, respectively (Fig. 2.6). If the plane rolls, pitches, and yaws, such as for a coordinated change in position,

$$\mathbf{R}_{\phi\theta\psi} = \mathbf{R}_{z,\psi}\mathbf{R}_{y,\theta}\mathbf{R}_{x,\phi} \tag{2.60}$$

or

$$\mathbf{R}_{\phi\theta\psi} = \begin{bmatrix} C\psi C\theta & C\psi S\theta S\phi - S\psi S\phi & C\psi S\theta C\phi + S\psi S\phi \\ S\psi C\theta & S\psi S\theta S\phi + C\psi C\phi & S\psi S\theta C\psi - C\psi S\phi \\ -S\theta & C\theta S\phi & C\theta C\phi \end{bmatrix}. \tag{2.61}$$

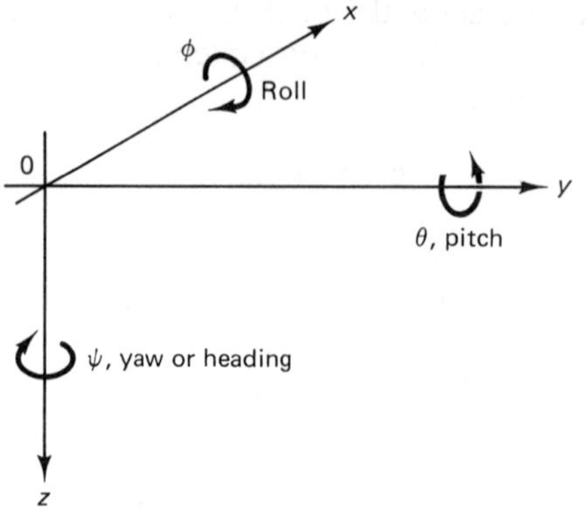

Figure 2.6 Traditional aircraft axes of rotation.

Certain geometric properties of rotation matrices are worth noting. First, it is apparent that the rotation matrix **R** geometrically represents the principal axis of the rotated coordinate system (X, Y, Z) relative to the reference system (x, y, z). That is, the column vectors of **R** represent the principal axes of (X, Y, Z) relative to (x, y, z). Also, with $\mathbf{R}^{-1} = \mathbf{R}^{\mathrm{T}}$, the row vectors represent the principal axis of the reference (x, y, z) with respect to (X, Y, Z), the rotated system. Also, recall that each row is a vector representation of orthonormal vectors such that the inner products of rows with other rows, and columns with other columns, are zero. It is seen that the determinant of the rotation for a right-hand coordinate system is $+1$ and for a left-hand system, -1. Rotation matrices and these properties are useful for such rigid-body dynamics as aircraft motion and multiple-link robots, which are discussed in Sections 2.2.3 and 2.2.4.

It should be noted that in the event that *rotation* of the *XYZ* frame takes place *about its own principal* axis, the appropriate basic rotation matrix must *postmultiply* the previous resultant rotation matrix (i.e., the reverse of the procedure above). Of course, with *XYZ* coordinates coincident with *xyz* coordinates initially, the starting resultant matrix of rotation is the 3×3 identity $\mathbf{I}_3$. The derivation of the equations of motion of a rigid body relative to an inertial coordinate system must include the basic laws of classic mechanics (i.e., Newton's laws, d'Alembert's principle, or the Euler–Lagrange equation) along with the kinematic relations above which constrain the motions to appropriate combinations of translations and rotations.

Consider a number of particles, points in a rigid body, or an assembly of rigid bodies such that from d'Alembert's principle (i.e., Newton's second law and a force conservation),

$$\sum_i (m_i\ddot{\mathbf{x}}_i - \mathbf{F}_i) \cdot \delta\mathbf{x}_i = 0, \tag{2.62}$$

where $\mathbf{F}_i$ is the external force acting on the ith particle, $\mathbf{x}_i$ is composed of appropriate x-y-z coordinates, and $\delta\mathbf{x}_i$ satisfy the kinematic relations. The latter requirement admits combinations of translations with $\delta\mathbf{x}_i = \boldsymbol{\epsilon}$ and rotations with $\delta\mathbf{x}_i = \boldsymbol{\epsilon} \times \mathbf{x}_i$. For the translation case with $\boldsymbol{\epsilon}$ an arbitrary vector, (2.62) yields

$$\overline{m}\ddot{\overline{\mathbf{x}}} = \overline{\mathbf{F}}, \tag{2.63}$$

where $\bar{m} = \Sigma_i m_i$ and $\bar{\mathbf{F}} = \Sigma_i \mathbf{F}_i$. $\bar{\mathbf{x}}$ is the center of mass position vector. For the rotational case with $\boldsymbol{\epsilon}$ again arbitrary, (2.62) becomes

$$\dot{\mathbf{M}} = \sum_i \mathbf{x}_i \times \mathbf{F}_i = \boldsymbol{\tau}. \tag{2.64}$$

Equations (2.63) and (2.64) are the basic rigid-body equations of motion. **M** is the total angular momentum

$$\mathbf{M} = \sum_i \mathbf{x}_i \times m_i\dot{\mathbf{x}}_i, \tag{2.65}$$

and the right-hand side of (2.65) is a moment summation. Both are taken with respect to the same point of reference. Equation (2.63) expresses the fact that the center of mass of a rigid body moves as if all the mass and forces were concentrated at that point.

For example, consider a pure rotation of a rigid body with rotational velocity $\boldsymbol{\omega}$ about its stationary point, so that

$$\dot{\mathbf{x}}_i = \boldsymbol{\omega} \times \mathbf{x}_i. \tag{2.66}$$

Then with $r_i^2 = \mathbf{x}_i \cdot \mathbf{x}_i$, (2.65) and (2.66) show that

$$\mathbf{M} = \sum_i m_i r_i^2 \boldsymbol{\omega} - \sum_i m_i \mathbf{x}_i(\mathbf{x}_i \cdot \boldsymbol{\omega}) \tag{2.67}$$

or

$$\mathbf{M} = \mathbf{J}\boldsymbol{\omega}, \tag{2.68}$$

where

$$\mathbf{J} = \begin{bmatrix} I_{xx} & I_{xy} & I_{xz} \\ I_{xy} & I_{yy} & I_{yz} \\ I_{xz} & I_{yz} & I_{zz} \end{bmatrix}, \tag{2.69}$$

$$I_{xx} = \sum_i m_i(r_i^2 - x_i^2) = \sum_i m_i(y_i^2 + z_i^2), \qquad I_{xy} = -\sum_i m_i x_i y_1, \qquad \text{etc.}$$

Here the ith–jth element of the inertia matrix **J**, with the sum over m_i approaching an integral over mass element dm, may be defined by

$$I_{ij} = \int \left[\delta_{ij}\left(\sum_k x_k^2\right) - x_i x_j \right] dm, \tag{2.70}$$

where i, j indicate the principal axes of the ith coordinate system, and the Kronecker delta $\delta_{ij} = 1$ for $i = j$ and $\delta_{ij} = 0$ for $i \neq j$.

If (2.68) refers to rotation about the principal axis of the rigid body, the inertia matrix **J** is diagonal. Therefore, the X, Y, Z elements of **M** become

$$\begin{aligned} M_x &= I_{xx}p, \\ M_y &= I_{yy}q, \\ M_z &= I_{zz}r, \end{aligned} \tag{2.71}$$

where **p**, **q**, **r** are the rotational velocities about X, Y, and Z axes. From (2.65), the kinetic energy of rotational motion is

$$T = \sum_i \tfrac{1}{2} m_i \dot{x}_i^2 = \tfrac{1}{2} \boldsymbol{\omega}^{\mathrm{T}} \mathbf{J} \boldsymbol{\omega}. \tag{2.72}$$

If the xyz axes are along the principal body axis,

$$T = \tfrac{1}{2} (I_{xx} p^2 + I_{yy} q^2 + I_{zz} r^2), \tag{2.73}$$

where p, q, r are the magnitudes of the **p**, **q**, **r** rotational velocities about the X, Y, Z axes, respectively.

If the rigid body moves with linear volocity **v** as well as rotational velocity $\boldsymbol{\omega}$,

$$\dot{\mathbf{x}}_i = \mathbf{v} + \boldsymbol{\omega} \times \mathbf{x}_i, \tag{2.74}$$

and with $|\mathbf{v}| = v$,

$$T = \tfrac{1}{2}\bar{m}v^2 + \tfrac{1}{2}\boldsymbol{\omega}^{\mathrm{T}} \mathbf{J} \boldsymbol{\omega} + \mathbf{v} \cdot (\boldsymbol{\omega} \times \bar{m}\bar{\mathbf{x}}) \tag{2.75}$$

where $\bar{m}$, $\bar{\mathbf{x}}$ are defined with (2.63). If, however, a moving coordinate system, with origin at the center of mass, is selected, then (2.75) simplifies with the last term zero. In this case,

$$T = \tfrac{1}{2} \bar{m}v^2 + \tfrac{1}{2} \boldsymbol{\omega}^{\mathrm{T}} \mathbf{J} \boldsymbol{\omega}. \tag{2.76}$$

Equation (2.76), along with an expression for the potential energy of the body as computed previously but with reference to the same coordinates as (2.76), can be useful for computing trajectories from the Euler–Lagrange equation.

Now, however, let us compute the so-called Euler equations for the p, q, r, the rotational speeds about the X, Y, Z axes, respectively. From (2.64) with an added motion relative to the fixed reference coordinates (x, y, z) due to $\boldsymbol{\omega}$, it is seen that

$$\dot{\mathbf{M}} + \boldsymbol{\omega} \times \mathbf{M} = \sum_i \mathbf{x}_i \times \mathbf{F}_i = \boldsymbol{\tau}. \tag{2.77}$$

then, from (2.77), or from body to inertia axis by (2.61), (2.64):

$$I_{xx}\dot{p} + (I_{zz} - I_{yy})qr = \tau_1 \tag{2.78}$$

$$I_{yy}\dot{q} + (I_{xx} - I_{zz})pr = \tau_2 \tag{2.79}$$

$$I_{zz}\dot{r} + (I_{yy} - I_{xx})pq = \tau_3. \tag{2.80}$$

Here τ_1, τ_2, τ_3 are body-axis components of the external torque vector τ. Only the rotational motion is given here in terms of p, q, r for rotated X, Y, Z body axes. The translational component is given by (2.63). If τ_1, τ_2, τ_3 are zero, the system above is conservative, and hense the energy is constant, that is,

$$\tfrac{1}{2} I_{xx} p^2 + \tfrac{1}{2} I_{yy} q^2 + \tfrac{1}{2} I_{zz} r^2 = E.$$

Similarly, the angular momentum is a constant of motion, so that

$$I_{xx}^2 p^2 + I_{yy}^2 q^2 + I_{zz}^2 r^2 = M^2.$$

Example 2.2

Consider a conservative system, such as is approximated by the earth's motion with $I_{xx} = I_{yy}$. Then, from (2.80), $I_{zz}\dot{r} = 0$ or $r = r_0$, constant. From (2.78) and (2.79),

$$I_{xx}\dot{p} + (I_{zz} - I_{xx})qr_0 = 0$$

and

$$I_{xx}\dot{q} + (I_{xx} - I_{zz})pr_0 = 0$$

Consequently, their solution yields the precession of the vector $\boldsymbol{\omega}$ around the Z axis such that

$$p = p_0 \cos \frac{(I_{zz} - I_{xx})r_0\, t}{I_{xx}}$$

and

$$q = q_0 \sin \frac{(I_{zz} - I_{xx})r_0\, t}{I_{xx}}.$$

For the earth, $(I_{zz} - I_{xx})/I_{xx}$ is about $\frac{1}{300}$, r_0 is 1 revolution per day, and the precessional period is of the order of 1 year.

2.2.3 Robotics

A robot arm or mechanical manipulator consists of a series of rigid bodies called *links*. These are connected together by revolute (rotary) or prismatic (sliding) joints as shown for the more popular revolute case by a typical robot in Fig. 2.7. Although not as popular, joints may be established by cylindrical, spherical, screw, and planar rubbing surfaces. For a manipulator with N joint–link pairs, there are N degrees of freedom (i.e., one degree of freedom per joint–link pair). It is assumed that the base line (link 0) establishes the inertial coordinate of the robot, and the last link is attached with a tool. To form a reference system, links are numbered outward from the base with each link connected to, at most, two others. Hence no closed link loops are formed. The treatment here follows the line of reasoning given in the excellent book on robotics [F2], which should be consulted for more details.

A joint axis is established as shown by Fig. 2.8 in order to determine the relative position of connecting links and eventually overall motion. Each joint has two normal vectors associated with it, one for each of the links. For example, in Fig. 2.8 joint i connects link $i - 1$ with link i. Here ψ_i, the joint angle, is measured in a plane that is normal to the joint axis. As shown in Fig. 2.8, ψ_i is the angle between normals x_i and x_{i-1} projected onto the (x_{i-1}, z_{i-1}) plane so that it is actually a measurement of the angle between adjacent links. Then the relative position of neighboring links is defined by ψ_i and a distance d_i between adjacent links. More precisely, as shown in Fig. 2.8, d_i is measured as the distance, along the ith-joint axis, between the origin of the $(i - 1\text{th})$ coordinate set to the intersection of the $\mathbf{z}_{i-1}$ axis with the $\mathbf{x}_i$ axis along the $\mathbf{z}_i$ axis.

The ith link is connected to the link $i - 1$ and link $i + 1$ in such a way that the relative orientation of joint i with respect to joint $i+1$ is described by a_i, a length along link i, and α_i, a twist angle that is associated with link i (see Fig. 2.8). More precisely, a_i is the shortest distance between the joint axes measured along the common normal.

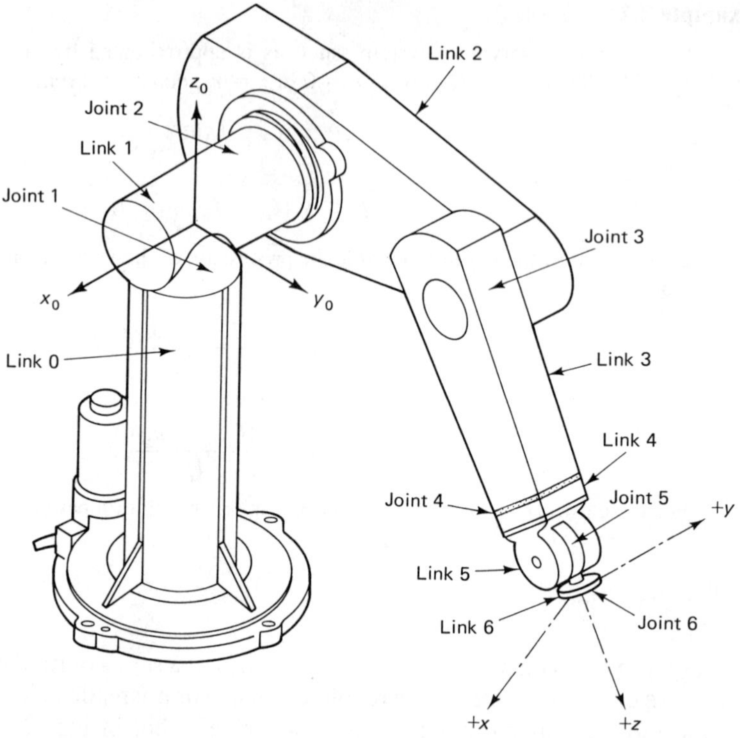

Figure 2.7 PUMA robot joints and links. (After Fu, et al. [F2].)

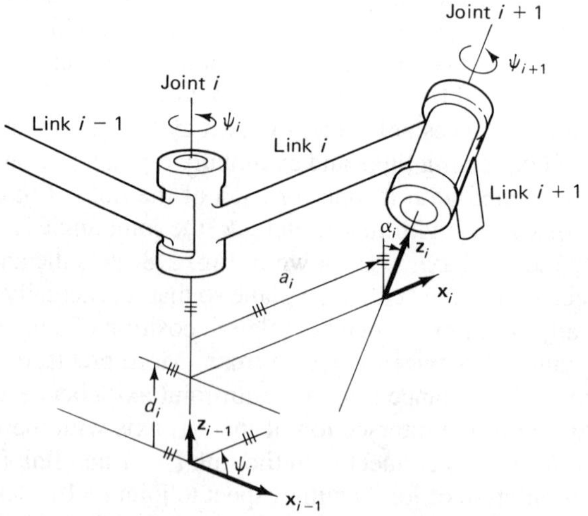

Figure 2.8 Link coordinate system. (After Fu, et al. [F2].)

For link i in Fig. 2.8, the joint axes are defined by the vectors $\mathbf{z}_{i-1}$ and $\mathbf{z}_i$, and the common normal by $\mathbf{x}_i$. The twist angle for link i measures the angle between the two opposite joint axes of link i projected onto a plane orthogonal to a_i (i.e., the offset angle from axis $\mathbf{z}_{i-1}$ to $\mathbf{z}_i$ about $\mathbf{x}_i$). Now, the link parameters, a_i and α_i, determine the link structure (i.e., relative position of link-i joints). Similarly, the joint parameters, d_i and ψ_i, define the relative position of link i to link $i - 1$.

A commonly accepted matrix method called the Denavit–Hartenberg (D-H) representation defines the translation and rotation of link i relative to link $i - 1$. Subsequently, a 4×4 transformation matrix represents each link's coordinate axes at the joint with respect to the previous link's coordinate axes. A sequence of transformations (matrix multiplications) leads to the representation of the end effector in inertial or "base coordinates." The end-effector (moving body) coordinates are referred to as "hand coordinates." For example, a six-axis PUMA robot arm has seven coordinate frames: $(\mathbf{x}_0, \mathbf{y}_0, \mathbf{z}_0), (\mathbf{x}_1, \mathbf{y}_1, \mathbf{z}_1), \ldots, (\mathbf{x}_6, \mathbf{y}_6, \mathbf{z}_6)$. Every coordinate frame is established with (1) the $\mathbf{z}_{i-1}$ axis along the motion axis of the ith joint, (2) the $\mathbf{x}_i$ axis normal to $\mathbf{z}_{i-1}$ but away from it, and (3) the $\mathbf{y}_i$ axis as the remaining axis of a right-hand coordinate system. For a rotary joint, d_i, a_i, and α_i are robot constants, while ψ_i is a variable that changes with the relative position of link i relative to link $i - 1$. ψ_i, a_i, and α_i are constant joint parameters, and d_i is a joint variable, for a prismatic joint.

A transformation matrix for a given sequence of rotations and translations can be derived according to the kinematic relations of the preceding section relative to the D-H coordinate system established above. The 3×3 rotation matrix, $\mathbf{R}$, which was derived in the preceding section, must be supplemented to include translation. Scaling could be included but will be assumed to be unity here. This is accomplished by addending the position vector of the preceding section with a unity element so that

$$\bar{\mathbf{p}} = [\mathbf{p}^{\mathrm{T}} 1]^{\mathrm{T}} = [p_x \quad p_y \quad p_z \quad 1]^{\mathrm{T}}. \tag{2.81}$$

This system is referred to as *homogeneous coordinates*. Then a homogeneous transformation for rotation and translation can be shown to have the following form, with four submatrices expanding the 3×3 $\mathbf{R}$ rotation matrix:

$$\mathbf{T} = \begin{bmatrix} \mathbf{R} & \mathbf{p} \\ \mathbf{0}^{\mathrm{T}} & 1 \end{bmatrix}, \tag{2.82}$$

where $\mathbf{p}$ represents the position vector of the origin of the rotated coordinates with respect to the reference (inertial) coordinates, and the zero row vector $\mathbf{0}^{\mathrm{T}}$ may be replaced by a prospective transformation such as may be used for computer vision. A homogeneous transformation 4×4 matrix corresponding to a single rotation ϕ about the x axis, given for a 3×3 transformation by (2.47), becomes

$$T_{x,\phi} = \begin{bmatrix} 1 & 0 & 0 & 0 \\ 0 & \cos\phi & -\sin\phi & 0 \\ 0 & \sin\phi & \cos\phi & 0 \\ 0 & 0 & 0 & 1 \end{bmatrix}. \tag{2.83}$$

Now suppose that it is desired to find the resultant transformation for the rotation given by (2.83) followed by a translation b along the rotated $\mathbf{y}$ axis (i.e., its own body-axis translation). Therefore, following the rule of postmultiplication of the previous (rotation) resultant (see Section 2.2.2), the transformation matrix becomes

$$\mathbf{T} = \mathbf{T}_{x,\phi}\mathbf{T}_{Y,b} = \begin{bmatrix} 1 & 0 & 0 & 0 \\ 0 & \cos\phi & -\sin\phi & 0 \\ 0 & \sin\phi & \cos\phi & 0 \\ 0 & 0 & 0 & 1 \end{bmatrix} \begin{bmatrix} 1 & 0 & 0 & 0 \\ 0 & 1 & 0 & b \\ 0 & 0 & 1 & 0 \\ 0 & 0 & 0 & 1 \end{bmatrix}$$

or

$$\mathbf{T} = \begin{bmatrix} 1 & 0 & 0 & 0 \\ 0 & \cos\phi & -\sin\phi & b\cos\phi \\ 0 & \sin\phi & \cos\phi & b\sin\phi \\ 0 & 0 & 0 & 1 \end{bmatrix}.$$

Note that premultiplication yields the same results here.

Now, let us examine the homogeneous transformation which relates the ith coordinate set to the $(i-1)^{\text{th}}$ coordinate set by the D-H representation for Fig. 2.8. Consequently, motion is constrained to the following sequence:

1. Rotate by an angle ψ_i about the $\mathbf{z}_{i-1}$ axis to align $\mathbf{x}_{i-1}$ with $\mathbf{x}_i$.
2. Translate by d_i along $\mathbf{z}_{i-1}$ axis to make $\mathbf{x}_{i-1}$ and $\mathbf{x}_i$ axes coincident.
3. Translate by a distance a_i along the $\mathbf{x}_i$ axis to bring two origins as well as $\mathbf{x}$ axis into coincidence.
4. Rotate about the $\mathbf{x}_i$ axis by an angle α_i, to bring the coordinate systems into coincidence. (Note the similarity of α_i to ϕ above.)

The order could be reversed and premultiplication convention followed. Let ${}^{i-1}\mathbf{A}_i$ denote the D-H transformation for adjacent, i and $i-1$, coordinate sets. Then the sequence above yields

$$ {}^{i-1}\mathbf{A}_i = \mathbf{T}_{z,\psi}\mathbf{T}_{z,d}\mathbf{T}_{x,a}\mathbf{T}_{x,\alpha}, \tag{2.84}$$

where

$$\mathbf{T}_{z,d} = \begin{bmatrix} 1 & 0 & 0 & 0 \\ 0 & 1 & 0 & 0 \\ 0 & 0 & 1 & d_i \\ 0 & 0 & 0 & 1 \end{bmatrix},$$

$$\mathbf{T}_{z,\psi} = \begin{bmatrix} \cos\psi_i & -\sin\psi_i & 0 & 0 \\ \sin\psi_i & \cos\psi_i & 0 & 0 \\ 0 & 0 & 1 & 0 \\ 0 & 0 & 0 & 1 \end{bmatrix},$$

$$\mathbf{T}_{x,a} = \begin{bmatrix} 1 & 0 & 0 & a_i \\ 0 & 1 & 0 & 0 \\ 0 & 0 & 1 & 0 \\ 0 & 0 & 0 & 1 \end{bmatrix},$$

and

$$\mathbf{T}_{x,\alpha} = \begin{bmatrix} 1 & 0 & 0 & 0 \\ 0 & \cos\alpha_i & -\sin\alpha_i & 0 \\ 0 & \sin\alpha_i & \cos\alpha_i & 0 \\ 0 & 0 & 0 & 1 \end{bmatrix}.$$

Then

$$^{i-1}\mathbf{A}_i = \begin{bmatrix} \cos\psi_i & -\cos\alpha_i \sin\psi_i & \sin\alpha_i \sin\psi_i & a_i \cos\psi_i \\ \sin\psi_i & \cos\alpha_i \cos\psi_i & -\sin\alpha_i \cos\psi_i & a_i \sin\psi_i \\ 0 & \sin\alpha_i & \cos\alpha_i & d_i \\ 0 & 0 & 0 & 1 \end{bmatrix}. \tag{2.85}$$

The inverse of (2.85) yields

$$[^{i-1}\mathbf{A}_i]^{-1} = {}^{i}\mathbf{A}_{i-1} = \begin{bmatrix} C\psi_i & S\psi_i & 0 & -a_i \\ -C\alpha_i S\psi_i & C\alpha_i C\psi_i & S\alpha_i & -d_i S\alpha_i \\ S\alpha_i S\psi_i & -S\alpha_i C\psi_i & C\alpha_i & -d_i C\alpha_i \\ 0 & 0 & 0 & 1 \end{bmatrix},$$

where C, S denote $\cos(\cdot)$, $\sin(\cdot)$, respectively; α_i, a_i, d_i are constants; and ψ_i is a variable for the revolute joint.

Now a homogeous matrix $^0\mathbf{T}_i$ may be defined to specify the location of the ith coordinate set relative to inertial (base) coordinates by a chain of coordinate transformations as follows:

$$^0\mathbf{T}_i = {}^0\mathbf{A}_1 {}^1\mathbf{A}_2 \cdots {}^{i-1}\mathbf{A}_i = \prod_{j=1}^{i} {}^{j-1}\mathbf{A}_j, \tag{2.86}$$

where $i = 1, 2, \ldots, N$. Again, the resultant $^0\mathbf{T}_i$ matrix is decomposed into a 3×3 rotation matrix $^0\mathbf{R}_i$, a 3×1 position vector $\mathbf{p}_i$, and unity. Consequently, this "arm matrix" is defined by

$$^0\mathbf{T}_i = \begin{bmatrix} ^0\mathbf{R}_i & \mathbf{p}_i \\ 0 & 1 \end{bmatrix}, \tag{2.87}$$

where $\mathbf{p}_i$ represents the ith-link origin with respect to the base coordinates. Then, in terms of the ith coordinate set at link i relative to the base coordinates, the rotation matrix is

$$^0\mathbf{R}_i = [\mathbf{x}_i \quad \mathbf{y}_i \quad \mathbf{z}_i]. \tag{2.88}$$

Now, consider a fixed point ${}^i\mathbf{r}_i$ at rest in link i with homogeneous coordinates

$$ {}^i\mathbf{r}_i = [x_i \quad y_i \quad z_i \quad 1]^{\mathrm{T}}. \tag{2.89}$$

Let ${}^0\mathbf{r}_i$ be the same point ${}^i\mathbf{r}_i$, but with respect to the fixed base coordinates; then

$$ {}^0\mathbf{r}_i = {}^0\mathbf{A}_i {}^i\mathbf{r}_i, \tag{2.90}$$

where ${}^0\mathbf{A}_i = {}^0\mathbf{T}_i$ from (2.86), and ${}^{i-1}\mathbf{A}_i$ is given by (2.85) if joint i is revolute. Since point ${}^i\mathbf{r}_i$ is at rest in link i, it has zero velocity with respect to the ith coordinate set (noninertial coordinates). Relative to the base (inertial) coordinate system, the *velocity* of ${}^i\mathbf{r}_i$ is given by

$$ {}^0\mathbf{v}_i \equiv \mathbf{v}_i = \frac{d}{dt}({}^0\mathbf{r}_i) = \frac{d}{dt}({}^0\mathbf{A}_i\, {}^i\mathbf{r}_i),$$

$$ \mathbf{v}_i = {}^0\dot{\mathbf{A}}_1 {}^1\mathbf{A}_2 \cdots {}^{i-1}\mathbf{A}_i {}^i\mathbf{r}_i + {}^0\mathbf{A}_1 {}^1\dot{\mathbf{A}}_2 \cdots {}^{i-1}\mathbf{A}_i {}^i\mathbf{r}_i + \cdots + {}^0\mathbf{A}_1 \cdots {}^{i-1}\dot{\mathbf{A}}_i {}^i\mathbf{r}_i + {}^0\mathbf{A}_i {}^i\dot{\mathbf{r}}_i $$

or

$$ \mathbf{v}_i = \left(\sum_{j=1}^{i} \frac{\partial\, {}^0\mathbf{A}_i}{\partial q_j} \dot{q}_j \right) {}^i\mathbf{r}_i, \tag{2.91}$$

since ${}^i\dot{\mathbf{r}}_i = \mathbf{0}$. Here q_j are appropriate generalized position coordinates. For convenience in the recursive computations that follow, the partial derivative of ${}^0\mathbf{A}_i$ with respect to q_i may be placed in the following form:

$$ \frac{\partial\, {}^{i-1}\mathbf{A}_i}{\partial q_j} = \mathbf{Q}_i({}^{i-1}\mathbf{A}_i), \tag{2.92}$$

where for a revolute joint,

$$ \mathbf{Q}_i = \begin{bmatrix} 0 & -1 & 0 & 0 \\ 1 & 0 & 0 & 0 \\ 0 & 0 & 0 & 0 \\ 0 & 0 & 0 & 0 \end{bmatrix}. \tag{2.93}$$

Example 2.3

A robot arm, with all rotary joints and $q_i = \psi_i$, has a coordinate transformation matrix ${}^{i-1}\mathbf{A}_i$ between the spatial displacement of the ith link and the $(i-1)$th link coordinate system which satisfies (2.85). Then, by direct computation,

$$ \frac{\partial\, {}^{i-1}\mathbf{A}_i}{\partial \psi_i} = \begin{bmatrix} -\sin\psi_i & -\cos\alpha_i \cos\psi_i & \sin\alpha_i \cos\psi_i & -a_i \sin\psi_i \\ \cos\psi_i & -\cos\alpha_i \sin\psi_i & \sin\alpha_i \sin\psi_i & a_i \cos\psi_i \\ 0 & 0 & 0 & 0 \\ 0 & 0 & 0 & 0 \end{bmatrix}.$$

By direct algebra, it is seen that this matrix is equal to $\mathbf{Q}_i({}^{i-1}\mathbf{A}_i)$.

It can thus be shown that

$$\mathbf{U}_{ij} \equiv \frac{\partial\, {}^0\mathbf{A}_i}{\partial q_j} = \begin{cases} {}^0\mathbf{A}_1\, {}^1\mathbf{A}_2 \cdots {}^{j-2}\mathbf{A}_{j-1}\mathbf{Q}_j\, {}^{j-1}\mathbf{A}_j \cdots {}^{i-1}\mathbf{A}_i, & j \le i \\ 0, & j > i, \end{cases} \tag{2.94}$$

$$i = 1, 2, \ldots, N,$$

which is a representation of the effect on all the points on link i caused by the motion of joint j. If the single transformations are collected into a single resultant,

$$\mathbf{U}_{ij} = \begin{cases} {}^0\mathbf{A}_{j-1}\mathbf{Q}_j\, {}^{j-1}\mathbf{A}_i, & j \le i, \\ \mathbf{0}, & j > i. \end{cases} \tag{2.95}$$

Then (2.91) can be rewritten as

$$\mathbf{v}_i = \left(\sum_{j=1}^{i} \mathbf{U}_{ij}\, q_j \right) {}^i\mathbf{r}_i. \tag{2.96}$$

One might question the wisdom of defining $\mathbf{U}_{ij}$, yet another matrix, in the maze of terms. However, its convenience will become more apparent in the chain computations such as that below for kinetic energy. Also note that $\partial^{i-1}\mathbf{A}_i/\partial q_i$ does not retain the structure of a homogeneous coordinate transformation matrix. However, it can be shown that premultiplying ${}^{i-1}\mathbf{A}_i$ by $\mathbf{Q}_i$ from (2.93) allows the use of ${}^{i-1}\mathbf{A}_i$ in a manner that is consistent with the algebraic rules above.

From the joint velocity of each link, (2.91) or (2.96), the kinetic energy of each link may be computed in terms of the base reference coordinate system. Rather than using the results of Section 2.2.2 directly, however, it is desired to formulate the kinetic energy in homogeneous coordinates using generalized position variables q_p, $p = 1, 2, \ldots, N$, where N is the number of links. Again, q_p may be an appropriate angle of rotation or a displacement, in general, for a link or joint. In terms of the fixed position ${}^i\mathbf{r}_i = [x \quad y \quad z \quad 1]^{\mathrm{T}}$ on the ith link, the 4×4 inertia matrix, from (2.70) for inertias about joint axes, becomes

$$\mathbf{J}_i = \int {}^i\mathbf{r}_i\, {}^i\mathbf{r}_i^{\mathrm{T}}\, dm \tag{2.97}$$

or

$$\mathbf{J}_i = \begin{bmatrix} \dfrac{-I_{xx} + I_{yy} + I_{zz}}{2} & I_{xy} & I_{xz} & m_i\bar{x}_i \\ I_{xy} & \dfrac{I_{xx} - I_{yy} + I_{zz}}{2} & I_{yz} & m_i\bar{y}_i \\ I_{xz} & I_{yz} & \dfrac{I_{xx} + I_{yy} - I_{zz}}{2} & m_i\bar{z}_i \\ m_i\bar{x}_i & m_i\bar{y}_i & m_i\bar{z}_i & m_i \end{bmatrix}, \tag{2.98}$$

where $\bar{x}_i$, $\bar{y}_i$, $\bar{z}_i$ are center of mass coordinates with respect to the ith-link reference coordinate system. Since a particle of differential mass dm in link i with velocity $\mathbf{v}_i$ has kinetic energy,

$$dT_i = \tfrac{1}{2}\,\mathrm{Tr}(\mathbf{v}_i\mathbf{v}_i^{\mathrm{T}})dm = \tfrac{1}{2}\,(\dot{x}_i^2 + \dot{y}_i^2 + \dot{z}_i^2)dm,$$

it is seen by integration over the mass of link i and substitution of (2.97) that

$$T_i = \frac{1}{2} \mathrm{Tr}\left[\sum_{p=1}^{i} \sum_{r=1}^{i} \mathbf{U}_{ip} \mathbf{J}_i \mathbf{U}_{ir}^{\mathrm{T}} \dot{q}_p \dot{q}_r \right], \tag{2.99}$$

where Tr[·] designates trace operation. Recall that Tr $\mathbf{M} = \sum_i m_{ii}$, the sum of the diagonal elements. Consequently, the total kinetic energy in an arm of N links becomes

$$T(\dot{\mathbf{q}}) = \sum_{i=1}^{N} T_i(\dot{\mathbf{q}}). \tag{2.100}$$

Note that the $\mathbf{U}_{ip}$ matrices are used conveniently to convert the velocity in the base coordinate system of the point ${}^0\mathbf{r}_i$ (i.e., fixed point of motion ${}^i\mathbf{r}_i$ with respect to link-i coodinates) to a generalized position variable q_p. Hence the kinetic energy is a function of generalized velocity components (or momenta) $\dot{q}_p$. The inertial matrices $\mathbf{J}_i$ are computed only once for each link, and of course in general the mass distribution for each link will result in different moments of inertia (I_{xx}, I_{yy}, I_{zz}) and products of inertia (I_{xy}, I_{xz}, I_{yz}).

The potential energy for each ith link results from gravity with

$$V_i = -\, m_i \mathbf{g}\, {}^0\bar{\mathbf{r}}_i = -\, m_i \mathbf{g}({}^0\mathbf{A}_i\, {}^i\bar{\mathbf{r}}_i), \tag{2.101}$$

where $i = 1, 2, \ldots, N$; again ${}^0\bar{\mathbf{r}}_i$ (${}^i\bar{\mathbf{r}}_i$) locates the center of mass with respect to the base coordinate system (link-i coordinate system), and $\mathbf{g}$ is a gravity row vector, $\mathbf{g} = [g_x \quad g_y \quad g_z \quad 0]$. Then the total potential energy of the robot arm with N links becomes

$$V = \sum_{i=1}^{N} V_i. \tag{2.102}$$

Recall the Euler–Lagrange equation with external torque $\boldsymbol{\tau}$ or force $\mathbf{F}$ on a particle from (2.19) so that

$$\frac{d}{dt}\left(\frac{\partial L}{\partial \dot{\mathbf{q}}}\right) - \frac{\partial L}{\partial \mathbf{q}} = \boldsymbol{\tau}, \tag{2.103}$$

where the Lagrangian $L(\mathbf{q}, \dot{\mathbf{q}}) = T(\dot{\mathbf{q}}) - V(\mathbf{q})$. Here $\mathbf{q} \in \mathbb{R}^N$ for an arm of N links. Obviously, (2.103) is a nonlinear system of equations and can be rewritten in the following form from (2.99) and (2.101):

$$\mathbf{D}(\mathbf{q}(t))\ddot{\mathbf{q}}(t) + \mathbf{h}(\mathbf{q}(t), \dot{\mathbf{q}}(t)) + \mathbf{c}(\mathbf{q}(t)) = \boldsymbol{\tau}(t), \tag{2.104}$$

where $\mathbf{D}(\mathbf{q})$ is an $N \times N$ symmetric matrix of elements, from (2.99) and (2.103),

$$D_{ik} = \sum_{j=\max(i,k)}^{n} \mathrm{Tr}(\mathbf{U}_{jk} \mathbf{J}_j \mathbf{U}_{ji}^{\mathrm{T}}); \tag{2.105}$$

$\mathbf{h}(\mathbf{q}, \dot{\mathbf{q}})$ is a Coriolis and centrifugal force vector with components

$$h_i = \sum_{k=1}^{N} \sum_{m=1}^{N} h_{ikm} \dot{q}_k \dot{q}_m \tag{2.106}$$

and

$$h_{ikm} = \sum_{j=\max(i,k,m)}^{N} \mathrm{Tr}(\mathbf{U}_{jkm}\,\mathbf{J}_j\,\mathbf{U}_{ji}^{\mathrm{T}}); \tag{2.107}$$

$\mathbf{c}(q)$ is a loading force (mass distribution) vector with elements

$$c_i = \sum_{j=1}^{N} -\, m_j \mathbf{g} \mathbf{U}_{ji}{}^{j}\bar{\mathbf{r}}_j; \tag{2.108}$$

in general, $i, j, k = 1, 2, \ldots, N$.

Example 2.4

Derive the equations of motion for the two-link manipulator shown in Fig. 2.9, where both joint rotations are about the **z** axis, $d_i = 0$, $\alpha_i = 0$, $a_i = l = l_i = l_2$, $i = 1, 2$[F2].

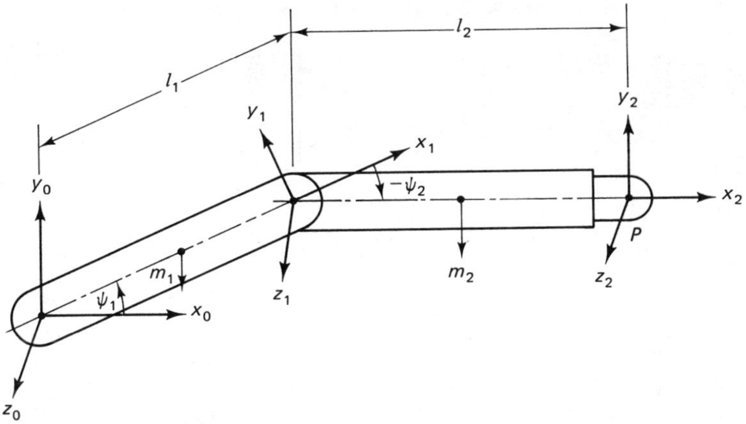

Figure 2.9 Two-link coordinate system for Example 2.4 and for Exercise 2.3.

Solution From (2.85) with $\alpha_i = 0$, the homogeneous transformation matrices become

$${}^{i-1}\mathbf{A}_i = \begin{bmatrix} C_i & -S_i & 0 & lC_i \\ S_i & C_i & 0 & lS_i \\ 0 & 0 & 1 & 0 \\ 0 & 0 & 0 & 1 \end{bmatrix},$$

where $i = 1, 2$; $C_i = \cos \psi_i$, $S_i = \sin \psi_i$. Then

$${}^{0}\mathbf{A}_2 = {}^{0}\mathbf{A}_1\,{}^{1}\mathbf{A}_2 = \begin{bmatrix} C_{12} & -S_{12} & 0 & l(C_{12} + C_1) \\ S_{12} & C_{12} & 0 & l(S_{12} + S_1) \\ 0 & 0 & 1 & 0 \\ 0 & 0 & 0 & 1 \end{bmatrix},$$

where $C_{12} = \cos(\psi_1 + \psi_2)$, $S_{12} = \sin(\psi_1 + \psi_2)$. Then with $\mathbf{Q}_i$ given by (2.93), substitution of ${}^{i-1}\mathbf{A}_i$ above into (2.95) yields

$$\mathbf{U}_{11} = \begin{bmatrix} -S_1 & -C_1 & 0 & -lS_1 \\ C_1 & -S_1 & 0 & lC_1 \\ 0 & 0 & 0 & 0 \\ 0 & 0 & 0 & 0 \end{bmatrix},$$

$$\mathbf{U}_{21} = \begin{bmatrix} -S_{12} & -C_{12} & 0 & -l(S_{12} + S_1) \\ C_{12} & -S_{12} & 0 & l(C_{12} + C_1) \\ 0 & 0 & 0 & 0 \\ 0 & 0 & 0 & 0 \end{bmatrix},$$

$$\mathbf{U}_{22} = \begin{bmatrix} -S_{12} & -C_{12} & 0 & -lS_{12} \\ C_{12} & -S_{12} & 0 & lC_{12} \\ 0 & 0 & 0 & 0 \\ 0 & 0 & 0 & 0 \end{bmatrix}.$$

From (2.98), with linkage symmetry so that $I_{xy} = I_{xz} = I_{yz} = 0$,

$$\mathbf{J}_i = \begin{bmatrix} \frac{1}{3}m_i l^2 & 0 & 0 & -\frac{1}{2}m_i l \\ 0 & 0 & 0 & 0 \\ 0 & 0 & 0 & 0 \\ -\frac{1}{2}m_i l & 0 & 0 & m_i \end{bmatrix},$$

$i = 1, 2$. Then, from (2.105),

$$\begin{aligned} D_{11} &= \mathrm{Tr}(\mathbf{U}_{11}\mathbf{J}_1\mathbf{U}_{11}^{\mathrm{T}}) + \mathrm{Tr}(\mathbf{U}_{21}\mathbf{J}_2\mathbf{U}_{21}^{\mathrm{T}}) \\ &= \tfrac{1}{3} m_1 l^2 + \tfrac{4}{3} m_2 l^2 + m_2 l^2 \cos \psi_2, \\ D_{12} &= D_{21} = Tr(\mathbf{U}_{22}\mathbf{J}_2\mathbf{U}_{21}^{\mathrm{T}}) \\ &= \tfrac{1}{3} m_2 l^2 + \tfrac{1}{2} m_2 l^2 \cos \psi_2, \\ D_{22} &= \mathrm{Tr}(\mathbf{U}_{22}\mathbf{J}_2\mathbf{U}_{22}^{\mathrm{T}}), \end{aligned}$$

or

$$D_{22} = \tfrac{1}{3} m_2 l^2.$$

From (2.106) and (2.107) with $N = 2$, the Coriolis and centrifugal (kinetic-energy-derived) forces are readily obtained with

$$\begin{aligned} h_1 &= -m_2 l^2 (\sin \psi_2)\dot{\psi}_2(\tfrac{1}{2}\dot{\psi}_2 + \dot{\psi}_1), \\ h_2 &= \tfrac{1}{2} m_2 l^2 (\sin \psi_2)\dot{\psi}_1^2. \end{aligned}$$

Finally, the gravity (potential-energy-derived) forces are computed from (2.108), so that

$$\begin{aligned} c_1 &= gl(\tfrac{1}{2} m_1 \cos \psi_1 + \tfrac{1}{2} m_2 \cos(\psi_1 + \psi_2) + m_2 \cos \psi_1), \\ c_2 &= \tfrac{1}{2} m_2 gl \cos(\psi_1 + \psi_2). \end{aligned}$$

Then the Euler–Lagrange equations of motion are found by substitution into (2.104), or

$$D_{11}(\psi_2)\ddot{\psi}_1 + D_{12}(\psi_2)\ddot{\psi}_2 + h_1(\psi_1, \psi_2, \dot{\psi}_1\dot{\psi}_2) + c_1(\psi_1, \psi_2) = \tau_1.$$

and

$$D_{12}(\psi_2)\ddot{\psi}_1 + D_{22}\ddot{\psi}_2 + h_2(\psi_2, \dot{\psi}_1) + c_2(\psi_1, \psi_2) = \tau_2.$$

Again, the role of nonlinear system theory in robotics appears to be significant indeed.

While the equations of motion for robotic manipulators are derived conveniently from the Euler–Lagrange equation as shown here, there are more computationally efficient methods. This has been an active area of research, with numerous assumptions and formulations suggested. The direct application of Newton's laws, d'Alembert's principle, and a Kane formulation generally are more computationally efficient. However, these methods require considerable derivation. Since the purpose here is merely to introduce robotic modeling and control rather than providing complete coverage, the reader with a particular interest in robotics is referred to [F2].

2.2.4 Airframe Dynamics

The rotation motion of a general rigid body was given by (2.78)–(2.80), which represents projection of the angular velocity vector onto the X,Y,Z body axis. For the airframe (aircraft or missile) p is the roll rate about the X axis, q the pitch rate about the Y axis, and r the yaw rate about the Z axis. Then if the X, Y, Z body axes are the principal axes of the airframe, the rotational dynamics (2.78)–(2.80) become

$$\dot{p} = \frac{L}{I_{xx}} - \frac{I_{zz} - I_{yy}}{I_{xx}} qr, \tag{2.109}$$

$$\dot{q} = \frac{M}{I_{yy}} - \frac{I_{xx} - I_{zz}}{I_{yy}} pr, \tag{2.110}$$

$$\dot{r} = \frac{N}{I_{zz}} - \frac{I_{yy} - I_{xx}}{I_{zz}} pq, \tag{2.111}$$

where following aerodynamic convention, L, M, N are the rolling moment (about the X axis), pitching moment (about the Y axis), and yawing moment (about the Z axis), respectively (Figs. 2.10 to 2.12).

To define the airframe translational motion, the velocity vector $\mathbf{v}$ is projected onto body fixed axes so that

$$\mathbf{v} = [u \quad v \quad w]^{\mathrm{T}}. \tag{2.112}$$

The translation dynamics are derived from the airframe linear momentum in an inertial x,y,z frame,

$$\mathbf{p} = m\mathbf{v}_{\mathrm{I}} = m\mathbf{R}_{IB}\mathbf{v},$$

so that

$$\frac{d}{dt}(m\mathbf{R}_{IB}\mathbf{v}) = \mathbf{F}_I,$$

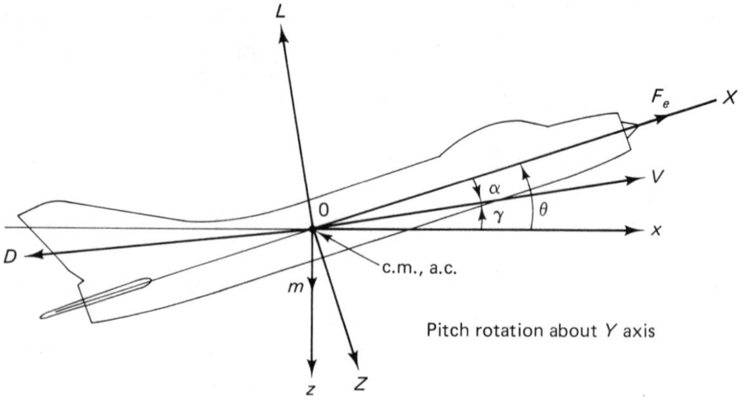

$\theta \equiv$ pitch angle
$\alpha \equiv$ angle of attack
$\gamma \equiv$ climb angle
$L \equiv$ lift force
$F_e \equiv$ thrust
$D \equiv$ drag force
a.c. $\equiv$ aerodynamic center
c.m. $\equiv$ mass center

Figure 2.10 Longitudinal airframe geometry.

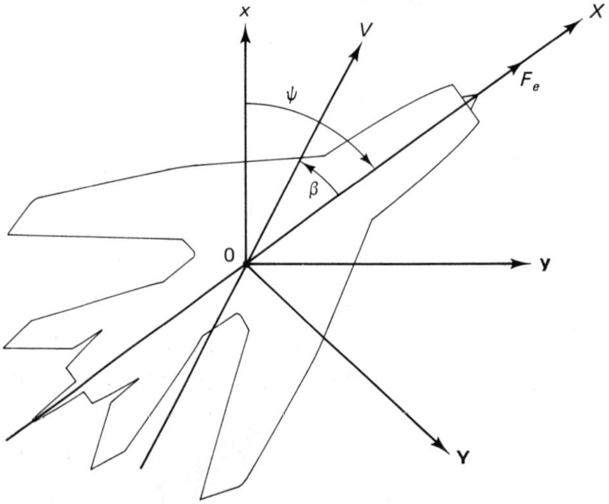

Figure 2.11 Yaw rotation about Z axis.

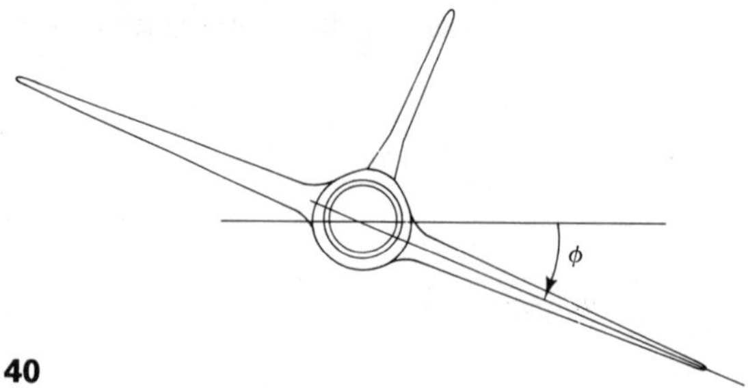

Figure 2.12 Roll rotation about X axis.

where $\mathbf{F}_I$ are the external translation forces referred to inertial axes and $\mathbf{R}_{IB}$ is the transformation from body to inertial axes. Then if m is constant,

$$\mathbf{R}_{IB}\frac{d\mathbf{v}}{dt} + \dot{\mathbf{R}}_{IB}\mathbf{v} = \mathbf{F}_I/m$$

or

$$\frac{d\mathbf{v}}{dt} = -\mathbf{R}_{BI}\dot{\mathbf{R}}_{IB}\mathbf{v} + \mathbf{F}_B/m, \tag{2.113}$$

where $\mathbf{F}_B = \mathbf{R}_{BI}\mathbf{F}_I$ is the external force vector resolved along body axes $\mathbf{R}_{BI} = \mathbf{R}_{IB}^{-1} = \mathbf{R}_{IB}^{\mathrm{T}}$. $\mathbf{R}_{IB}$ involves the product of (2.47), (2.55), and (2.54) from left to right for, first, rotation ψ about the z axis; second, θ about the y axis; and last, ϕ about the x axis. At the beginning, the body axes are aligned with inertial axes, so that the precise order of rotations yields the body axes' orientation relative to the inertial axes. Consequently, (2.61) yields $\mathbf{R}_{IB} = \mathbf{R}_{\phi\theta\psi}$. It should be noted that aeronautical engineers generally use a right-hand coordinate system, with the z axis pointed down instead of up. Then after some algebra,

$$\mathbf{R}_{BI}\dot{\mathbf{R}}_{IB} = \begin{bmatrix} 0 & r & -q \\ -r & 0 & p \\ q & -p & 0 \end{bmatrix}$$

or

$$\dot{u} = rv - qw + F_{xB}/m, \tag{2.114}$$

$$\dot{v} = -ru + pw + F_{yB}/m, \tag{2.115}$$

$$\dot{w} = qu - pv + F_{zB}/m, \tag{2.116}$$

where F_{xB}, F_{yB}, F_{zB} are the total of engine, aerodynamic, and gravity forces acting on the body and resolved in body axes. The engine thrust is usually assumed to act along the body X axis. The aerodynamic force acts through the center of pressure (aerodynamic center) and the gravity force along the z inertial axis. The products ru, and so on, are Coriolis accelerations due to body-axis rotation. In summary, the airframe six-degree-of-freedom dynamics is expressed by (2.109)–(2.111) and (2.114)–(2.116). These can be expressed in terms of 12 first-order state differential equations.

The aerodynamic forces and moments depend on the dynamic pressure

$$Q = \tfrac{1}{2}\rho V^2,$$

where ρ is air density (primarily a function of altitude, h) and V is the aircraft speed,

$$V = (u^2 + v^2 + w^2)^{1/2}.$$

Then the aerodynamic forces and moments may be related as follows:

$$F_{xA} = QAC_x, \qquad F_{yA} = QAC_y, \qquad F_{zA} = QAC_z,$$

$$L = l_L QAC_L, \qquad M = l_M QAC_M, \qquad N = l_N QAC_N,$$

where C_x, and so on, are dimensionless aerodynamic coefficients; A is an airframe aerodynamic reference area; and l_L, l_M, l_N are reference aerodynamic moment arms.

While these are convenient for model formulations, it is usually more convenient to get the coefficients in terms of coefficients of lift and drag so that

$$F_D = QAC_D, \qquad F_L = QAC_L,$$

where F_D and F_L are drag and lift forces in terms of the drag and lift coefficients, C_D and C_L, respectively. F_D acts along the aircraft velocity vector and through the aerodynamic center (a.c.). F_L acts orthogonal to the velocity vector and through the a.c. Then the forces may be resolved appropriately into the desired body or other convenient axis systems, and the appropriate moment arms computed. The aerodynamic coefficients in any form (C_L, C_D, C_X, etc.) are primarily functions of aircraft Mach number (speed/acoustic velocity), angle of attack (α), sideslip angle (β), altitude, and to some extent, control surface angles. The angle of attack (α) and sideslip angle (β) are given by

$$\alpha = \tan^{-1}\frac{w}{(u^2 + v^2)^{1/2}} \approx \frac{w}{u},$$

$$\beta = \tan^{-1}\frac{v}{u} \approx \frac{v}{u}.$$

Typically, aircraft pitch motion is controlled by the elevators (with angle δ_e), the roll by ailerons (δ_a), and the yaw by the rudder (δ_r).

Figure 2.10 shows a typical longitudinal airframe configuration of forces. Here it is assumed that the center of mass (c.m.) and the aerodynamic center (a.c.) are at the same point. Usually, the c.m. is a little in front of the a.c. for static (or trim) stability. In designing longitudinal controllers, such as pitch control or airspeed control, traditionally it is assumed that the lateral mode is in its trimstate with no change in yaw or roll and no sideslip. Then the longitudinal dynamics are nearly decoupled from the other modes. This motion is studied by Example 3.6 together with linearized equations for small perturbations from the trimstate.

It is common practice to derive such perturbation transfer functions. Two second-order modes with complex poles characterize these transfer functions with damping coefficients and natural frequencies as functions of trim Mach number (M), altitude (h), angle of attack (α), and others to a lesser extent. The so-called short-period mode appears predominantly in pitch (θ) motion, while the phugoid (long-period mode) predominates in airspeed changes. Typically, the phugoid is lowly damped and the short period a bit better damped. Then it is the task of typical aircraft or missile controllers to provide improved response in returning the object to desired trim after disturbance. Although linear control may be sufficient for small perturbations, nonlinear analysis and/or simulation is necessary, and nonlinear control may improve general trajectory control.

2.3 NUCLEAR AND THERMAL PROCESSES

An electric power generation station may consist of a nuclear reactor (eventually, a fast-fission breeder type), a direct heat exchanger that transfers fission heat to the coolant of a primary cycle, and an indirect heat exhanger to transfer coolant heat to cycling water. The latter, in turn, may drive a steam turbine. For space stations, nuclear power sources are ideal because of their long lifetime and high power density. Indeed, space stations must be reliable and lightweight. Consequently, nuclear power stations may use thermal-electric energy converters such as thermionic units, thermoelectric devices, fuel cells, or magnetohydrodynamic (MHD) systems. Any optimum allocation of these converters to a plant would depend on overall system requirements. Magnetohydrodynamic systems and fuel cells may be quite efficient, but they are also generally heavy. This usually leads to a compromised design according to overall specifications. An ultrahigh-heat-transport device called a heat pipe seems to be most amenable to electric power generation in space. Nuclear-propelled rockets, with their large specific impulses (thrust per unit of coolant mass flow rate), will be very efficient for interplanetary space missions [M4]. For example, with present technology nuclear rockets can obtain specific impulses of about 1000 sec, compared to about 400 sec at the most for chemical systems.

The fission process, carefully regulated in the form of a nuclear chain reactor, produces heat as a useful output. Heat exchangers, which are required to cool the reactor and to provide the thermal energy to any required converter, may be quite complex in design. The basic reactor may have a solid, liquid, or gaseous core with direct or indirect heat transfer arranged by convection, conduction, and/or radiation. Probably the most common design involves a solid core with direct transfer of heat to a moving fluid. Within the reactor itself, fissions of the fuel nucleus take place mainly in the core. The core normally is surrounded by two concentrically surrounding shells which form a neutron reflector and a radiation-absorbing shield. The fission process usually is made more efficient by adding a moderator to the core so that the energy of the fission neutrons is decreased by inelastic scatterings. This in turn increases the probability of neutrons fissioning fuel nuclei. For some applications however, the added weight and design complexity due to the moderation are not justified and the configuration is called a fast reactor. Fertile material such as ^{238}U can be added to certain fast-fission reactors to breed more fuel, yet augment reactor power by as much as 25 percent. Such reactors, which produce more fissionable material than they deplete, are termed breeder systems. Unfortunately, breeder reactors are quite expensive and require advanced material technology.

2.3.1 Nuclear Fission

When a fertile nucleus undergoes fission, an average of two or three neutrons are emitted together with nuclear radiation and a relatively large amount of energy. The energy causes rapid motion of fision fragments, which produces heat. A rigorous derivation of the neutronic state equations based on the time-dependent monoenergetic transport equation is given in [K2, pp. 161–166]. The heuristic derivation given here follows that of [M4, pp. 3–6]. The net change in neutron population over one generation by neutron conservation is

$$\frac{dn}{dt} = (k - 1)\frac{n}{l} \tag{2.117}$$

where k, the average number of first-generation offspring per neutron death, is called the multiplication constant; l is the mean prompt neutron generation time; and all neutrons are produced in t time $t \ll l$. Here l may be a millisecond for thermal reactors, or might be a microsecond for fast reactors. This model assumes that all neutrons are produced promptly. It is common knowledge, however, that a small portion of neutrons are derived from unstable fission products called precursors. Six distinct groups of delayed neutrons have been observed, and their characteristics are observed in Table 2.1 for ^{235}U fuel.

TABLE 2.1 ^{235}U DELAYED NEUTRON PARAMETERS FOR FAST FISSION

Group i	Half-life (sec)	Decay constant λ_i (sec^{-1})	Portion* $\beta i/\beta$
1	54.51	0.0127	0.038
2	21.84	0.0317	0.213
3	6.00	0.115	0.188
4	2.23	0.311	0.407
5	0.496	1.40	0.128
6	0.179	3.87	0.026

Source: After Keepin [K2].

$*\beta = \sum_{i=1}^{6} \beta_i \approx 0.0065.$

Equation (2.117) for prompt neutrons may be modified to account for delayed neutrons merely by subtracting $k\beta n/l$, and for a neutron source rate s, by adding s. Then the additional delayed neutrons emitted by the six precursors cause the rate of neutron change to be

$$\frac{dn}{dt} = \frac{k(1 - \beta) - 1}{l} n + \sum_{i=1}^{6} \lambda_i c_i + s, \tag{2.118}$$

where λ_i is the decay constant for the ith group of precursors and c_i is the population of the ith precursor group. It is assumed that delayed neutrons have the same effect on the process as that of prompt fission neutrons. Here source s is usually a relatively small constant.

The rate of precursor population change equals birth rate minus death rate:

$$\dot{c}_i = \frac{dc_i}{dt} = \frac{k\beta_i}{l} n - \lambda_i c_i, \qquad i = 1, \cdots, 6, \tag{2.119}$$

where β_i is the portion of neutrons generated from the ith precursor. The total neutron population is at constant value n_1 if

$$k = \left(l \sum_{i=1}^{6} \frac{\dot{c}_i}{n_1} \right) + 1,$$

and $s = 0$. At this "delayed critical condition," the neutron kinetics, (2.118) and (2.119), are in an equilibrium state, $n(t) = n_1$ and $c_i(t) = c_{1i}(i = 1, \ldots, 6)$, if $k = 1$ and $c_{1i} = (\beta/l\lambda_i)n_1$.

Ordinarily, around the design level the system is operated near delayed critical with k approximately 1 and s negligible. Then the kinetics are approximated by

$$\frac{dn}{dt} = \frac{\delta k - \beta}{l} n + \sum_{i=1}^{6} \lambda_i c_i \tag{2.120}$$

and

$$\frac{dc_i}{dt} = \frac{\beta_i}{l} n - \lambda_i c_i \qquad i = 1, \cdots, 6, \tag{2.121}$$

where the reactivity δk is $k-1$. Frequently, the neutron kinetics are accurately approximated by a single-precursor model of the form

$$\frac{dn}{dt} = \frac{\delta k - \beta}{l} n + \lambda c \tag{2.122}$$

and

$$\frac{dc}{dt} = \frac{\beta}{l} n - \lambda c \tag{2.123}$$

where λ is an average decay constant for an average precursor of population c.

Control of this fission process is affected by means of reactivity δk or multiplication k; either may be utilized as a control variable, depending on which equation is used. Again, the system model is nonlinear and superposition cannot be invoked generally. With $u = k$ or $u = \delta k$, the neutron fission process is a bilinear system, as defined by (1.10).

2.3.2 Heat Transfer

Heat exchanges are naturally distributed processes described by partial differential equations. In practice, particularly for control purposes, however, it is common to approximate the system to some desired accuracy by finite difference models or "compartmented" models. Furthermore, solid reactor cores with direct heat transfer to a moving fluid such as those considered here are carefully designed so that radial flux and temperature distributions are quite uniform. Consequently, an average radial element is utilized to approximate the process. A qualitative description of the heat transfer is provided by a single axial lump as well.

Assume that the nuclear reactor with multiple coolant flow passages is modeled by a single average cylinder. If axial heat transfer is neglected, due to small temperature gradients in the direction of coolant velocity, a thermal energy balance yields: Fission heat q (proportional to n) equals heat stored in fuel and moderator plus heat transferred to coolant. In mathematical terms,

$$q = c_m \frac{dT_u}{dt} + hA(T_u - T_g), \tag{2.124}$$

where T_u is the average core temperature, T_g the average coolant temperature, c_m the effective heat capacity of core mass, h an effective heat-transfer coefficient, and A an effective heat-transfer area. An energy balance on the coolant yields

$$hA(T_u - T_g) = c_{mg}\frac{dT_g}{dt} + c_g w(T_e - T_i), \tag{2.125}$$

where w is the coolant mass flow rate, T_e the coolant exit temperature, T_i the coolant entrance temperature, c_{mg} the coolant mass heat capacity, and c_g the coolant specific heat. The average coolant temperature is a weighted average of inlet and outlet temperatures,

$$T_g = \frac{T_i + \theta T_e}{1 + \theta}, \tag{2.126}$$

where θ is a weighting constant (sometimes equal to 1 or 2).

Generated from convection, conduction, and radiation, heat-transfer coefficients are generally nonlinear functions of appropriate temperatures. Of course, the effective heat-transfer areas are geometrically dependent. For a solid reactor core with a fluid coolant, radiation is quite small and conduction impedance is frequently neglected compared to convective impedance. If conduction impedance is considered, usually it is assumed to be constant dependent on thermal conduction of the material and effective thickness. The convective coefficient is strongly dependent on coolant flow rate. If the flow rate is turbulent and the conduction term is neglected, the coefficient is approximated by

$$h = kw^{0.8}, \tag{2.127}$$

where k is a geometrically dependent constant. For a nuclear rocket test reactor it was found that $k \approx c_n/n^{0.8}d^{1.8}$, where n is the number of coolant passages (cylindrical) and d is the average passage diameter; c_n was approximately $6.2(10)^3$ Btu/(in. sec)$^{0.2}$ °R lb$^{0.8}$ in this case [M4]. The remaining coefficients (c_m, c_g, c_{mg}) usually have a weak temperature dependence over a broad range of high temperatures. Therefore, if (2.127) is approximated by a linear function, $h = kw$, the process is bilinear, with w and q as input variables.

The model frequently may be further simplified by neglecting coolant mass heat capacity and coolant temperature at the core inlet. Then (2.124) and (2.125) may be approximated by

$$\frac{dT_u}{dt} = \frac{q}{c_m} - awT_u \tag{2.128}$$

where a is a heat-transfer constant. Again, the natural model is a bilinear system of form (1.10). In all cases, the system is linear and time-variant if the controls, in this case power, $q(t)$, and coolant flow rate, $w(t)$, are specified functions of time. In the next section it will be seen that coupling the heat exchanger to the fission process, as for a nuclear reactor, results in an overall nonlinear (and nonbilinear) system which has a sometimes-convenient coupled bilinear structure. Of course, it should be noted that similar heat-exchange processes are a significant part of many engineering designs.

Solar plants and heat-pump-supplemented heating, ventilating, and air-conditioning (HVAC) systems exemplify well where nonlinear dynamic analysis and control can yield big payoffs (see [R1] and [M2]).

2.3.3 Reactor Control

The neutronic and heat-exchange state equations (2.122), (2.123), and (2.128) are coupled together in a physical reactor by coefficients of reactivity. The most common reactivity coefficient couples core temperature T_u and neutron level n or power q by means of a temperature reactivity ([A1, pp. 79–86] gives a detailed analysis of temperature reactivity), which is commonly approximated by

$$\delta k_t = \alpha_t T_u, \tag{2.129}$$

where α_t usually may be assumed constant within the accuracy of which it may be estimated. The temperature coefficient is usually negative as a consequence of thermal expansion, which reduces core density, and thereby neutron moderation and reactivity. Also, core temperature affects neutron capture cross sections, fission cross sections, absorption cross sections, and Doppler broadening associated with the resonance region of the neutron energy spectrum. For certain fast reactors, the Doppler coefficient is positive and so large that the overall temperature coefficient can be positive. Fortunately, this is not usually the case. For KIWI A prime, one of the first nuclear rocket test reactors, $\alpha_t = -1.52(10)^5/°R$.

A negative coefficient of reactivity almost always has a stabilizing effect on the reactor dynamics. Even for a linear heat-exchange model (h = constant), however, the multiplicative term in the neutron kinetics, (2.120) and (2.121), causes the reactor to be nonlinear and indeed nonbilinear. In addition to temperature-reactivity feedback, some reactors use a coolant that is also a neutron moderator. For example, liquid-hydrogen propellant generally used with nuclear rockets is a very good moderator. Even as a dense gas at the reactor core entrance, it has a strong moderating effect on intermediate energy (thermal) neutrons and on fast neutrons. Such coolant reactivity is approximated by

$$\delta k_p = k_p \rho_g V, \tag{2.130}$$

where k_p is a positive constant, V the coolant volume in the core, and ρ_g the coolant density. While this reactivity feedback is positive, it is apparent that again the temperature effect is negative according to the gas equation,

$$\rho_g = \frac{p}{zRT_g}, \tag{2.131}$$

where p is the average coolant pressure in the core, R the universal gas constant, and z the compressibility factor (usually of value 1). The coolant density reactivity also couples the neutron kinetics to the coolant feed system. The total reactivity δk, which is the input to the neutron kinetics, (2.120) and (2.121), is a summation of reactivity, described by

$$\delta k = \delta k_t + \delta k_p + \delta k_c, \tag{2.132}$$

where δk_c is the reactivity generated usually by neutron absorbers or reflectors in the form of control rods or control vanes. Reactor control will be described by examples in later chapters and in Vol. 2.

2.4 BODY-TEMPERATURE REGULATION

The discussion here follows [M1] and [M5]. Briefly, heat produced by biochemical reactions of cellular metabolism is transported by circulation of blood to the skin. There some of it is transferred to the environment normally by convection, by radiation, and by energy of vaporization of sweat. The parametric control in this case is by vasomotor control on the cutaneous arterioles and by sweat glands. As for water balance, which is closely regulated, feedback control is accomplished through the hypothalamus. Although cellular metabolism is increased by secretion of thyroid hormone in order to add heat to a body placed in a cold environment, the major source of control heat is due to shivering and increased general activity of skeletal muscle. Feedback signals to the hypothalamus are obtained mainly from deep central and skin receptors [H2]. Other receptors are strategically located at various points, such as in the respiratory tract.

Thermoregulatory models have been derived by conveniently lumping or discretizing the naturally distributed process in several different ways. For example, an energy balance may be performed on each lumped element. Anatomically, it is convenient and has been found to be surprisingly accurate for many studies to divide the body into three lumps: core, skeletal muscle, and skin. Again, it is assumed that the temperature and the properties of a given compartment are the same throughout that compartment. The core includes such internal organs as the lungs, heart, kidney, liver, and intestines.

As in Section 2.3, a heat balance on each compartment is used to derive a dynamical model. It is assumed that geometrically the system consists of three concentric cylinders with axial conduction negligible compared to radial conduction. The latter assumption is a common one for systems of this sort. The model is formulated so that surface-to-volume ratios of each compartment conform with those of human beings themselves. As for the body, the outer compartment corresponds to skin and the inner compartment to core.

Vasomotor activity is used to vary blood-flow patterns through core and skin, thus coupling skin to core. The blood has negligible mass heat capacity, and its heat transfer usually takes place in the form of convection, For simple modeling purposes, however, it may be assumed that this effect may be represented by variable tissue conductivity, Such representation is actually quite accurate since the blood circulation time within the body is very short (on the order of 1 minute) compared to the temperature response (on the order of several hours), and since the rate of heat transfer is nearly proportional to the temperature difference between core and skin. Consequently, vasomotor control is recognized as a core–skin conductivity term.

A heat balance on the core yields

$$c_{mc} \frac{dT_c}{dt} = q_b - q_{re} - q_{cm} - q_{cs}, \tag{2.133}$$

where T_c is the core temperature, q_b the heat generated from basal metabolism, q_{re} the heat loss due to respiration and excretion, q_{cm} the heat transferred from the core to muscle, q_{cs} the effective heat transferred from core to skin, and c_{mc} the core mass heat capacity. Heat transferred from core to the muscle and from the core to the skin is described by

$$q_{cm} = \frac{kA_{cm}}{l_{cm}}(T_c - T_m) \tag{2.134}$$

and

$$q_{cs} = \frac{k_v A_{cs}}{l_{cs}}(T_c - T_s), \tag{2.135}$$

where k is effective conductivity, l_{cm} and l_{cs} the effective heat transfer lengths, A_{cm} and A_{cs} the effective heat-transfer areas for appropriate compartments, $k_v(t)$ the equivalent vasomotor conductivity term (which is a control variable), T_c the average core temperature, and T_m the average muscle temperature. While (2.134) and (2.135) couple the core with other compartments, basal metabolism and respiration–excretion allow for some additive control, and vasomotor conductivity provides bilinear control.

A heat balance on muscles yields

$$c_{mm}\frac{dT_m}{dt} = q_s + q_{ex} + q_{cm} - q_{ms}, \tag{2.136}$$

where T_m is the average muscle temperature, c_{mm} the muscle core heat capacity, q_s the muscle shivering metabolism, q_{ex} the heat from muscle exercise, and q_{ms} the heat transferred to muscle from skin according to

$$q_{ms} = \frac{kA_{ms}}{l_{ms}}(T_m - T_s). \tag{2.137}$$

Here A_{ms} and l_{ms} are effective heat-transfer areas and lengths, respectively.

Similarly, a heat balance taken at the skin shows that

$$c_{ms}\frac{dT_s}{dt} = q_{cs} + q_{ms} - (q_c + q_e + q_r), \tag{2.138}$$

where T_s is the average skin temperature; c_{ms} the skin mass heat capacity; and q_c, q_e, and q_r the heat-loss terms due to convection, evaporation, and radiation, respectively. Again the system is described by a set of compartmental bilinear differential equations (2.133)–(2.138) with multiplicative and additive control terms. Compartmental temperatures T_c, T_m, and T_s are appropriate state variables for this thermoregulatory plant.

Numerous attempts have been made to model the feedback control equations which again make the closed-loop system more highly nonlinear or nonbilinear. Evaporation rate q_e may be approximated by [M5]

$$q_e = \begin{cases} k_1(T_c - T_c^*), & T_c > T_c^*, \\ q_{e0}, & T_c < T_c^*, \end{cases}$$

where T_c^* is the desired core temperature (normal body temperature), q_{e0} the basal evaporation rate (almost zero), and k_1 is a positive constant. This model assumes that the corresponding response time is negligible compared to core temperature response. This equilibrium assumption is usually valid, but a more accurate function of system temperature is proposed in [C2] and has the form

$$q_e = \begin{cases} q_{e0} + k_4[k_2(T_b - T_b^*) + k_3(T_b - T_b^*)^4], & T_b > T_b^*, \\ q_{e0}, & T_b < T_b^*, \end{cases} \tag{2.139}$$

where k_2 and k_3 are appropriate empirical constants, k_4 is an exercise constant, T_b^* is a desired weighted body temperature, and

$$T_b = \frac{l_sT_s + l_mT_m + l_cT_c}{l_s + l_m + l_c},$$

where l_s, l_m, and l_c are effective thicknesses of the appropriate compartments.

Muscle shivering heat q_s and basal metabolic heat q_b are also nonlinear functions of core temperature and skin temperature, which may be approximated by linear equations; q_b is nearly constant, and q_s is approximated by [C2]

$$q_s = \begin{cases} k_5(T_b^* - T_b), & T_b < T_b^*, \\ 0, & T_b > T_b^*, \end{cases} \tag{2.140}$$

where k_5 is a positive constant. There seems to be some anticipation in the vasomotor control mechanism, which suggests a proportional plus derivative feedback controller. Such controllers for a slab model have been derived from physiological data.

Of course, humans may readily add heat, q_{ex}, by exercising and by varying the quantity and color of their clothing. The latter directly affects the q_c and q_r terms in (2.138). These terms are also nonlinear functions of skin temperature and environmental conditions. The dynamics of this system is studied in more detail in Section 5.3.2. Other nonlinear (including bilinear) biomedical processes are studied in Vol. 2. This includes carbon dioxide and cardiovascular-flow regulation ([G3], [G4], and [M1]), along with compartmental dynamical analysis ([H3], [H4], [L2], [M6], and [M7]). Mathematical models in immunology, cancer, and AIDS are also analyzed in Vol. 2 ([E1], [D1], [B2], [C3], [H5], [M6]–[M10], [P2], and [R2]).

2.5 MISCELLANEOUS PROCESSES AND COMMENTS

Only a small representative sample of physical systems and their mathematical models have been studied here. A complete treatment of this topic is beyond the scope of this book. Still it is hoped that at this point, the reader will have a better appreciation of this most significant and most complicated problem. Also, although linear systems are convenient for analysis, they are not all that common in the real world. Only for academic problems in general can one expect to have plant transfer functions or plant state and control coefficient matrices provided. Chemical process control, electric power networks, ecology and socioeconomics are the most significant areas that have not been studied here due to space limitations. However, the same fundamental concepts of conservation, the Euler–Lagrange equation, and others are still useful. Some of these processes will be introduced along with examples throughout the book.

Compartmental models can be useful in chemical process control and socioeconomic regional modeling as well as in biological systems, which were noted above. The same structure may arise with tracer experiments, including appropriate polls as well as chemical or isotopic tracers. These topics and immunology are studied in more detail in Vol. 2. Bilinear models arise in agriculture, predator–prey population dynamics, and socioeconomics, as studied in [M1]. For an industrial plant analysis, for example, bilinear control may take the form of allocation of different portions of profit to new equipment, expanded work force, new research, salary increases, or stock dividends. In the government sector, tax rates on earnings may be multiplicative controls to stimulate activity of a particular sector or compartment. Even if the control engineer were to be faced with a linear plant, nonlinear control might still hold the possibility of improved performance. For examples see [M2] and [R3]. Such aspects are investigated throughout this book and in Vol. 2.

EXERCISES

2.1. **(a)** Derive the nonlinear differential equations of motion for m_1 and m_2 in Fig. 2.13.
(b) Draw the lossless electrical network equivalence with inductances analogous to m_1, m_2 and capacitance analogous to $k_1(\cdot)$, $k_2(\cdot)$.
(c) Derive the network equations for capacitor charges analogous to x_1, x_3.

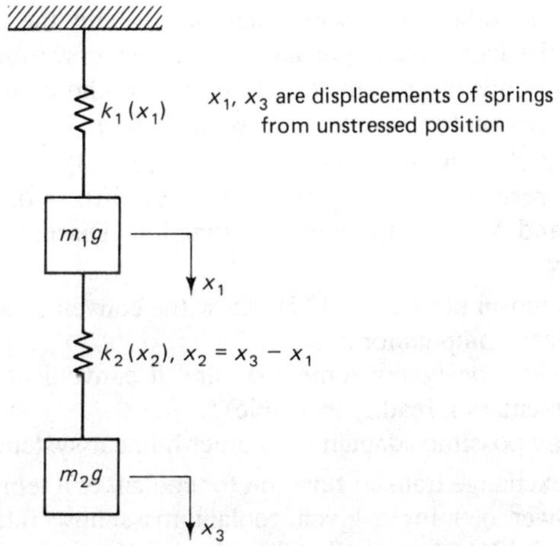

Figure 2.13 Rigid-body configuration for Exercise 2.1.

2.2. Consider Fig. 2.14 with a bead sliding on a wire that rotates about the y axis with angular speed ω. The wire has parabolic shape with $y = x^2/2x_0$. Here x_0 is the initial position of the bead. Compute the dynamic equations of motion for the bead of point mass m by **(a)**

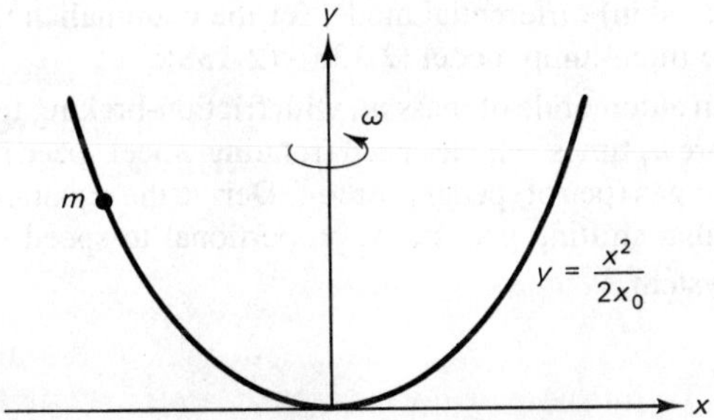

Figure 2.14 Rotating bead for Exercise 2.2.

a force summation and (**b**) the Euler–Lagrange equation. (**c**) Express your answer in phase-variable form. (**d**) How many degrees of freedom does this system have?

2.3. For the two-link arm shown in Fig. 2.9 ($l_1 \neq l_2$), calculate the link angles ψ_1, ψ_2 according to specified coordinates p_x, p_y for work point p. This is the classic robot inverse problem. Specify inaccessible regions for p (i.e., coordinates outside the workspace of the robot.) Are the accessible coordinates generally realized by a unique ψ_1, ψ_2?

2.4. Show that by appropriate measurements of link angles (e.g., Fig. 2.9) the nonlinear robot dynamics have a conditionally linear structure as defined by (1.11). Is there any reason to expect that such a structure may be computationally convenient?

2.5. Consider the PUMA robot arm of Fig. 2.7, where the lengths of the links are a_0, a_1, a_2, a_3, and $a_4 = 0$.

(a) Compute, in D-H representation, matrix ${}^0A_4$, relating the set of coordinates at the basis of link 0 to link 4. Link 0 rotates about the z_0 axis and link 4 rotates about the z_4 axis.

(b) Solve the inverse robot problem; that is, find for a point p the corresponding link angles $\psi_0 \cdots \psi_4$. For $p = (x_0, y_0, 0)$ with x_0, y_0 on the circle of radius r_1.

2.6. Consider the first four joints of the PUMA robot arm in Fig. 2.7, where the lengths of the links are a_0, a_1, a_2, a_3. Link masses are neglected for links 0 and 1 and are equal to m for links 2 and 3 such that the inertia matrix of links 2 and 3 is similar to J_i in Example 2.4. Compute the equations of motion for this four-link robot, that is, compute ${}^0A_3$ and matrices U_{ip}, D_{ik} to express the Euler–Lagrange equation (2.104).

2.7. For a one-link telescopic robot arm, with rotation ψ about the z axis and length $a \in [\alpha_1, \alpha_2]$, compute the accessible region and solve the inverse robot problem; that is, for a work point p, compute the parameters ψ, a. Is the choice unique? Derive the inertialess equations of motion for controls $w = \dot{\psi}$ and $v = \dot{a}$.

2.8. For the second-order nuclear fission process (2.122)–(2.123), compute the Laplace transform $n(s)$ and time response $n(t)$ for given $n(0) = n_0$, $\dot{n}(0) = 0$, and $\delta k = \gamma\delta_{-1}(t)$, where γ is a constant and $\delta_{-1}(t)$ is the unit-step function. Discuss the relevance of γ relative to system safety.

2.9. **(a)** For the "prompt" neutron kinetics (2.117), show the convenience of replacing $n(t)$ by $\ln n(t)$ for response computation.

(b) Why do reactor control designers sometimes find it convenient to control $\ln n(t)$ (with $\ln n(t)$ instrumentation readily available)?

(c) Does this lead to any possible adaptation to other bilinear systems?

2.10. **(a)** Compute the heat-exchange transfer function for coolant exit temperature $T_e(t)$ with respect to heat power $q(t)$ for a given coolant mass flow rate w according to equations (2.124)–(2.126). Assume that $T_i = 0$.

(b) Compute an expression for the heat-exchange time constant τ_h in terms of appropriate system parameters in (2.124)–(2.126). Assume that coolant mass heat capacity c_{mg} is zero (i.e., negligible relative to c_m).

2.11. Derive a two-lump (core, skin) differential model for the mammalian thermal regulator as a simplification of the three-lump model (2.133)–(2.138).

2.12. Consider the forces on an automobile of mass m, with friction-braking force proportional to applied control pressure u_1 times vehicle speed (rotating wheel speed) and acceleration control u_2 proportional to gas (petrol) pedal position. Derive the equations of automobile motion. Further, show that shifting gear ratios proportional to speed result in a piecewise-constant bilinear system.

CHAPTER

3

PERTURBATION THEORY

More likely than not, the system that an engineer or scientist must design, analyze, and/or control is nonlinear. That is, the dynamical process is approximated by some appropriate set of nonlinear differential equations such as those derived for specific processes in Chapter 2. Unfortunately, relatively few methods are available for the study of nonlinear processes. Certainly, one of the most common assumptions in engineering practice is the assumption that signal variations about some reference are so small that the process is linear. Therefore, the process is analyzed conveniently by linear system theory, for which there exists a wealth of methodologies. It is obvious that such an assumption, although useful under appropriate conditions, does have its limitations. However, let us study this topic in more detail.

3.1 PERTURBATION DYNAMICS

3.1.1 Reference Trajectories and Equilibria

Consider Fig. 3.1, where an object moves along some trajectory $\mathbf{x}^*(t)$ which is a particular solution to the following dynamical equation of motion for the object:

$$\frac{d\mathbf{x}}{dt} = \mathbf{f}(\mathbf{x}, \mathbf{u}), \tag{3.1}$$

where the state vector $\mathbf{x} \in \mathbb{R}^n$, the control vector $\mathbf{u} \in \mathbb{R}^m$ and $\mathbf{f}(\cdot)$ is differentiable with respect to both arguments. Actually, $\mathbf{f}(\cdot)$ need only be differentiable in some neighborhood of $\mathbf{x}^*(t)$ for our purposes. Hence

$$\frac{d\mathbf{x}^*}{dt} = \mathbf{f}(\mathbf{x}^*, \mathbf{u}^*), \tag{3.2}$$

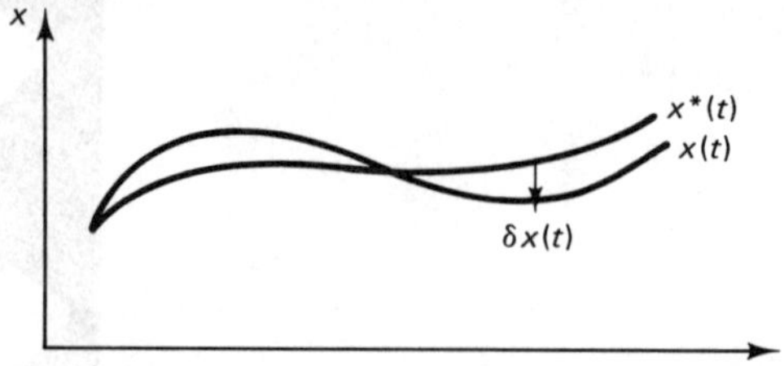

Figure 3.1 Trajectory variation.

describes the reference trajectory $\mathbf{x}^*(t)$. Since the vector function $\mathbf{f}(\cdot)$ is differentiable, $\mathbf{f}(\cdot)$ satisfies the Lipschitz condition, and $\mathbf{x}^*(t)$ and more generally $\mathbf{x}(t)$ are unique solutions to (3.1) and (3.2) which correspond to given initial states $\mathbf{x}(t_0)$, $\mathbf{x}^*(t_0)$ and specified controls $\mathbf{u}(t)$, $\mathbf{u}^*(t)$ (see Appendix B and [C1]). As examples, $\mathbf{x}^*(t)$ may represent a nominal path of robotic links for a particular task, or an altitude- and Mach-hold trajectory for an aircraft, or the healthy state in the immune system or the design state of a nuclear reactor. Then consider a small trajectory change $\delta\mathbf{x}(t)$ that results from a small input change $\delta\mathbf{u}(t)$. Of course, more generally, $\delta\mathbf{u}(t)$ might include appropriately modeled system disturbances.

Now, as shown in Fig. 3.1, the general trajectory $\mathbf{x}(t)$ caused by $\mathbf{u}(t)$ is comprised of the reference and perturbation portions, so that

$$\mathbf{x}(t) = \mathbf{x}^*(t) + \delta\mathbf{x}(t) \tag{3.3}$$

and

$$\mathbf{u}(t) = \mathbf{u}^*(t) + \delta\mathbf{u}(t). \tag{3.4}$$

Then, subtraction of (3.2) from (3.1) yields the perturbation differential equation

$$\delta\dot{\mathbf{x}} = \mathbf{f}(\mathbf{x}, \mathbf{u}) - \mathbf{f}(\mathbf{x}^*, \mathbf{u}^*). \tag{3.5}$$

If $\mathbf{f}(\cdot,\cdot)$ is analytic, it can be expanded in an infinite Taylor series in terms of both arguments. For our purposes let

$$\mathbf{f}(\mathbf{x}, \mathbf{u}) = \mathbf{f}(\mathbf{x}^*, \mathbf{u}^*) + \frac{\partial \mathbf{f}}{\partial \mathbf{x}}(\mathbf{x}^*, \mathbf{u}^*)\, \delta\mathbf{x} + \frac{\partial \mathbf{f}}{\partial \mathbf{u}}(\mathbf{x}^*, \mathbf{u}^*)\, \delta\mathbf{u} + \mathbf{r}(\mathbf{x}, \mathbf{u}), \tag{3.6}$$

where $\mathbf{r}(\mathbf{x}, \mathbf{u}) = \mathbf{r}(\mathbf{x}^* + \delta\mathbf{x}, \mathbf{u}^* + \delta\mathbf{u}) = \mathbf{r}(\delta\mathbf{x}, \delta\mathbf{u})$ represents a remainder term beyond the first-order approximation. Next, assume that as $\delta\mathbf{x}$, $\delta\mathbf{u}$ contract to the origin, $\mathbf{r}(\delta\mathbf{x}, \delta\mathbf{u})$ approaches $\mathbf{0}$ faster than the linear terms in $\delta\mathbf{x}$, $\delta\mathbf{u}$ for (3.6). That is,

$$\lim_{\substack{\|\delta\mathbf{x}\| \to 0 \\ \|\delta\mathbf{u}\| \to 0}} \frac{\mathbf{r}(\delta\mathbf{x}, \delta\mathbf{u})}{\|\delta\mathbf{x}\|} = \lim_{\substack{\|\delta\mathbf{x}\| \to 0 \\ \|\delta\mathbf{u}\| \to 0}} \frac{\mathbf{r}(\delta\mathbf{x}, \delta\mathbf{u})}{\|\delta\mathbf{u}\|} = \mathbf{0}.$$

Then, in sufficiently small neighborhoods of $\mathbf{x}^*(t)$ and $\mathbf{u}^*(t)$, the perturbation equation or locally linearized equation approximates the nonlinear system. Consequently,

$$\delta\dot{\mathbf{x}} = \frac{\partial \mathbf{f}}{\partial \mathbf{x}}(\mathbf{x}^*, \mathbf{u}^*)\, \delta\mathbf{x} + \frac{\partial \mathbf{f}}{\partial \mathbf{u}}(\mathbf{x}^*, \mathbf{u}^*)\, \delta\mathbf{u}. \tag{3.7}$$

If $\mathbf{x}^* = \mathbf{x}^*(t)$ and $\mathbf{u}^* = \mathbf{u}^*(t)$ are appropriate time-variant reference vector trajectories, (3.7) is a linear time-variant system so that

$$\delta\dot{\mathbf{x}} = \mathbf{A}(t)\,\delta\mathbf{x} + \mathbf{B}(t)\,\delta\mathbf{u}, \tag{3.8}$$

where

$$\mathbf{A}(t) = \frac{\partial \mathbf{f}}{\partial \mathbf{x}}(\mathbf{x}^*(t), \mathbf{u}^*(t)), \tag{3.9}$$

$$\mathbf{B}(t) = \frac{\partial \mathbf{f}}{\partial \mathbf{u}}(\mathbf{x}^*(t), \mathbf{u}^*(t)). \tag{3.10}$$

Furthermore, if $\mathbf{x}^* = \mathbf{x}_e$ is an equilibrium state, which corresponds to an equilibrium $\mathbf{u}_e$ such that

$$\dot{\mathbf{x}} = \mathbf{0} = \mathbf{f}(\mathbf{x}_e, \mathbf{u}_e), \tag{3.11}$$

then the linearized system is time invariant. Therefore, (3.8) becomes

$$\delta\dot{\mathbf{x}} = \mathbf{A}\,\delta\mathbf{x} + \mathbf{B}\,\delta\mathbf{u}, \tag{3.12}$$

where

$$\mathbf{A} = \frac{\partial \mathbf{f}}{\partial \mathbf{x}}(\mathbf{x}_e, \mathbf{u}_e), \tag{3.13}$$

$$\mathbf{B} = \frac{\partial \mathbf{f}}{\partial \mathbf{u}}(\mathbf{x}_e, \mathbf{u}_e). \tag{3.14}$$

Example 3.1

Recall the nonlinear (bilinear) nuclear fission process (2.122)–(2.123) with reactivity $\delta k(t) = u(t)$; then with $\mathbf{x}(t) = [n(t) \quad c(t)]^T$,

$$\frac{dn}{dt} = \frac{u - \beta}{l}n + \lambda c, \tag{3.15}$$

$$\frac{dc}{dt} = \frac{\beta}{l}n - \lambda c. \tag{3.16}$$

At equilibrium, (3.16) requires that

$$c_e = \frac{\beta}{l\lambda}n_e$$

and from (3.15), $u_e = 0$. Therefore, from (3.12)–(3.14),

$$\delta\dot{n} = \frac{n^*}{l}\,\delta u - \frac{\beta}{l}\,\delta n + \lambda\delta c, \tag{3.17}$$

$$\delta\dot{c} = \frac{\beta}{l}\,\delta n - \lambda\,\delta c,$$

or

$$\delta\dot{\mathbf{x}} = \mathbf{A}\,\delta\mathbf{x} + \mathbf{b}(n^*)\,\delta u, \tag{3.18}$$

where

$$\mathbf{A} = \begin{bmatrix} \dfrac{-\beta}{l} & \lambda \\ \dfrac{\beta}{l} & -\lambda \end{bmatrix}, \tag{3.19}$$

$$\mathbf{b}(n^*) = \begin{bmatrix} \dfrac{n^*}{l} \\ 0 \end{bmatrix}. \tag{3.20}$$

Example 3.2

Suppose that $\mathbf{x}^*(t) = [n_e \quad c^*(t)]^T$. That is, $\dot{n}^* = 0$ but $\dot{c}^*(t) \neq 0$ such that from (3.15) and (3.16),

$$u_e(t) = \frac{l\dot{c}^*(t)}{n_e} = \beta - \frac{l\lambda}{n_e} c^*(t). \tag{3.21}$$

Then, from (3.8)–(3.10),

$$\delta\dot{n} = \frac{u_e - \beta}{l}\,\delta n + \lambda\,\delta c + \frac{n_e}{l}\,\delta u, \tag{3.22}$$

$$\delta\dot{c} = \frac{\beta}{l}\,\delta n - \lambda\,\delta c, \tag{3.23}$$

where $u_e(t)$ is given by (3.21). In this case the linearized system is time variant as given by (3.8) or by (3.18) with $\mathbf{A} = \mathbf{A}(t)$, so that

$$\mathbf{A}(t) = \begin{bmatrix} \dfrac{u_e(t) - \beta}{l} & \lambda \\ \dfrac{\beta}{l} & -\lambda \end{bmatrix}. \tag{3.24}$$

In practice, both cases may be important. The first condition, equilibrium state (n_e, c_e) corresponds to the case whereby neutron level (or power level) and precursor level are both at equilibrium, say operating at design level for some period of time. On the other hand, it may be desired to maintain a constant neutron level n_e as accurately as possible even though the slower-responding precursor is not yet at equilibrium (sometimes referred to as steady-state neutron or power level as a distinction from equilibrium). Stability problems of this terminal control in practice as part of an optimal control policy will be discussed in Vol. 2.

Example 3.3

Consider the pnuematic control of flow rate $\dot{m}(t)$ in Fig. 3.2. It is conceivable that such a system might regulate the coolant flow rate through a heat exchanger as part of a nuclear (or other) power plant. It is assumed that upstream pressure p_a and downstream pressure

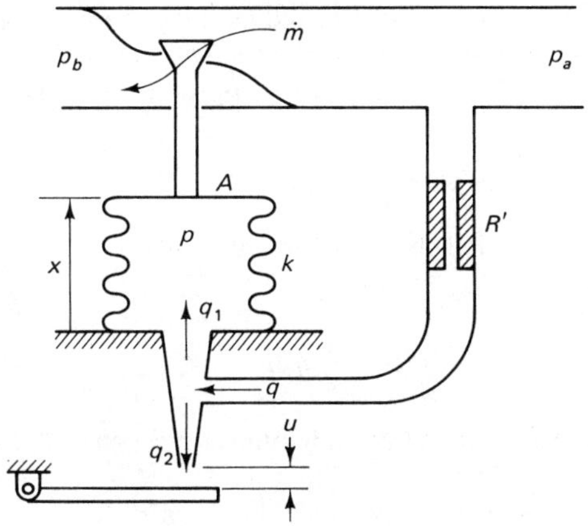

Figure 3.2 Pneumatic flow controller.

p_b are constant. A small quantity of coolant of volumetric flow rate $q(t)$ is bled from the main line for control purposes. The control line resistance is a constant R'. A portion, with flow rate $q_1(t)$, of the bled coolant enters the bellows chamber, which has a pressure $p(t)$, a cross-sectional area A, and a stiffness k. Another portion $q_2(t)$ is manipulated directly by the flapper valve of position $u(t)$. The main control valve position or bellows head position is given by $x(t)$. Find the system equations and the transfer function that relate the perturbations $\delta x(t)$ and $\delta u(t)$ from equilibrium conditions.

Solution First consider the control line resistance,

$$q(t) = \frac{p_a - p(t)}{R'}, \tag{3.25}$$

and the flow conservation

$$q(t) = q_1(t) + q_2(t), \tag{3.26}$$

both of which are linear relationships. The exit flow rate $q_2(t)$ at the flapper valve is defined by Bernoulli's flow equation

$$q_2(t) = c_f u(t)\sqrt{p(t)}. \tag{3.27}$$

Here c_f is a constant, and pressure $p(t)$ is taken relative to the atmospheric (exit) pressure (with no appreciable pressure drop between the bellows and the flapper valve). Similarly, Bernoulli's equation at the main line valve yields the following for mass flow rate:

$$\dot{m}(t) = c_v x(t)\sqrt{p_a - p_b}, \tag{3.28}$$

where c_v is a valve constant. While (3.27) is a nonlinear algebraic relation of system dependent variables, $u(t)$ and $p(t)$, (3.28) is linear with p_a and p_b assumed constant.

Next, a conservation of forces on the bellows head provides a linear algebraic equation,

$$Ap(t) = kx(t), \tag{3.29}$$

between bellows spring force and internal pressure. Finally the perfect gas law

$$p(t)v(t) = RTm_b(t), \tag{3.30}$$

where R is the universal gas constant and T, the bellows chamber gas temperature, is assumed constant. $v(t)$ and $m_b(t)$ are the bellows chamber volume and mass of coolant in the chamber. Then (3.30) is differentiated with respect to time so that

$$\dot{m}_b(t) = \frac{v}{RT}\dot{p}(t) + \frac{p}{RT}\dot{v}(t) = \gamma q_1(t),$$

where γ is the coolant density. If it is further assumed that $v(t)$ is nearly a constant v_e, $\dot{v} = 0$, and

$$\dot{p}(t) = \frac{\gamma}{c} q_1(t), \tag{3.31}$$

where capacitance $c = v_e/RT$. While (3.30) is nonlinear, the constant-volume assumption makes (3.31) linear.

The perturbation equations for (3.25)–(3.31) become

$$\delta q(t) = \frac{p_a - \delta p(t)}{R'} = \delta q_1(t) + \delta q_2(t),$$

$$\delta q_2(t) = \frac{c_f u_e}{2\sqrt{p_e}}\,\delta p(t) + c_f\sqrt{p_e}\,\delta u(t),$$

$$\delta \dot{m}(t) = c_v\sqrt{p_a - p_b}\,\delta x(t),$$

$$\delta x(t) = \frac{A}{k}\,\delta p(t),$$

$$\delta \dot{p}(t) = \frac{\gamma}{c}\,\delta q_1(t).$$

Then, by Laplace transformation of the pertubation variables with zero initial conditions, the desired system transfer function is given by

$$\frac{\delta x/x_e}{\delta u/u_e}(s) = \frac{\delta \dot{m}/\dot{m}_e}{\delta u/u_e}(s) = \frac{k_v}{\tau_v s + 1}, \tag{3.32}$$

where the gain and time constant are

$$k_v = \frac{-2R'Ap_e q_{2e}}{kx_e(2p_e + R'q_{2e})},$$

$$\tau_v = \frac{2R'cp_e}{\gamma(2p_e + R'q_{2e})},$$

and s is the Laplace transform complex variable. Again e designates equilibrium values ($\dot{m} = 0 = \dot{p}$) which are related by (3.25)–(3.31).

Example 3.4 [H6]

Consider the dynamical equations for a boiler–turbine–generator–transmission system with following definitions:

ϕ = rotor angle	Ψ = flux linkage
T_e = airgap torque	T_m = generator shaft torque
v = voltage	i = current
R = resistance	x_T = transmission line reactance

J = inertia
τ = time constant
u = activating signal to governor on inlet valve
ω_0 = infinite busbar frequency
θ_v = inlet valve position
F = fraction of total power from turbine stage (subscript H for high-pressure stage and L for low-pressure stage).

Subscripts are: d and q for each axis, m for stator, f for field, k for damper, t for machine terminal, b for busbar, and T for turbine. Then, flux linkages for the field, stator, and damper (referred to the direct d, and quadrature q axes, respectivly, are given by

$$\dot{\Psi}_{fd} = \omega_0(v_{fd} - R_{fd}\, i_{fd}),$$

$$\text{(NL)} \quad \dot{\Psi}_d = \omega_0[v_b \sin\phi + (R_m + R_T)i_d + \Psi_q] - \Psi_q\phi,$$

$$\text{(NL)} \quad \dot{\Psi}_q = \omega_0[v_b \cos\phi + (R_m + R_T)i_q + \Psi_d] - \Psi_d\phi,$$

$$\dot{\Psi}_{kd} = \omega_0 R_{kd} i_{kd},$$

$$\dot{\Psi}_{kq} = \omega_0 R_{kq} i_{kq}.$$

The rotor swing angle ϕ is given by

$$J\ddot{\phi} = \omega_0(T_m - T_e),$$

where

$$T_e = \Psi_d i_q - \Psi_q i_d,$$

$$\text{(NL)} \quad \tau_m \dot{T}_m = F_H\theta_v - F_L\theta_v u - T_m,$$

$$\tau_v \dot{\theta}_v = u - \theta_v.$$

and where τ_v and τ_m are the inlet-valve and turbine-generator inertial time constants.

Currents are related to fluxes by linear equations of form $\mathbf{i} = \mathbf{X}^{-1}\Psi$, where

$$\mathbf{i} = [i_{fd} \quad i_d \quad i_{kd} \quad i_q \quad i_{kq}]^{\mathrm{T}},$$

$$\mathbf{\Psi} = [\Psi_{fd} \quad \Psi_d \quad \Psi_{kd} \quad \Psi_q \quad \Psi_{kq}]^{\mathrm{T}},$$

and

$$\mathbf{X} = \begin{bmatrix} x_{fd} & -x_{ad} & x_{ad} & 0 & 0 \\ x_{ad} & -(x_d + x_T) & x_{ad} & 0 & 0 \\ x_{ad} & -x_{ad} & x_{kd} & 0 & 0 \\ 0 & 0 & 0 & -(x_q + x_T) & x_{aq} \\ 0 & 0 & 0 & -x_{aq} & x_{kq} \end{bmatrix}.$$

Machine terminal voltages become

$$\text{(NL)} \quad v_{td} = v_b \sin\phi + R_T i_d - x_T i_q,$$

$$\text{(NL)} \quad v_{tq} = v_b \cos\phi + R_T i_q + x_T i_d,$$

$$\text{(NL)} \quad v_t = (v_{td}^2 + v_{tq}^2)^{1/2}.$$

Here six equations (NL) are nonlinear. Note that the third (NL) is bilinear. Again, equilibrium is found with all time derivatives set to zero. The linear equations remain linear for the perturbation equations (only replace i_{fd} by δi_{fd}, etc.). When linearized the (NL) equations become

$$\delta\dot{\Psi}_d = \omega_0[v_{be}\cos\phi_e\,\delta\phi + \sin\phi_e\,\delta v_b + (R_m + R_T)\,\delta i_d + \delta\Psi_q] - \Psi_{qe}\,\delta\phi - \phi_e\,\delta\Psi_q,$$

$$\delta\dot{\Psi}_q = \omega_0[-v_{be}\sin\phi_e\,\delta\phi + \cos\phi_e\,\delta v_b + (R_m + R_T)\,\delta i_q + \delta\Psi_d] - \Psi_{de}\,\delta\phi - \phi_e\,\delta\Psi_d,$$

$$\tau\,\delta\dot{T} = F_H\,\delta\theta_v - F_L(u_e\,\delta\theta_v + \theta_{ve}\,\delta u) - \delta T_m,$$

$$\delta v_{td} = v_{be}\cos\phi_e\,\delta\phi + \sin\phi_e\,\delta v_b + R_T\,\delta i_d - x_T\,\delta i_q,$$

$$\delta v_{tq} = -v_{be}\sin\phi_e\,\delta\phi + \cos\phi_e\,\delta v_b + R_T\,\delta i_q + x_T\,\delta i_d,$$

$$\delta v_t = \tfrac{1}{2}(v_{tde}^2 + v_{tqe}^2)^{-1/2}\,(\delta v_{td} + \delta v_{tq}).$$

A typical control scheme would utilize the excitation signal to the valve governor shown above as well as an exciter to affect generator field current.

In summary, consider the following nonlinear system:

$$\frac{d\mathbf{x}}{dt} = \mathbf{f}(\mathbf{x}, \mathbf{u}), \tag{3.33}$$

$$\mathbf{y} = \mathbf{g}(\mathbf{x}, \mathbf{u}), \tag{3.34}$$

where $\mathbf{y}(t) \in \mathbb{R}^p$ is an output vector. Furthermore, suppose that we have a reference trajectory [i.e., solution to (3.33)–(3.34)] given by $\mathbf{x}^*(t)$, $\mathbf{y}^*(t)$ for a reference $\mathbf{u}^*(t)$. Then the linearized perturbation model of (3.33)–(3.34) about the reference becomes

$$\delta\dot{\mathbf{x}} = \frac{\partial\mathbf{f}}{\partial\mathbf{x}}(\mathbf{x}^*, \mathbf{u}^*)\,\delta\mathbf{x} + \frac{\partial\mathbf{f}}{\partial\mathbf{u}}(\mathbf{x}^*, \mathbf{u}^*)\,\delta\mathbf{u}, \tag{3.35}$$

$$\delta\mathbf{y} = \frac{\partial\mathbf{g}}{\partial\mathbf{x}}(\mathbf{x}^*, \mathbf{u}^*)\,\delta\mathbf{x} + \frac{\partial\mathbf{g}}{\partial\mathbf{u}}(\mathbf{x}^*, \mathbf{u}^*)\,\delta\mathbf{u}, \tag{3.36}$$

or

$$\delta\dot{\mathbf{x}} = \mathbf{A}^*\,\delta\mathbf{x} + \mathbf{B}^*\,\delta\mathbf{u},$$

$$\delta\mathbf{y} = \mathbf{C}^*\,\delta\mathbf{x} + \mathbf{D}^*\,\delta\mathbf{u},$$

where $\mathbf{A}^*$, $\mathbf{B}^*$, $\mathbf{C}^*$, $\mathbf{D}^*$ are appropriately defined constant matrices according to (3.35) and (3.36) if $\mathbf{x}^*$, $\mathbf{u}^*$ are constants. In general, of course, the system would be time variant and linear.

Once the linearized (perturbation) model is derived for a nonlinear system, it would be convenient to apply linear system techniques to the analysis and design of the nonlinear system with some confidence. Formally, an argument can be made for such a procedure in many practical cases as long as the perturbations $\delta\mathbf{x}(t)$ are quite small relative to reference $\mathbf{x}^*(t)$. This formally includes concepts of local stability, local controllability, and local observability. Rigorously derived results, however, are somewhat scarce except for very specific cases. The most popular result in this regard probably is Liapunov's first method for the evaluation of stability for an equilibrium point, which is studied in the next section.

3.2 STABILITY OF AN EQUILIBRIUM POINT

In Sections 3.2 and 3.3 the discussion is limited to autonomous systems. First, in this section equilibrium-point stability is studied for nonlinear systems by means of linear system stability. Then, with this motivation, autonomous linear system theory is reviewed in Section 3.3. Prior to a statement of Liapunov's first method for local stability at an equilibrium point, consider several relevant definitions.

3.2.1 Definitions

The nonlinear *autonomous system* is described by

$$\frac{d\mathbf{x}}{dt} = \mathbf{f}(\mathbf{x}), \tag{3.37}$$

where the equilibrium states $\mathbf{x}_e$ are given by

$$\mathbf{f}(\mathbf{x}_e) = \mathbf{0}. \tag{3.38}$$

In general, (3.38) may yield one or more and even an infinite number of equilibrium points. Again, if $\mathbf{f}(\mathbf{x})$ satisfies the Lipschitz condition (see Appendix B and [C1]), the solution is unique for a given initial state $\mathbf{x}(t_0) = \mathbf{x}_0$. Then the equilibrium solution points cannot be reached (nor can they be left), in finite time from (to) any other state. Again, this is an apparent requirement of a proper model for a physical system.

Consider an *isolated equilibrium point* (state) in state space, that is, a point $\mathbf{x}_e$ such that no other equilibrium points lie within its infinitesimal neighborhood. Then stability of $\mathbf{x}_e$ is defined conveniently by Fig. 3.3. Roughly, the system (3.37) is stable at $\mathbf{x}_e$ if for every initial state $\mathbf{x}_0$ that is sufficiently close to $\mathbf{x}_e$, the solution $\mathbf{x}(\mathbf{x}_0, t)$ remains near $\mathbf{x}_e$. The following definition is more precise: The equilibrium point $\mathbf{x}_e$ is *stable* if for every real number $\epsilon > 0$, there exists a real number $\delta(\epsilon) > 0$ such that $\|\mathbf{x}_0 - \mathbf{x}_e\| < \delta$ implies that $\|\mathbf{x}(\mathbf{x}_0, \mathbf{t}) - \mathbf{x}_e\| < \epsilon$ for all $t \geq t_0$. If $\mathbf{x}_e$ is not stable, it is *unstable*. From the definition of uniform continuity (see Section A.2) it is apparent that stability here is nothing more than uniform continuous dependence on $\mathbf{x}_0$.

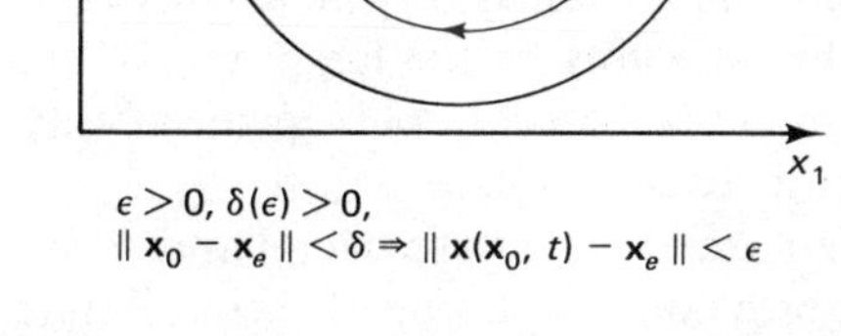

$\epsilon > 0$, $\delta(\epsilon) > 0$,
$\| \mathbf{x}_0 - \mathbf{x}_e \| < \delta \Rightarrow \| \mathbf{x}(\mathbf{x}_0, t) - \mathbf{x}_e \| < \epsilon$

Figure 3.3 Illustration of equilibrium point stability.

Formally, the system (3.37) is asymptotically stable at $\mathbf{x}_e$ if it is stable, and furthermore, $\mathbf{x}(t)$ approaches $\mathbf{x}_e$ as $t \to \infty$. Or more precisely, the following definition can be made from Fig. 3.4: The equilibrium state $\mathbf{x}_e$ is *asymptotically stable* if it is *stable and convergent*. "Convergent" here means that there exists a real number $\delta_1 > 0$, and for every $\epsilon_1 > 0$ there exists a $T(\epsilon_1) > 0$ such that $\|\mathbf{x}(t_0) - \mathbf{x}_e\| < \delta_1$ implies that $\|\mathbf{x}(x_0, t) - \mathbf{x}_e\| < \epsilon_1$ for all $t \geq T + t_0$. It should be noted, however, that more commonly, T is dropped from the definition and convergence is given by

$$\lim_{t\to\infty} \|\mathbf{x}(\mathbf{x}_0, t) - \mathbf{x}_e\| = 0.$$

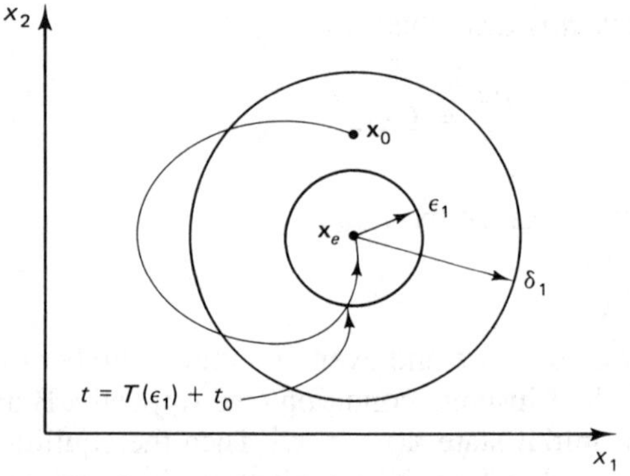

$\epsilon_1 > 0,\ \delta_1 > 0,\ T(\epsilon_1) > 0$
such that
$\| \mathbf{x}_0 - \mathbf{x}_e \| < \delta_1 \Rightarrow \| \mathbf{x}(\mathbf{x}_0, t) - \mathbf{x}_e \| < \epsilon_1 \ \forall\ t \geqslant T(\epsilon_1) + t_0$

Figure 3.4 Asymptotically stable equilibrium point.

For a system (3.37) with only a single equilibrium, we can ask a stronger question: Is the system asymptotically stable in the large, or is $\mathbf{x}_e$ globally asymptotically stable? It follows that $\mathbf{x}_e$ is *globally asymptotically stable* if it is asymptotically stable for all initial states (i.e., $\delta_1 \to \infty$ in Fig. 3.4).

In all of these defintions $\|\cdot\|$ refers to the appropriate vector norm, such as

$$\|\mathbf{x}\| = \sum_{i=1}^{n} |x_i|$$

or the Euclidean norm

$$\|\mathbf{x}\| = \left(\sum_{i=1}^{n} x_i^2\right)^{1/2}.$$

See Appendix A and [C2] for a more detailed discussion. It should be noted that for a particular case, one norm may be mathematically more convenient or physically more meaningful than others. While stability in the large will be studied by Liapunov's direct method or second method in Chapter 6, only local stability at $\mathbf{x}_e$ in terms of the linear perturbation equations is discussed here.

For some reason, students tend to confuse the relative meaning of δ and ϵ in this definition of stability. It is worth noting again that the equilibrium point is stable if for every (1) real $\epsilon > 0$ there exists (2) a real $\delta(\epsilon) > 0$ such that motion started interior to the region defined by (2) never leaves the region defined by (1), as exemplified by Fig. 3.3. Note that as ϵ is selected smaller, $\delta(\epsilon)$ generally is smaller. Then asymptotic stability requires further that the motion proceeds to $\mathbf{x}_e$ in the limit as $t \to \infty$. Furthermore, it is required that the motion converges asymptotically so that the longer it lasts, the closer it gets to $\mathbf{x}_e$. Of course, the regions of asymptotic stability generally are smaller than for stability—hence the use of ϵ_1 and δ_1 for regions of convergence. As will be seen in Chapter 6, these concepts are generalized to include regions of attraction to "invariant sets," which include equilibrium points.

3.2.2 Liapunov's First Method

Local Stability Theorem. Given the autonomous system (3.37) with a perturbation equation at an equilibrium state $\mathbf{x}_e$ given by

$$\delta\dot{\mathbf{x}} = \frac{\partial \mathbf{f}}{\partial \mathbf{x}}(\mathbf{x}_e)\,\delta\mathbf{x} + \mathbf{r}(\mathbf{x}_e, \delta\mathbf{x}) \tag{3.39}$$

so that

$$\lim_{\|\delta\mathbf{x}\|\to 0} \frac{\mathbf{r}(\mathbf{x}_e, \delta\mathbf{x})}{\|\delta\mathbf{x}\|} = \mathbf{0},$$

the following conclusions can be made:

1. If all the eigenvalues of $\partial\mathbf{f}(\mathbf{x}_e)/\partial\mathbf{x}$ have only negative real parts, $\mathbf{x}_e$ is *asymptotically stable*.
2. If one or more of the eigenvalues of $\partial\mathbf{f}(\mathbf{x}_e)/\partial\mathbf{x}$ have positive real parts, $\mathbf{x}_e$ is *unstable*.
3. If one or more eigenvalues of $\partial\mathbf{f}(\mathbf{x}_e)/\partial\mathbf{x}$ have zero real parts and no eigenvalues with positive real parts, stability of $\mathbf{x}_e$ cannot be ascertained by perturbation theory.

Here it should be recalled that the eigenvalues, $\lambda_i, i = 1, \ldots, n$ of the $n \times n$ matrix $\mathbf{A}$, which is $\partial\mathbf{f}(\mathbf{x}_e)/\partial\mathbf{x}$ in (3.39), are simply the solution to the algebraic equation

$$\det[\mathbf{I}\lambda_i - \mathbf{A}] = \mathbf{0}, \tag{3.40}$$

where det[·] refers to the matrix determinant. Such important aspects of linear system theory will be reviewed in more detail in Section 3.3, now that their need has been shown even for nonlinear systems. Of course, for linear autonomous systems, eigenvalues in the left-half complex plane define asymptotic stability, and eigenvalues in the right-half complex plane define instability. Simple eigenvalues on the imaginary axis correspond to stable isolated equilibrium points at the origin. These three conclusions would be expected, but a proof will be deferred until further results are available in Chapter 6. One might question the third conclusion, but realizing that even for linear cases the system is really only marginally stable, and unstable in the case of repeated eigenvalues with zero-real parts, it would seem reasonable. If the perturbation equations yield any eigenvalues with zero-real parts, the system (3.37) is said to be *structurally unstable*.

Example 3.5

Consider the neutronic perturbation equations (3.18)–(3.20) for the nonlinear systems (3.15)–(3.16), from Example 3.1. Since det $\mathbf{A} = 0$ from (3.19), it is immediately known that $\lambda_1\lambda_2 = 0$. Consequently, the process is structurally unstable, and no conclusion by Liapunov's first method can be made about stability of the nonlinear system (3.15)–(3.16) at its equilibrium point (n_e, c_e) with $u_e = 0$.

Actually, the eigenvalues of the matrix $\mathbf{A}$ [equation (3.19)] are given by

$$\det \begin{bmatrix} \lambda_i + \dfrac{\beta}{l} & -\lambda \\ -\dfrac{\beta}{l} & \lambda_i + \lambda \end{bmatrix} = 0, \qquad i = 1, 2,$$

or

$$\lambda_1 = 0, \qquad \lambda_2 = -\lambda - \frac{\beta}{l}.$$

Although (3.15) is not generally autonomous, it can be assumed so for a constant u that satisfies the class of equilibrium controls here.

Example 3.6

Consider the nonlinear autonomous system

$$\dot{x}_1 = x_2,$$

$$\dot{x}_2 = -\ ax_2 - \sin x_1,$$

which has multiple equilibrium points at

$$\mathbf{x}_{ek} = [k\pi \quad 0]^{\mathrm{T}},$$

where $k = \ldots, -1, 0, 1, 2, \ldots$. Then the perturbation model is

$$\delta\dot{x}_1 = \delta x_2,$$

$$\delta\dot{x}_2 = -\ a\ \delta x_2 - (-1)^k\ \delta x_1.$$

Therefore,

1. If $a < 0$, $\mathbf{x}_{ek}$ are unstable.
2. If $a > 0$, $\mathbf{x}_{ek} = [2k\pi \quad 0]^{\mathrm{T}}$ are asymptotically stable equilibrium states, where k are the integers; the other $\mathbf{x}_{ej}$ are unstable.
3. If $a = 0$, $\mathbf{x}_{ek} = [(2k + 1)\pi \quad 0]^{\mathrm{T}}$ are unstable; for other $\mathbf{x}_e$ no conclusion on stability can be made based on the first method.

Example 3.7

Consider the dynamics of the simple longitudinal aircraft dynamics shown in Fig. 3.5. A summation of forces in the direction orthogonal to the velocity vector (called the velocity axis) and a sum of moments about the center of mass yields (see Section 2.2.4) $\mathbf{v} \times \boldsymbol{\omega} = V\dot{\gamma}$, or

$$mV\dot{\gamma} = AQC_L(\cdot), \tag{3.41}$$

$$I\ddot{\theta} = AlQC_M(\cdot), \tag{3.42}$$

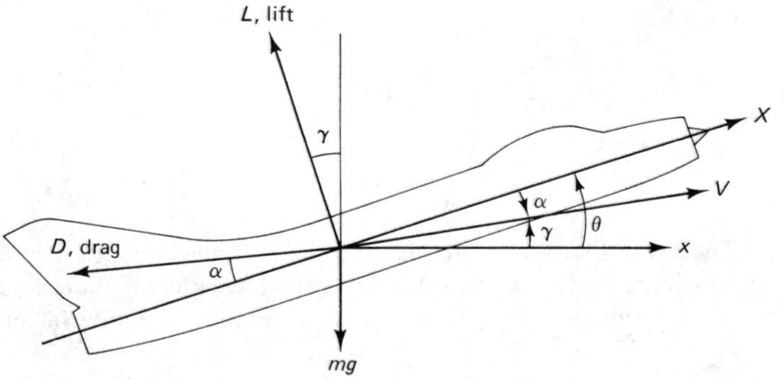

Figure 3.5 Longitudinal airframe geometry.

where V is the aircraft velocity magnitude, m the aircraft mass, I its pitch moment of inertia, γ the climb angle, θ the pitch angle, Q the dynamic pressure, A an aircraft characteristic area (for which aerodynamic pressure is integrated over to obtain the resultant force), l the resultant pitch force moment arm, $C_L(\cdot)$ the aerodynamic lift coefficient, and $C_M(\cdot)$ the aerodynamic pitch moment coefficient. Here V, m, I, Q, A, and l are assumed constants. $C_L(\cdot)$, $C_M(\cdot)$ in general are functions of angle of attack α, control fin angle u, Mach number, and altitude. But the lattter two are not important here (being assumed constant in the same way that Q is assumed constant.) By use of a moving coordinate system, the lateral force equation may be neglected here. Consequently, we are modeling only one degree of freedom, the pitch motion about the wing axis. Since the nonlinear terms, $C_L(\alpha, u)$ and $C_M(\alpha, u)$, are functions of angle of attack, α, and control deflection u, let α be a state variable (replacing γ) with $\gamma = \theta - \alpha$. Also, for convenience let

$$C_M'(\alpha, u) = AlQC_M(\alpha, u)$$

and

$$C_L'(\alpha, u) = AQC_L(\alpha, u).$$

Now (3.41) and (3.42) are replaced by

$$\ddot{\theta} = \frac{C_M'(\alpha, u)}{I}, \tag{3.43}$$

$$\dot{\alpha} = \dot{\theta} - \frac{C_L'(\alpha, u)}{mV}. \tag{3.44}$$

An aerodynamic trim condition establishes the desired equilibrium here, so that

$$C_M'(\alpha_e, u_e) = 0, \qquad \dot{\theta}_e = 0, \qquad \text{and} \qquad C_L'(\alpha_e, u_e) = 0.$$

Then linearization about this trim condition yields

$$\delta\ddot{\theta} = \frac{1}{I}[C_{M\alpha}'(\alpha_e, u_e)\,\delta\alpha + C_{Mu}'(\alpha_e, u_e)\,\delta u] \tag{3.45}$$

$$\delta\dot{\alpha} = \frac{-1}{m\mathrm{V}}[C_{L\alpha}'(\alpha_e, u_e)\,\delta\alpha + C_{Lu}'(\alpha_e, u_e)\,\delta u] + \delta\dot{\theta}, \tag{3.46}$$

where

$$C'_{M\alpha}(\cdot) = \frac{\partial C'_M}{\partial \alpha}, \qquad C'_{Mu}(\cdot) = \frac{\partial C'_M}{\partial u},$$

$$C'_{L\alpha}(\cdot) = \frac{\partial C'_L}{\partial \alpha}, \qquad C'_{Lu}(\cdot) = \frac{\partial C'_L}{\partial u}.$$

The latter four terms are classical aerodynamic stability derivaties (normalized here) which are evaluated from standard wind tunnel and flight test data.

From (3.45) and (3.46), the transfer functions are evaluated as follows:

$$\frac{\Delta\dot{\theta}}{\Delta u}(s) = \frac{a_\theta(s + \sigma_\theta)}{D(s)}, \tag{3.47}$$

$$\frac{\Delta\alpha}{\Delta u}(s) = \frac{a_\alpha(s + \sigma_\alpha)}{D(s)}, \tag{3.48}$$

where

$$a_\theta = \frac{C'_{Mu}(\alpha_e, u_e)}{I}, \qquad a_\alpha = \frac{-C'_{Lu}(\alpha_e, u_e)}{mV},$$

$$\sigma_\theta = \frac{1}{mV}C'_{L\alpha}(\alpha_e, u_e)\left[1 - \frac{C'_{M\alpha}(\alpha_e, u_e)C'_{Lu}(\alpha_e, u_e)}{C'_{Mu}(\alpha_e, u_e)C'_{L\alpha}(\alpha_e, u_e)}\right],$$

$$\sigma_\alpha = \frac{-m\mathrm{V}}{I}\frac{C'_{Mu}(\alpha_e, u_e)}{C'_{Lu}(\alpha_e, u_e)},$$

and the characteristic equation is given by

$$D(s) = s^2 + bs + c = 0. \tag{3.49}$$

Here

$$b = \frac{C'_{L\alpha}(\alpha_e, u_e)}{m\mathrm{V}}$$

$$c = \frac{-C'_{M\alpha}(\alpha_e, u_e)}{I}.$$

Again the first method of Liapunov provides stability conditions with the roots of (3.49), that is, eigenvalues of the perturbation state matrix or poles of the transfer function, in the left-half complex plane. Here $b > 0$ (i.e., $C'_{L\alpha}$ positive) and $c < 0$ (i.e., $C_{M\alpha}$ negative) ensure a stable equilibrium (trim state) for the pitch mode.

Although $C'_{L\alpha}(\cdot)$ normally is positive, $C'_{M\alpha}$ may be of either sign, depending on the angle of attack and the relative location of the aerodynamic center (center of pressure) relative to the center of mass. Normally, the center of mass of the aircraft is forward of the aerodynamic center in order to guarantee so-called static stability of this pitch mode. Of course, it is assumed that the higher-order terms of the Taylor series for the stability derivatives are sufficiently small. It can be shown that this is generally a valid assumption for normal flight envelopes (excluding Mach 1).

3.3 AUTONOMOUS LINEAR SYSTEMS

In Section 3.2 we have seen that within its limitations, linear system theory may be used to analyze certain aspects of nonlinear systems for small perturbations about some reference state or trajectory. In particular, it was seen by Liapunov's first method that structually stable autonomous systems may be analyzed for stability at isolated equilibrium points by means of its linearized autonomous equations if the higher-order terms of the Taylor series are sufficiently small. Consequently, it is important to recall certain relevant aspects of linear autonomous systems. Many of these results are used throughout this book.

3.3.1 Definitions and Solutions

A linear autonomous system here is defined again by state differential equation

$$\frac{d\mathbf{x}}{dt} = \mathbf{A}\mathbf{x}, \tag{3.50}$$

where $\mathbf{x} \in \mathbb{R}^n$; $\mathbf{A}$ is an $n \times n$ matrix with ij^{th} element given by a_{ij}. Obviously, (3.50) has an equilibrium state at the origin, $\mathbf{x}_e = \mathbf{0}$. If the nonhomogeneous system

$$\frac{d\mathbf{x}}{dt} = \mathbf{A}\mathbf{x} + \mathbf{b} \tag{3.51}$$

has an equilibrium state $\mathbf{x}_e$, then a simple translation, $\mathbf{z} = \mathbf{x} - \mathbf{x}_e$, converts (3.51) to the form of (3.50). Consequently, the stability properties of $\mathbf{x}_e$ for (3.51) are identical to those for $\mathbf{0}$ in (3.50), or for that matter for any other $\mathbf{x}_e$ in (3.51). Indeed, if $\mathbf{A}$ is a nonsingular matrix (i.e., $\det|\mathbf{A}| \neq 0$ so that $\mathbf{A}^{-1}$ does exist), it can be shown that (3.50) has a unique equilibrium state at the origin.

For an initial state $\mathbf{x}(t_0) = \mathbf{x}_0$, the solution to (3.50) is given by

$$\mathbf{x}(t, \mathbf{x}_0, t_0) = \exp\,[\mathbf{A}(t - t_0)]\mathbf{x}_0, \tag{3.52}$$

or in briefer notation with $t_0 = 0$,

$$\mathbf{x}(t) = \mathbf{e}^{\mathbf{A}t}\mathbf{x}_0. \tag{3.53}$$

Here $\mathbf{e}^{\mathbf{A}t}$ is the exponential matrix or the state-transition matrix of system (3.50). In general, the state-transition matrix is a particular fundamental matrix, $\mathbf{X}(t)$, solution to matrix equation $\dot{\mathbf{X}} = \mathbf{A}\mathbf{X}$, with $\mathbf{X}(t_0) = \mathbf{I}$. That is, $\mathbf{e}^{\mathbf{A}(t-t_0)}$ is a matrix of n linearly independent solutions to (3.50) with $\mathbf{x}^1(t_0) = [1\ \ 0\ \ \cdots\ \ 0]^T$, $\mathbf{x}^2(t_0) = [0\ \ 1\ \ 0\ \ \cdots\ \ 0]^T, \ldots, \mathbf{x}^n(t_0) = [0\ \ \cdots\ \ 0\ \ 1]^T$.

In terms of its Taylor series,

$$\begin{aligned} \mathbf{e}^{\mathbf{A}(t-t_0)} &= \mathbf{I} + \mathbf{A}(t - t_0) + \frac{\mathbf{A}^2(t - t_0)^2}{2!} + \cdots \\ &= \sum_{k=0}^{\infty} \frac{\mathbf{A}^k(t - t_0)^k}{k!}. \end{aligned} \tag{3.54}$$

Consider the Laplace transform of (3.50) which is given by

$$s\mathbf{x}(s) - \mathbf{x}_0 = \mathbf{A}\mathbf{x}(s)$$

or

$$\mathbf{x}(s) = [\mathbf{I}s - \mathbf{A}]^{-1}\mathbf{x}_0. \tag{3.55}$$

Then, since $\mathbf{x}(t) = \mathbf{e}^{\mathbf{A}t}\mathbf{x}_0$ if $\mathbf{x}(0) = \mathbf{x}_0$, the exponential state-transition matrix may be computed from the inverse Laplace transform of (3.55) as follows:

$$\mathbf{e}^{\mathbf{A}t} = \mathcal{L}^{-1}\{[\mathbf{I}s - \mathbf{A}]^{-1}\}. \tag{3.56}$$

Example 3.8

Compute the state-transition matrix, $\mathbf{e}^{\mathbf{A}t}$, for the following:

$$\dot{x}_1 = x_2,$$

$$\dot{x}_2 = x_3,$$

$$\dot{x}_3 = -\sigma_1\sigma_2 x_2 - (\sigma_1 + \sigma_2)x_3.$$

Solution

$$[\mathbf{I}s - \mathbf{A}] = \begin{bmatrix} s & -1 & 0 \\ 0 & s & -1 \\ 0 & \sigma_1\sigma_2 & s + (\sigma_1 + \sigma_2) \end{bmatrix}$$

$$[\mathbf{I}s - \mathbf{A}]^{-1} = \begin{bmatrix} \frac{1}{s} & \frac{s + (\sigma_1 + \sigma_2)}{sD(s)} & \frac{1}{sD(s)} \\ 0 & \frac{s + (\sigma_1 + \sigma_2)}{D(s)} & \frac{1}{D(s)} \\ 0 & \frac{-\sigma_1\sigma_2}{D(s)} & \frac{s}{D(s)} \end{bmatrix},$$

where $D(s) = (s + \sigma_1)(s + \sigma_2)$. Then from (3.56),

$$\mathbf{e}^{\mathbf{A}t} = \begin{bmatrix} 1 & \frac{\sigma_1 + \sigma_2}{\sigma_1\sigma_2} + \frac{\sigma_2 e^{-\sigma_1 t}}{\sigma_1(\sigma_1 - \sigma_2)} + \frac{\sigma_1 e^{-\sigma_2 t}}{\sigma_2(\sigma_2 - \alpha_1)} & \frac{1}{\sigma_1\sigma_2} + \frac{e^{-\sigma_1 t}}{\sigma_1(\sigma_1 - \sigma_2)} + \frac{e^{-\sigma_2 t}}{\sigma_2(\sigma_2 - \sigma_1)} \\ 0 & \frac{\sigma_2 e^{-\sigma_1 t}}{\sigma_2 - \sigma_1} + \frac{\sigma_1 e^{-\sigma_2 t}}{\sigma_1 - \sigma_2} & \frac{e^{-\sigma_1 t}}{\sigma_2 - \sigma_1} + \frac{e^{-\sigma_2 t}}{\sigma_1 - \sigma_2} \\ 0 & \frac{\sigma_1\sigma_2}{\sigma_1 - \sigma_2}(e^{-\sigma_1 t} - e^{-\sigma_2 t}) & \frac{\sigma_1 e^{-\sigma_1 t}}{\sigma_1 - \sigma_2} + \frac{\sigma_2 e^{-\sigma_2 t}}{\sigma_2 - \sigma_1} \end{bmatrix} \tag{3.57}$$

Example 3.9

Compute $\mathbf{x}(t)$ for Example 3.8 if $\mathbf{x}(0) = [1 \quad 0 \quad 1]^{\mathrm{T}}$, $\sigma_1 = 1$, and $\sigma_2 = 2$.

Solution From (3.53) and (3.57),

$$\mathbf{x}(t) = \mathbf{e}^{\mathbf{A}t}\mathbf{x}_0$$

or

$$x_1(t) = \frac{3 - 2e^{-t} + e^{-2t}}{2},$$
$$x_2(t) = e^{-t} - e^{-2t},$$
$$x_3(t) = 2e^{-2t} - e^{-t}.$$

3.3.2 Eigenvalues and Eigenvector Transformations

In some cases it is possible and convenient to transform $\mathbf{x}(t)$ to a new variable $\mathbf{z}(t)$ so that (3.50) takes a diagonal form

$$\dot{\mathbf{z}} = \mathbf{\Lambda z} \tag{3.58}$$

or

$$\dot{z}_1 = \lambda_1 z_1, \quad \dot{z}_2 = \lambda_2 z_2, \quad \ldots, \quad \dot{z}_n = \lambda_n z_n.$$

Then with $z_i(0) = z_{i0}$, $i = 1, \ldots, n$, $z_i(t) = e^{\lambda_i t} z_{i0}$, or

$$\mathbf{z}(t) = \mathbf{e}^{\mathbf{\Lambda} t}\mathbf{z}_0. \tag{3.59}$$

With this in mind assume a constant matrix transformation of (3.50):

$$\mathbf{P}\dot{\mathbf{z}} = \mathbf{APz} \tag{3.60}$$

or

$$\dot{\mathbf{z}} = \mathbf{P}^{-1}\mathbf{APz} = \mathbf{\Lambda z}. \tag{3.61}$$

Any transformation of the form $\mathbf{P}^{-1}\mathbf{AP}$ is called a similarity transformation, and the related $\mathbf{A}$ and $\mathbf{\Lambda}$ are similar matrices (with the same λ_i). If we rewrite $\mathbf{P}^{-1}\mathbf{AP} = \mathbf{\Lambda}$ as $\mathbf{AP} = \mathbf{P\Lambda}$, it is seen that a convenient way to compute $\mathbf{P}$ is to find the n eigenvectors, $\mathbf{p}^i$, $i = 1, \ldots, n$, such that

$$\mathbf{Ap}^i = \lambda_i \mathbf{p}^i,$$

or

$$[\mathbf{I}\lambda_i - \mathbf{A}]\mathbf{p}^i = \mathbf{0}, \qquad i = 1, \ldots, n. \tag{3.62}$$

Then (3.62) has a nonsingular solution if

$$\det[\mathbf{i}\lambda_i - \mathbf{A}] = 0, \qquad i = 1, \ldots, n, \tag{3.63}$$

which is the characteristic equation for which the zeros are the eigenvalues of matrix $\mathbf{A}$. Then for each eigenvalue one can compute from (3.62) a corresponding eigenvector (actually, a unique direction only). It can be seen that such a transformation is possible as long as the eigenvectors are unique (i.e., linearly independent) so that the resulting eigenvector matrix, or modal matrix $\mathbf{P}$, is of rank n, the system order. Therefore, $\mathbf{P}^{-1}$ exists. The $\mathbf{z}$ coordinates are sometimes called *canonical normal coordinates*.

If the original state equation $\dot{\mathbf{x}} = \mathbf{Ax}$ is in canonical phase-variable form $\dot{x}_1 = x_2$, $\dot{x}_2 = x_3, \ldots, \dot{x}_n = -a_n x_1 - \cdots - a_1 x_n$,

$$\mathbf{A} = \begin{bmatrix} 0 & 1 & 0 & \cdot & \cdot & \cdot & 0 \\ 0 & 0 & 1 & 0 & \cdot & \cdot & 0 \\ \cdot & & & & & & \\ \cdot & & & & & & \\ \cdot & & & & & & \\ -a_n & -a_{n-1} & \cdot & \cdot & \cdot & \cdot & -a_1 \end{bmatrix}.$$

Then it can be shown from (3.62) and (3.63) that the Vandermonde matrix

$$\mathbf{V} = \begin{bmatrix} 1 & 1 & \cdot & \cdot & \cdot & 1 \\ \lambda_1 & \lambda_2 & \cdot & \cdot & \cdot & \lambda_n \\ \lambda_1^2 & \lambda_2^2 & \cdot & \cdot & \cdot & \lambda_n^2 \\ \cdot & \cdot & & & & \cdot \\ \cdot & \cdot & & & & \cdot \\ \cdot & \cdot & & & & \cdot \\ \lambda_1^{n-1} & \lambda_2^{n-1} & \cdot & \cdot & \cdot & \lambda_n^{n-1} \end{bmatrix} \tag{3.64}$$

is an eigenvector matrix $\mathbf{P} = \mathbf{VC}$, where $\mathbf{C}$ is an appropriate $n \times n$ matrix since the modal matrix only establishes directions. For example, it may be desirable to maintain a vector length or norm in the new space by proper selection of $\mathbf{C}$.

It is apparent that such a transformation also may be a convenient way to compute the general exponential state-transition matrix. To see this, it is convenient to examine (3.54) with $\mathbf{\Lambda} = \mathbf{P}^{-1}\mathbf{AP}$ or $\mathbf{A} = \mathbf{P\Lambda P}^{-1}$ and in general $\mathbf{A}^k = \mathbf{P\Lambda}^k\mathbf{P}^{-1}$. Consequently,

$$\mathbf{e}^{\mathbf{A}(t-t_0)} = \mathbf{Pe}^{\mathbf{\Lambda}(t-t_0)}\mathbf{P}^{-1}. \tag{3.65}$$

This of course checks with the $e^{\mathbf{A}(t-t_0)}$ solution to the matrix equation $\dot{\mathbf{X}} = \mathbf{AX}, \mathbf{X}(t_0) = \mathbf{I}$ or now $\dot{\mathbf{Z}} = \mathbf{\Lambda Z}$, so that $\mathbf{e}^{\mathbf{A}(t-t_0)} = \mathbf{PZ}(t)$ with $\mathbf{Z}(t_0) = \mathbf{P}^{-1}$.

Example 3.10

The characteristic equation $\det[\mathbf{I}\lambda - \mathbf{A}] = 0$ for Example 3.8 yields $\lambda_1 = -\sigma_1$, $\lambda_2 = -\sigma_2$, $\lambda_3 = 0$. Therefore,

$$\mathbf{V} = \begin{bmatrix} 1 & 1 & 1 \\ -\sigma_1 & -\sigma_2 & 0 \\ \sigma_1^2 & \sigma_2^2 & 0 \end{bmatrix}.$$

Select $\mathbf{P} = \mathbf{V}$ since norm preservation is not required (i.e., $\mathbf{C} = \mathbf{I}$ above). Then

$$\mathbf{P}^{-1} = \frac{\mathbf{V}^{\text{Adj}}}{\det \mathbf{V}} = \frac{\begin{bmatrix} 0 & \sigma_2^2 & \sigma_2 \\ 0 & -\sigma_1^2 & -\sigma_1 \\ \sigma_1\sigma_2(\sigma_1 - \sigma_2) & \sigma_1^2 - \sigma_2^2 & \sigma_1 - \sigma_2 \end{bmatrix}}{\sigma_1\sigma_2(\sigma_1 - \sigma_2)}.$$

It is readily seen that (3.65) is satisfied so that with $t_0 = 0$,

$$\mathbf{Pe}^{\mathbf{\Lambda}t}\mathbf{P}^{-1} = \mathbf{e}^{\mathbf{A}t}$$

yields (3.57) with $\mathbf{e}^{\Lambda t} = \text{diag}[e^{-\sigma_1 t} \quad e^{-\sigma_2 t} \quad 1]$, where diag $[\cdot]$ designates nonzero elements of a diagonal matrix.

Then with $\mathbf{x}(0) = [1 \quad 0 \quad 1]^T$, $\sigma_1 = 1$, $\sigma_2 = 2$ as in Example 3.9, $\mathbf{z}(0) = \mathbf{P}^{-1}\mathbf{x}(0) = [-1 \quad \frac{1}{2} \quad \frac{3}{2}]^T$, where

$$\mathbf{P} = \begin{bmatrix} 1 & 1 & 1 \\ -1 & -2 & 0 \\ 1 & 4 & 0 \end{bmatrix}, \qquad \mathbf{P}^{-1} = \begin{bmatrix} 0 & -2 & -1 \\ 0 & \frac{1}{2} & \frac{1}{2} \\ 1 & \frac{3}{2} & \frac{1}{2} \end{bmatrix}.$$

Then

$$\mathbf{z}(t) = \mathbf{e}^{\Lambda t}\mathbf{z}_0,$$

where $\mathbf{z}(0) = \mathbf{z}_0$. Or $z_1(t) = -e^{-t}$, $z_2(t) = \frac{1}{2}e^{-2t}$, $z_3(t) = \frac{3}{2}$, $t \geq 0$. Again, $\mathbf{x}(t) = \mathbf{P}\mathbf{z}(t)$, or $x_1(t) = \frac{3}{2} - e^{-t} + \frac{1}{2}e^{-2t}$, $x_2(t) = e^{-t} - e^{-2t}$, $x_3(t) = 2e^{-2t} - e^{-t}$. Hence, the solution checks with Example 3.9. Note the very unlikely initial condition along an eigenvector, say $\mathbf{z}(0) = [-1 \quad 0 \quad 0]$ or $\mathbf{x}(0) = [-1 \quad 1 \quad -1]$. Then $z_1(t) = -e^{-t}$, $z_2(t) = z_3(t) = 0$, and $x_1(t) = -e^{-t}$, $x_2(t) = e^{-t}$, $x_3(t) = -e^{-t}$. In general, of course, the solution $\mathbf{x}(t)$ is a linear combination of eigenvector solutions or sometimes called normal modes. In other words, the n eigenvectors span the $\mathbb{R}^n$ vector space.

As unlikely as it is in practice (since physical components are not likely to create nonisolated eigenvalues with random-size variations from admissible tolerances), mathematically there may exist repeated eigenvectors (a sufficient condition for repeated eigenvalues). In such a mathematical case, the procedure described above is not valid for that portion of the decomposition with repeated eigenvectors. (Note the singularity of the modal **P** matrix in this case.) However, an "off-diagonal" canonical form, or Jordon canonical form, exists. For example, suppose that λ_k is a repeated eigenvalue of **A** of order r. That is the matrix **A** has r eigenvalues of value λ_k. Furthermore, suppose that **A** has m such repeated eigenvalues in general (with $r = 1$, the nonrepeated special case is as above). Then there exists a Jordan form $\mathbf{J} = \mathbf{T}^{-1}\mathbf{A}\mathbf{T}$ such that

$$\mathbf{J} = \text{diag}[\mathbf{J}_1 \quad \mathbf{J}_2 \quad \cdots \quad \mathbf{J}_m] = \begin{bmatrix} \mathbf{J}_1 & & & 0 \\ & \cdot & & \\ & & \cdot & \\ 0 & & & \mathbf{J}_m \end{bmatrix},$$

where

$$\mathbf{J}_k = \begin{bmatrix} \lambda_k & 1 & & & & 0 \\ & \lambda_k & \cdot & & & \\ & & \cdot & \cdot & & \\ & & & \cdot & \cdot & \\ & & & & \cdot & 1 \\ & & & & \lambda_k & 1 \\ 0 & & & & & \lambda_k \end{bmatrix}$$

The order of $\mathbf{J}_k$ is no larger than the multiplicity r of λ_k. From (3.65),

$$\mathbf{e}^{\mathbf{A}t} = \mathbf{T}\mathbf{e}^{\mathbf{J}t}\mathbf{T}^{-1},$$

where

$$\mathbf{e}^{\mathbf{J}t} = \text{diag}[\mathbf{e}^{\mathbf{J}_1 t} \quad \cdots \quad \mathbf{e}^{\mathbf{J}_m t}]$$

and

$$\mathbf{e}^{\mathbf{J}_k t} = \begin{bmatrix} 1 & t & t^2/2! & \cdots & t^{r-1}/(r-1)! \\ 0 & 1 & t & \cdots & t^{r-2}/(r-2)! \\ \cdot & \cdot & \cdot & & \cdot \\ \cdot & \cdot & \cdot & & \cdot \\ \cdot & \cdot & \cdot & & \cdot \\ 0 & 0 & 0 & \cdots & 1 \end{bmatrix} e^{\lambda_k t}.$$

Then the solution to (3.50) generally may be expressed by

$$\mathbf{x}(t) = \mathbf{T}\mathbf{e}^{\mathbf{J}t}\mathbf{T}^{-1}\mathbf{x}_0, \tag{3.66}$$

where $\mathbf{x}_0 = \mathbf{x}(0)$.

From these canonical forms, it is seen readily that the autonomous linear system is stable at the origin (i.e., null solution is stable) if the eigenvalues are not in the right-hand complex plane(i.e, they have nonpositive real parts) and are not repeated (i.e., $r = 1$) if on the imaginary axis. Otherwise, it is unstable. Again, left-half plane eigenvalues correspond to asymptotic stability in any case. For linear systems such asymptotic stability is global.

In the case of complex eigenvalues, the modal matrix is complex, and the canonical normal variables would be complex. However, it is possible to make another transformation (complex in this case) so that the overall transformation is real. Then, $\mathbf{z} = \mathbf{P}^{-1}\mathbf{x}$ results in complex $\mathbf{P}$ and complex $\mathbf{z}$. If $\mathbf{z}_R = \mathbf{Q}\mathbf{z} = \mathbf{Q}\mathbf{P}^{-1}\mathbf{x}$, it is possible to find such a $\mathbf{Q}$ so that $\mathbf{Q}\mathbf{P}^{-1}$(and $\mathbf{P}\mathbf{Q}^{-1}$) are real so that both $\mathbf{x}$ and $\mathbf{z}_R$ are real. Again, as for the repeated λ_k case, the new form has off-diagonal elements for the complex eigenvalues. Suppose that $n = 4$ with real λ_1, λ_2 but complex

$$\lambda_3 = \alpha + j\beta, \qquad \lambda_4 = \alpha - j\beta.$$

In this case, $\dot{\mathbf{x}} = \mathbf{A}\mathbf{x}$, $\dot{\mathbf{z}} = \mathbf{\Lambda}\mathbf{z}$, and $\dot{\mathbf{z}}_R = \mathbf{Q}\mathbf{P}^{-1}\mathbf{A}\mathbf{P}\mathbf{Q}^{-1}\mathbf{z}_R$, or

$$\dot{\mathbf{z}}_R = \begin{bmatrix} \lambda_1 & 0 & 0 & 0 \\ 0 & \lambda_2 & 0 & 0 \\ 0 & 0 & \alpha & \beta \\ 0 & 0 & -\beta & \alpha \end{bmatrix} \mathbf{z}_R. \tag{3.67}$$

Here $\mathbf{P}$ is the modal matrix and $\mathbf{Q}$ has the form

$$\mathbf{Q} = \begin{bmatrix} 1 & 0 & 0 & 0 \\ 0 & 1 & 0 & 0 \\ 0 & 0 & jk & -k \\ 0 & 0 & jk & k \end{bmatrix}.$$

Therefore, $\dot{z}_{R1} = \lambda_1 z_{R1}$, $\dot{z}_{R2} = \lambda_2 z_{R2}$, $\dot{z}_{R3} = \alpha z_{R3} + \beta z_{R4}$, $\dot{z}_{R4} = -\beta z_{R3} + \alpha z_{R4}$, and we have two eigenvectors and an "eigenplane" such that motion starting along an eigenvector never leaves it, and motion starting in the eigenplane never leaves it.

Then by a simple transformation of the z_{R3}, z_{R4} components to polar coordinates, with $r = (z_{R3}^2 + z_{R4}^2)^{1/2}$ and $\phi = \tan^{-1} z_{R4}/z_{R3}$,

$$\dot{r} = \alpha r \tag{3.68}$$

and

$$\dot{\phi} = -\beta. \tag{3.69}$$

Therefore, the new simple canonical normal solutions are

$$r(t) = r_0 e^{\alpha t},$$

$$\phi(t) = \phi_0 - \beta t,$$

$$z_1(t) = z_{10} e^{\lambda_1 t},$$

and

$$z_2(t) = z_{20} e^{\lambda_2 t},$$

where $r_0 = r(0)$, $\phi_0 = \phi(0)$, $z_{10} = z_1(0)$, and $z_{20} = z_2(0)$.

3.3.3 Characteristic State-Plane Portraits

Consider the second-order autonomous linear system

$$\frac{d\mathbf{x}}{dt} = \mathbf{A}\mathbf{x}, \tag{3.70}$$

where

$$\mathbf{A} = \begin{bmatrix} a_{11} & a_{12} \\ a_{21} & a_{22} \end{bmatrix} \tag{3.71}$$

Then, the characteristic equation, $\det[\mathbf{I}\lambda - \mathbf{A}] = 0$ becomes

$$\det\begin{bmatrix} \lambda - a_{11} & -a_{12} \\ -a_{21} & \lambda - a_{22} \end{bmatrix} = 0$$

or

$$\lambda^2 + b\lambda + c = 0, \tag{3.72}$$

where $b = -a_{11} - a_{22}$, $c = a_{11}a_{22} - a_{12}a_{21}$. Then it is seen that (3.70) is asymptotically stable at the origin if $b > 0$, $c > 0$. The roots are complex if $b^2 \leq 4c$. Figure 3.6 shows a decomposition by type of dynamic response in terms of parameters b and c.

From (3.62), the eigenvectors of $\mathbf{A}$ in (3.70) are computed by $[\mathbf{I}\lambda_i - \mathbf{A}]\mathbf{p}^i = \mathbf{0}$,

or

$$\frac{p_2^i}{p_1^i} = \frac{a_{21}}{\lambda_i - a_{22}} = \frac{\lambda_i - a_{11}}{a_{12}}, \tag{3.73}$$

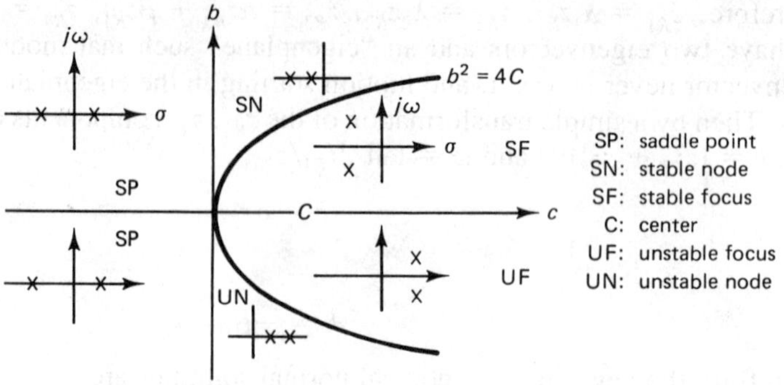

Figure 3.6 Second-order parameter plane.

where $\mathbf{p}^i = [p_1^i \quad p_2^i]^T$, $i = 1, 2$. Again, for canonical phase variables ($\dot{x}_1 = x_2, \ldots$), we can select $\mathbf{P} = \mathbf{V}$, the Vandermonde matrix. Here

$$\mathbf{V} = \begin{bmatrix} 1 & 1 \\ \lambda_1 & \lambda_2 \end{bmatrix}. \tag{3.74}$$

Example 3.11

For

$$\dot{x}_1 = x_2,$$
$$\dot{x}_2 = -\sigma_1\sigma_2 x_1 - (\sigma_1 + \sigma_2)x_2,$$
$$\mathbf{A} = \begin{bmatrix} 0 & 1 \\ -\sigma_1\sigma_2 & -(\sigma_1 + \sigma_2) \end{bmatrix}, \qquad \lambda_1 = -\sigma_1, \lambda_2 = -\sigma_2,$$

and

$$\mathbf{V} = \begin{bmatrix} 1 & 1 \\ -\sigma_1 & -\sigma_2 \end{bmatrix}.$$

If $\sigma_1 = 1$ and $\sigma_2 = 2$, then $\lambda_1 = -1$, $\lambda_2 = -2$, and the isolated equilibrium point at the origin is asymptotically stable, as shown by its state-plane portrait in Fig. 3.7. Note here that the motion is always to the right for increasing time in the upper half-plane. That is,

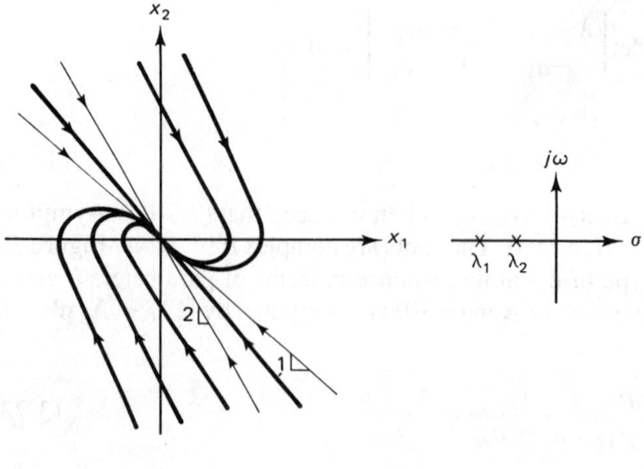

Figure 3.7 Phase-plane portrait (stable node) for Example 3.11.

$\dot{x}_1 > 0$ if $x_2 > 0$. Also, motion is to the left in the lower half-plane and crosses the x_1 axis orthogonal to it. Note that $dx_1 = x_2\,dt$ indicates that the infinitesimal change in x_1 for an infinitesimal change in t is proportional to x_2, so that the motion proceeds faster to the right for motion starting with large $x_2 = \dot{x}_1$.

With the eigenvectors for **z** coordinates, the motion is shown in Fig. 3.8. The equilibrium point corresponding to such motion, which is typical of that for second-order autonomous systems with two real negative eigenvalues, is called a stable node. The state-plane portrait in **z** coordinates is easily found from (3.58), with $n = 2$, to be $z_2 = z_{20}(z_1/z_{10})^{\lambda_2/\lambda_1}$. This is a general expression for state-plane portraits in **z** coordinates.

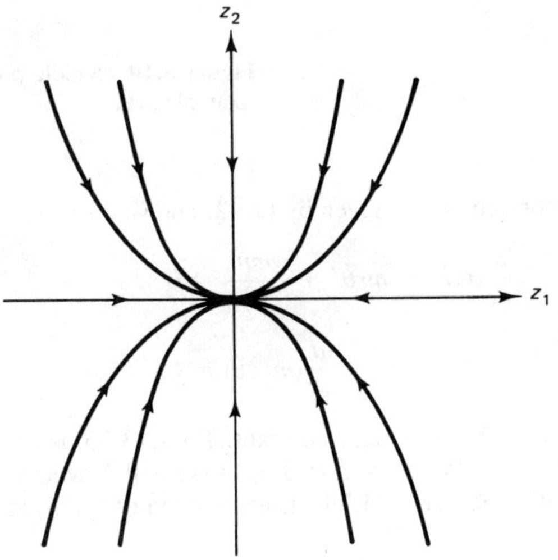

Figure 3.8 Normal coordinate portrait for Example 3.11.

Example 3.12

In Example 3.11, let $\lambda_1 = -\sigma_1 = 1$ and $\lambda_2 = -\sigma_2 = 2$. Then the origin is unstable and the corresponding unstable-node portrait is shown in Fig. 3.9. Also, with $\lambda_1 = -\sigma_1 = 1$ and $\lambda_2 = -\sigma_2 = -2$, the origin is unstable (saddle point) and the corresponding motion is shown in Fig. 3.10.

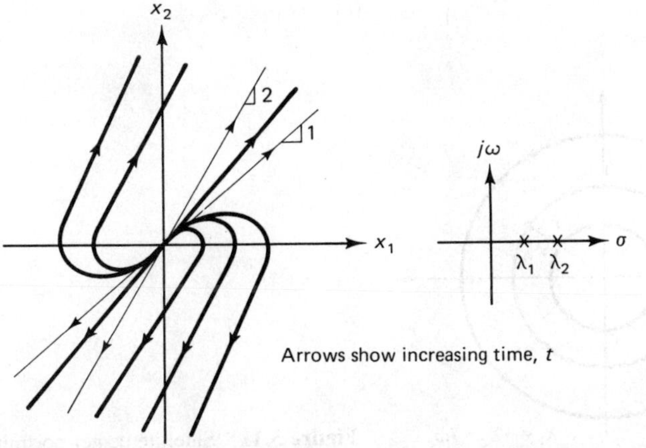

Figure 3.9 Unstable-node portrait for Example 3.12.

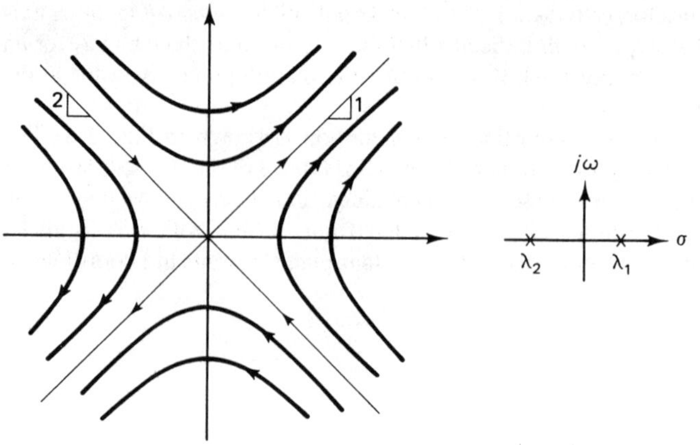

Figure 3.10 Saddle-point portrait for Example 3.12.

Example 3.13

Recall the satellite trajectory equations given by (2.42) and (2.43):

$$\frac{d}{dt}(m\dot{r}) - mr\dot{\theta}^2 + \frac{\gamma m m_e}{r^2} = 0, \tag{3.75}$$

$$\frac{d}{dt}(mr^2\dot{\theta}) = 0. \tag{3.76}$$

If it is assumed that $\ddot{\theta}_e = 0$ so that $\dot{\theta}_e = \omega_e$, a constant. Then, (3.75) is analyzed for small perturbations in δr about r_e, ω_e. With $\ddot{r} = \dot{r} = 0$, $r_e = (k/\omega_e^2)^{1/3}$, where $k = \gamma m_e$ and m is assumed constant and divided out of (3.75). Linearization of (3.75) at r_e, ω_e yields

$$\delta\ddot{r} - \left(\omega_e^2 - \frac{2k}{r_e^3}\right)\delta r = 0$$

or

$$\delta\ddot{r} + \omega_e^2 \ \delta r = 0. \tag{3.77}$$

Let $x_1 = \delta r$, $x_2 = \dot{x}_1 = \delta\dot{r}$, so that $\dot{\mathbf{x}} = \mathbf{A}\mathbf{x}$, where

$$\mathbf{A} = \begin{bmatrix} 0 & 1 \\ -\omega_e^2 & 0 \end{bmatrix}.$$

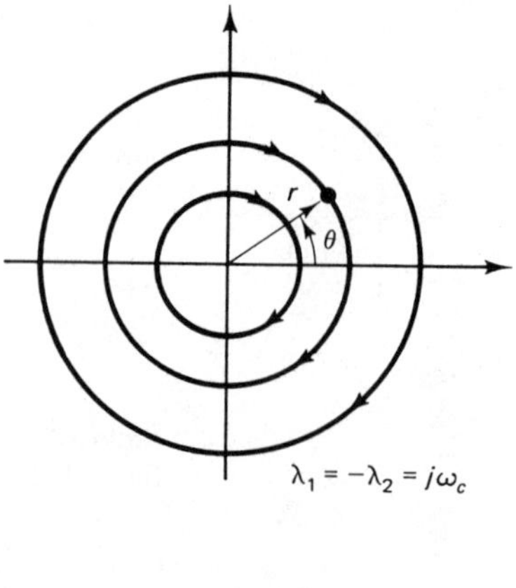

Figure 3.11 Satellite center portrait.

For $x_1(0) = 1$, $x_2(0) = 0$, $x_1(t) = \cos \omega_e t$ and $x_2(t) = \sin \omega_e t$. Here the eigenvalues $\lambda_1 = j\omega_e$, $\lambda_2 = -j\omega_e$, and the origin is stable for the perturbation equations, but not asymptotically stable. Hence no conclusion can be drawn about stability of the nonlinear equations from Liapunov's first method. The perturbation equations yield simple harmonic motion for any nonzero initial state. This results in the portrait shown in Fig. 3.11, for which the equilibrium state is called a *center point*. If the satellite were placed in a 24-hour communication orbit (fixed relative to a point on the revolving earth), then $\omega_e = 1$ per day. If it were in a low trajectory for surveillance purposes, ω_e would be larger and some atmospheric damping may eventually take place that would yield a dissipation of energy.

A stable focus and an unstable focus portrait for complex eigenvalues are shown in Fig. 3.12. Again, with $\lambda_{1,2} = \alpha \pm j\beta$, a transformation to canonical form can be made so that $\dot{z}_{1R} = \alpha_1 z_{1R} + \beta z_{2R}$ and $\dot{z}_{2R} = -\beta z_{1R} + \alpha z_{2R}$; or in polar coordinates $\dot{r} = \alpha r$, $\dot{\phi} = -\beta$. Then the solution is given simply as $r(t) = r_0 e^{\alpha t}$, $\phi(t) = \phi_0 - \beta t$. Degenerate state-plane portraits are shown in Figs. 3.13 to 3.16. Stable and unstable degenerate nodes are shown in Fig. 3.13 where one eigenvalue is zero. In this case, there exists an equilibrium line. If $\lambda_1 = \lambda_2 \neq 0$, two distinct cases result. Figure 3.14 shows the case where the modal matrix is nonsingular (i.e., linearly independent eigenvectors) so that every vector is an eigenvector. In the more general case of Fig. 3.15, there is really only one eigenvector, and again a degenerate mode results. Figure 3.16 shows the case with two zero eigenvalues, which results in an equilibrium line. Perhaps this is best visualized as a center portrait (Fig. 3.10) as $\beta \rightarrow 0$, or as a degenerate node (Fig. 3.15) as $\sigma \rightarrow 0$.

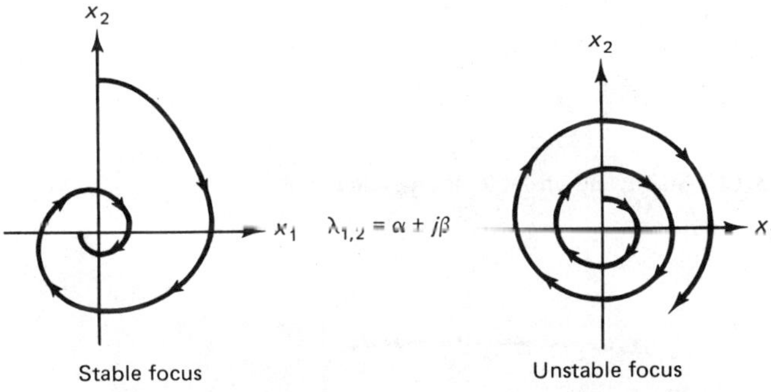

Figure 3.12 Stable and unstable foci.

It should be noted here that the state-plane trajectories result from the solution to (3.70) with t eliminated; that is

$$\frac{dx_2}{dx_1} = \frac{a_{21}x_1 + a_{22}x_2}{a_{11}x_1 + a_{12}x_2}. \tag{3.78}$$

Therefore, the solutions to (3.78) do not uniquely define a set of state equations since there may be common terms such as $\dot{x}_1 = f_1(\mathbf{x})\alpha(\mathbf{x})$, $\dot{x}_2 = f_2(\mathbf{x})\alpha(\mathbf{x})$. Points in time, expressed as a parameter, would need to be indicated on given trajectories to define unique time trajectories. Also, it should be noted that $\mathbf{x}_e$ an equilibrium state for $\dot{\mathbf{x}} = \mathbf{f}(\mathbf{x})$ is more correctly called a singular point $\mathbf{x}_s$ of (3.78). However, we have not made this distinction thus far.

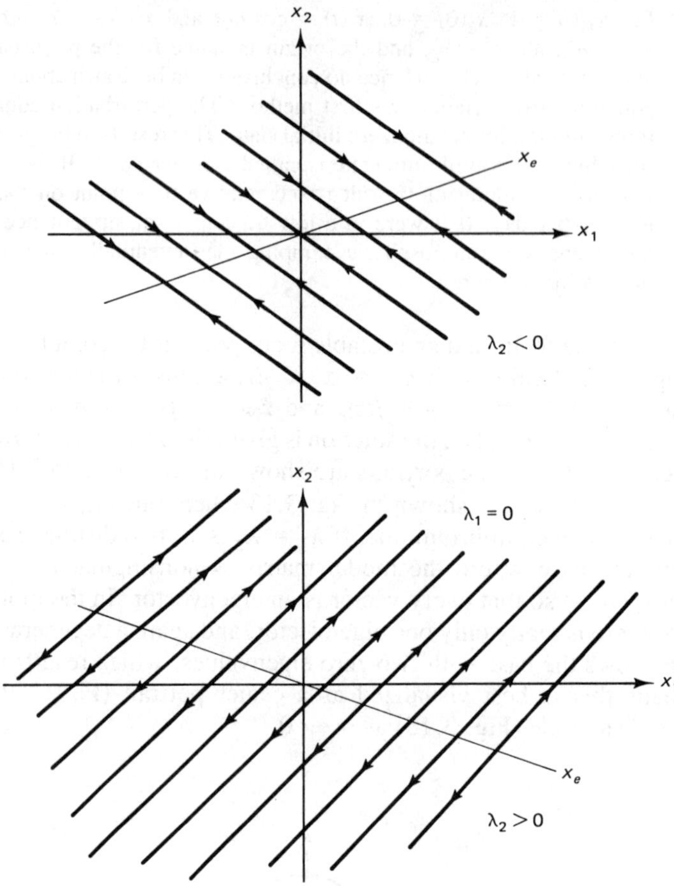

Figure 3.13 Stable and unstable degenerate nodes.

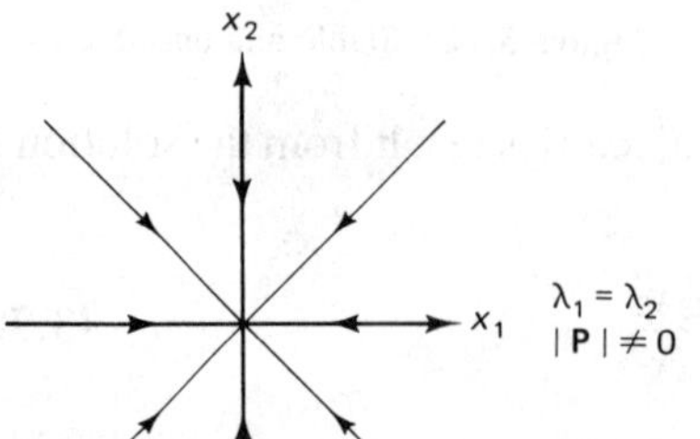

Figure 3.14 Stable portrait with equal eigenvalues but linearly independent eigenvectors.

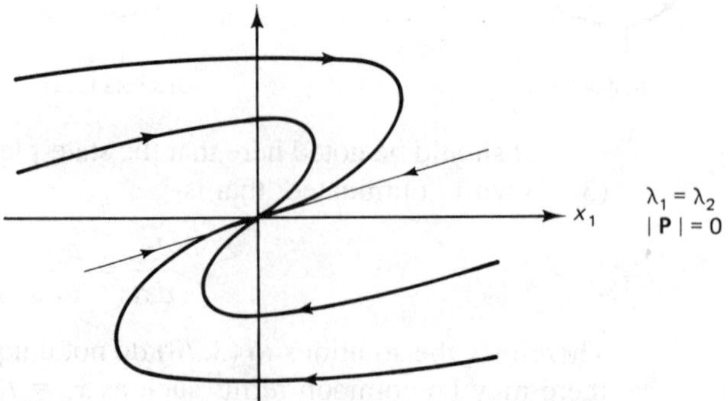

Figure 3.15 Stable-node portrait with equal eigenvalues and linearly dependent eigenvectors.

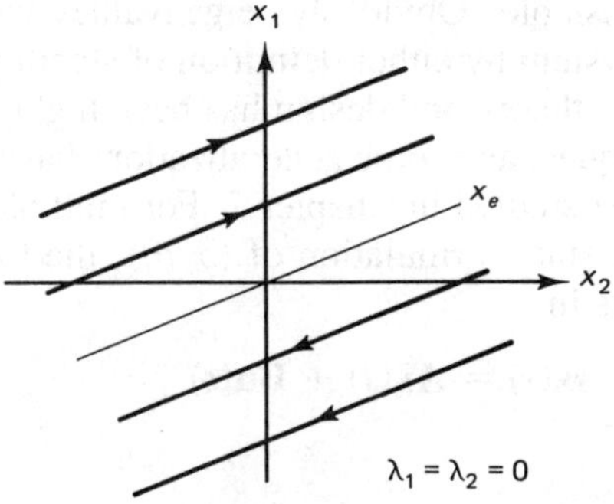

Figure 3.16 Zero-eigenvalues portrait.

3.4 FORCED LINEAR SYSTEMS

Here the discussion is restricted to time-invariant (constant **A** and **B** matrices) linear systems with time-variant inputs on controls. Thus

$$\frac{d\mathbf{x}}{dt} = \mathbf{Ax} + \mathbf{Bu}, \tag{3.79}$$

where again $\mathbf{x} \in \mathbb{R}^n$ is the state vector, $\mathbf{u} \in \mathbb{R}^m$ the control vector, and **A**, **B** are appropriate constant matrices. As noted previously, for constant inputs, a translation of state variables can be made so that (3.79) is represented by an autonomous linear system of form (3.70) with new equilibrium points.

The initial-state solution to (3.79) obviously is identical to the *zero-input solution* of (3.79). It is readily seen that the *zero-state solution* to (3.79) is given by

$$\mathbf{x}(t) = \int_{t_0}^{t} \mathbf{e}^{\mathbf{A}(t-\sigma)}\mathbf{Bu}(\sigma)\, d\sigma,$$

where $\mathbf{e}^{\mathbf{A}t}$ is the state-transition matrix given in (3.53). Then, by superposition (linear systems), the general solution is given by

$$\mathbf{x}(t) = \mathbf{e}^{\mathbf{A}(t-t_0)}\mathbf{x}_0 + \int_{t_0}^{t} \mathbf{e}^{\mathbf{A}(t-\sigma)}\mathbf{Bu}(\sigma)\, d\sigma. \tag{3.80}$$

Recall from Section 3.1.1 that the perturbation model (3.12) at an equilibrium state and control for the nonlinear system (3.1) results in the linear nonautonomous system structure of (3.79). The linear nonautonomous system (3.79) generally is termed asymptotically stable if the eigenvalues of **A** all have negative real parts. This is true whether we refer to the unforced system (zero-input response) or to the bounded-input system—the latter being termed bounded-input, bounded-output (BIBO) stability. Again, however, for eigenvalues on the imaginary axis we must be careful. Unique imaginary eigenvalues correspond to a stable unforced response. But note that a sinusoidal input of the same frequency as the imaginary eigenvalue will cause a divergence, and hence (3.79) would not be BIBO stable. And, of course, with mathematically repeated imaginary eigenvalues, as shown by (3.66), even the unforced system would be unstable. Obviously, eigenvalues with positive real parts correspond to an unstable system by either definition of stability.

Classical linear control theory and design has been highly successful in the use of frequency-domain techniques, and some generalizations have been found useful for nonlinear systems, as will be studied in Chapter 7. For multiple-input, multiple-state linear systems given by the state formulation of (3.70), the Laplace transform with zero initial conditions results in

$$s\mathbf{x}(s) = \mathbf{A}\mathbf{x}(s) + \mathbf{B}\mathbf{u}(s)$$

or

$$\mathbf{x}(s) = [\mathbf{I}s - \mathbf{A}]^{-1}\mathbf{B}\mathbf{u}(s). \tag{3.81}$$

There the state-control transfer function is given by

$$\mathbf{H}(s) = [\mathbf{I}s - \mathbf{A}]^{-1}\mathbf{B}, \tag{3.82}$$

so that

$$\mathbf{x}(s) = \mathbf{H}(s)\mathbf{u}(s), \tag{3.83}$$

where $\mathbf{H}(s)$ is an $n \times m$ matrix function of the complex variable s, and $\mathbf{x}(s)$, $\mathbf{u}(s)$ are the Laplace transforms of the state and control with zero initial conditions. Since $[\mathbf{I}s - \mathbf{A}]^{-1}$ is the Laplace transform of the state transition matrix, $\mathbf{e}^{\mathbf{A}t}$, $\mathbf{e}^{\mathbf{A}t}\mathbf{B}$ represents an impulse-response matrix and the transfer function matrix

$$\mathbf{H}(s) = \mathcal{L}[\mathbf{e}^{\mathbf{A}t}]\mathbf{B}. \tag{3.84}$$

If the output $\mathbf{y}(t)$ is given as a linear combination of state and possibly control,

$$\mathbf{y}(t) = \mathbf{C}\mathbf{x}(t) + \mathbf{D}\mathbf{u}(t), \tag{3.85}$$

where $\mathbf{y}(t) \in \mathbb{R}^p$, and $\mathbf{C}$, $\mathbf{D}$ are appropriate constant matrices. Then the input–output transfer function for (3.85) and (3.79) is given by

$$\mathbf{G}(s) = \mathbf{C}[\mathbf{I}s - \mathbf{A}]^{-1}\mathbf{B} + \mathbf{D}, \tag{3.86}$$

such that

$$\mathbf{y}(s) = \mathbf{G}(s)\mathbf{u}(s). \tag{3.87}$$

At this point it should be apparent that the system poles generally correspond to the eigenvalues of $\mathbf{A}$. However, it should be noted that the input–output relationship could mathematically result in pole–zero cancellation or so-called system hidden modes. Consequently, we see that asymptotic system stability, from (3.79) or (3.83), and input–output stability from (3.87) are equivalent only if there are no hidden modes with positive- or zero-real parts. The designer must be cautioned about the practical problems that may arise from pole–zero cancellation. While it is obvious that such cancellations will result in instability with positive-real modes due to inexact components causing the existence of finite residues, it can still be detrimental to performance when an ever-so-annoying small residue for a slow response mode never reaches equilibrium.

Example 3.14

Given an open-loop transfer function

$$G(s) = \frac{k}{s(s + \alpha)},$$

where $\alpha = -1$, compute a stabilizing linear controller.

Solution Mathematically, it would be very convenient to simply cancel the pole with a zero and perhaps compensate further for response time, and so on. In reality, the compensation or the plant will not have precise parametric values, and the unstable mode will persist despite its small residue, as shown in Fig. 3.17. Consider a small nonzero ϵ error in zero placement. Even if we use negative unity feedback with $G_c(s) = k_c(s - 1 - \epsilon)$ as shown in Fig. 3.18, the root locus of Fig. 3.17 shows that the closed-loop system is still unstable. Of course, simple negative gain or lag feedback will not stabilize it, as can be shown from the root locus. However, a conventional proportional plus derivative (PD) controller or lead compenstion stabilizes the system very nicely, as shown in Fig. 3.19. In this case $G_c(s) = k_c(s + 1)$. Since the closed-loop system is only of second order, it is quite easy to solve for the closed-loop transfer function poles and zeros. The closed-loop zeros are simply the open-loop zeros since there are no feedback (parallel) elements. The closed-loop transfer function is

$$\frac{C(s)}{R(s)} = \frac{kk_c(s + 1)}{s(s - 1) + kk_c(s + 1)} = \frac{kk_c(s + 1)}{s^2 + (kk_c - 1)s + kk_c}.$$

If $kk_c = 4$, $\omega_n = 2$ and damping coefficient $\zeta = \frac{3}{4}$ for the closed-loop transfer function, which results in relatively good response characteristics without further specifications. Note here that even a small error in the traditional zero placement relative to the plant pole has minor effect.

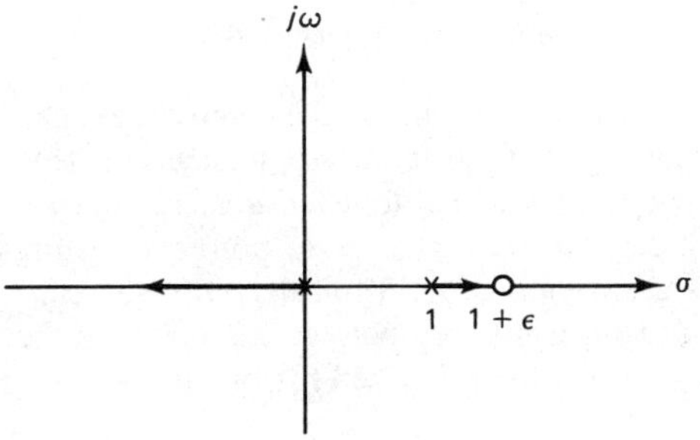

Figure 3.17 Effect of improper pole cancellation.

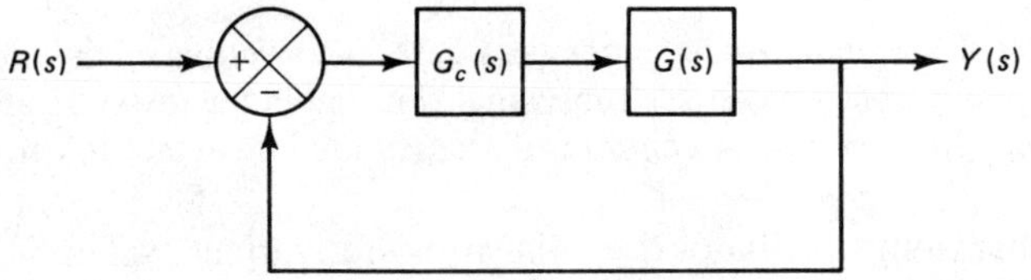

Figure 3.18 Block diagram for Example 3.14.

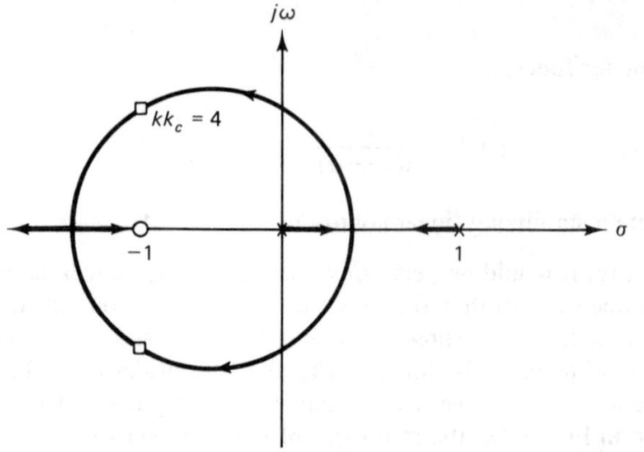

Figure 3.19 Simple proportional plus derivative control.

Example 3.15

Consider the nonlinear heat-transfer dynamics given by (2.124)–(2.127). Linearization about equilibrium conditions with $c_{mg} = 0$ (i.e., negligible coolant mass heat capacity relative to the heat exchanger) and $T_i = 0$ yields the following transfer function:

$$\frac{\delta T_u(s)}{\delta q(s)} = \frac{k_1}{s + \tau_h^{-1}}, \tag{3.88}$$

where $k_1 = c_m^{-1}$ and

$$\tau_h = c_m\left[\frac{\theta}{(1+\theta)c_p w_e} + \frac{1}{Ah_e}\right]. \tag{3.89}$$

If the heat is generated from a nuclear fission process with transfer function, derived from (2.122),

$$\frac{\delta q(s)}{\delta u(s)} = \frac{q_e(s+\lambda)}{s(s+\beta/l+\lambda)}. \tag{3.90}$$

Here w_e is the equilibrium coolant flow rate, h_e the equilibrium heat-transfer coefficient, and q_e the equilibrium heat power, proportional to neutron density. c_m is the heat-exchanger mass heat capacity, A the heat transfer area, c_p the specific heat at constant pressure, λ the average precursor decay rate, β the portion of neutrons from precursors, and l the mean-neutron generation time. Typically, $\tau_h = 5$ sec, $\lambda = 0.1$ sec^{-1}, and $(\beta/l + \lambda) = 50$ sec^{-1}, which results in a pole at 0.2 sec^{-1} for the linearized heat exchanger and at 0 and 50 sec^{-1} for the linearized fission process along with a zero at 0.1. Consequently,

$$\frac{\delta T_u(s)}{\delta u(s)} = \frac{k_1 q_e(s+0.1)}{s(s+50)(s+0.2)},$$

which is obviously a "stiff" system. Again, attempted zero cancellation of the slow heat-exchange mode in practice may produce a substantial slow tail in the response after the fast transients subside. Nuclear reactor control is discussed in more detail in Vol. 2.

It is apparent that the harmonic oscillator (i.e., linear time-invariant system with imaginary eigenvalues) presents complications in linear system theory; for example, note the stability complications above. But even further complications can arise that give its response to sinusoidal inputs some of the characteristics of nonlinear systems. For example, consider

$$\dot{x}_1 = x_2,$$
$$\dot{x}_2 = -\beta^2 x_1 + F \cos \omega t, \tag{3.91}$$

where β and F are constants, and $x_1(0) = x_{10}$, $x_2(0) = 0$. Then

$$x_1(t) = B \cos \beta t - C \cos \omega t,$$

where

$$C = \frac{F}{\omega^2 - \beta^2},$$
$$B = x_{10} + C.$$

All the following cases of periodic motion can be demonstrated[S1]:

1. $F = 0$, period $T = \dfrac{2\pi}{\beta}$.
2. $B = 0$, no free oscillation, $T = \dfrac{2\pi}{\omega}$.
3. $\omega = k\beta$, k is a nonunity integer, $B = 0$, $T = \dfrac{2\pi}{\beta} = \dfrac{2\pi k}{\omega}$, subharmonic oscillation.
4. $B \neq 0$, $\omega = \dfrac{\beta}{m}$, m is a nonunity integer, $T = \dfrac{2m\pi}{\beta} = \dfrac{2\pi}{\omega}$; as case (2) but here there is a superposition of two different periods which produce ultraharmonics.
5. $B \neq 0$, $\omega = \dfrac{k\beta}{m}$; k, m are relative prime integers (i.e., no common divisors except unity); least period, $T = \dfrac{2k\pi}{\omega} = \dfrac{2m\pi}{\beta}$ with ultrasubharmonics.

Again, it is noted that $k = 1$ in the above corresponds to a linearly increasing enveloped cosine wave ($t \cos \beta t$ term).

3.5 CONTROLLABILITY AND OBSERVABILITY

Controllability is a basic property of systems which is indicative of its ability to control. More precisely, a system is *controllable* if it is possible to reach any state from any other state in finite time with some control. In practice, it is sometimes necessary to add further stipulations, such as admissible controls which may include magnitude constraints. This will be investigated in Vol. 2.

For linear systems, the definition above is equivalent to drive it to or from the origin. That is, the *linear system is controllable* if it is possible to transfer any given state from the origin to any finite final state in finite time by some $\mathbf{u}(t)$. One test for controllability of the linear time-invariant system (3.79) is to transform it to canonical normal form so that with $\mathbf{x} = \mathbf{Pz}$, where $\mathbf{P}$ is the modal matrix, (3.79) becomes $\mathbf{P\dot{z}} = \mathbf{APz} + \mathbf{Bu}$, or

$$\frac{d\mathbf{z}}{dt} = \mathbf{\Lambda z} + \mathbf{P}^{-1}\mathbf{Bu}, \tag{3.92}$$

again with

$$\mathbf{\Lambda} = \text{diag}[\lambda_1 \quad \cdots \quad \lambda_n].$$

where $\lambda_1, \ldots, \lambda_n$ are unique and real.

While the definition of controllability requires that every state be directly or indirectly affected by a control variable, in the canonical normal form every state variable (i.e., every mode of the system) must be manipulated directly by a control variable. In (3.92) there must be at least one nonzero element in every row of $\mathbf{P}^{-1}\mathbf{B}$.

If, however, the eigenvalues are complex, another transformation to the off-diagonal form of (3.67) shows that a control need only affect one of the two rows, corresponding, for example, to either $\dot{x}_3 = \alpha x_3 + \beta x_4$ or $\dot{x}_4 = -\beta x_3 + \alpha x_4$, where $\beta \neq 0$. But in the polar canonical form, $\dot{r} = \alpha r$, $\dot{\theta} = -\beta$, a control must affect both rows. If there are repeated eigenvalues, in general, a transformation can be made to a Jordan canonical form (3.66). Then it is only required that the last row of each J_i must have one control affecting it. A simple algebraic test is available for the general case of (3.79) by the following theorem [F1].

Theorem. System (3.79) is controllable if and only if the $n \times nm$ matrix

$$\mathbf{M} = [\mathbf{B}\ \mathbf{AB}\ \mathbf{A}^2\mathbf{B} \quad \cdots \quad \mathbf{A}^{n-1}\mathbf{B}] \tag{3.93}$$

has rank n.

Example 3.16

Consider

$$\dot{x}_1 = x_2,$$

$$\dot{x}_2 = -2x_1 - x_2 + u.$$

$\mathbf{B} = \mathbf{b} = [0 \quad 1]^T$ and

$$\mathbf{A} = \begin{bmatrix} 0 & 1 \\ -2 & -1 \end{bmatrix},$$

so, from (3.93),

$$\mathbf{M} = \begin{bmatrix} 0 & 1 \\ 1 & -1 \end{bmatrix}$$

which has rank $n = 2$, so that the system is controllable.

Example 3.17

Consider Example 3.8 again with added control so that

$$\dot{x}_1 = x_2,$$

$$\dot{x}_2 = x_3,$$

$$\dot{x}_3 = -\sigma_1\sigma_2 x_2 - (\sigma_1 + \sigma_2)x_3 + u.$$

Then $\mathbf{B} = [0 \quad 0 \quad 1]^T$,

$$\mathbf{A} = \begin{bmatrix} 0 & 1 & 0 \\ 0 & 0 & 1 \\ 0 & -\sigma_1\sigma_2 & -\sigma_1 - \sigma_2 \end{bmatrix},$$

$$\mathbf{M} = \begin{bmatrix} 0 & 0 & 1 \\ 0 & 1 & -\sigma_1 - \sigma_2 \\ 1 & -\sigma_1 - \sigma_2 & \sigma_1\sigma_2 + (\sigma_1 + \sigma_2)^2 \end{bmatrix}.$$

Again, rank of $\mathbf{M}$, $\rho[\mathbf{M}] = n = 3$, and the system is controllable. In the canonical normal form $\dot{\mathbf{z}} = \mathbf{\Lambda z} + \mathbf{P}^{-1}\mathbf{b}u$,

$$\dot{z}_1 = -\sigma_1 z_1 + \sigma_1^{-1}(\sigma_1 - \sigma_2)^{-1}u,$$

$$\dot{z}_2 = -\sigma_2 z_2 - \sigma_2^{-1}(\sigma_1 - \sigma_2)^{-1}u,$$

$$\dot{z}_3 = (\sigma_1\sigma_2)^{-1}u,$$

and the system controllability is checked.

Several conclusions can be derived from the discussion above that are quite relevant to control system design. These are summarized here without proof.

1. The controllable linear constant-coefficient system (3.79) has the pole-placement property. That is, the closed-loop system poles (eigenvalues of $\mathbf{A} + \mathbf{BK}$) may be arbitrarily assigned by constant-gain linear feedback $\mathbf{u} = \mathbf{Kx}$.[follows (3.92)].
2. If (3.79) undergoes scalar control, with $\mathbf{b}$ replacing $\mathbf{B}$, a specific selection of poles implies a unique $\mathbf{k}$, such that $u = \mathbf{k}^T\mathbf{x}$.
3. The controllable subspace, $C \subseteq \mathbb{R}^n$, generally is spanned by the column vectors of the controllability matrix $\mathbf{M}$ given by (3.93). In fact, $\mathbf{M}$ was generated in this manner.
4. Although nothing has been said about the class of controls required here, it should be noted that in one sense it is a quite realistic class. The class of square integrable controls may be used in the proofs. That is, $\int_{t_0}^{t_1} \mathbf{u}^T(t)\mathbf{u}(t)dt < \infty$ for the appropriate control period $[t_0, t_1]$.

Roughly, observability is the ability to determine the state of a system from certain output observations or measurements. More precisely, the linear time-invariant system (3.79) is *observable* if the initial state $\mathbf{x}(0)$ can be determined uniquely from the output $\mathbf{y}(t)$ and the input (control) $\mathbf{u}(t)$ which are given for a finite interval of time, $t \in [0, T]$. It is assumed that the output is given by (3.85). Again, observability for linear systems means that every normal mode must be observed directly in the canonical normal form

$$\frac{d\mathbf{z}}{dt} = \mathbf{\Lambda z} + \mathbf{P}^{-1}\mathbf{Bu}$$

and

$$\mathbf{y} = \mathbf{CPz} + \mathbf{Du}, \tag{3.94}$$

where again **P** is the modal matrix. As for controllability, the determination is simplest in this form if $\mathbf{\Lambda}$ has no off-diagonal elements such as from real simple (nonrepeated) eigenvalues. In this case, it is required only that the output matrix, **CP**, has at least one nonzero element in each column. Similar to the controllability test above for the case with off-diagonal elements, only one column of each such combination must have a nonzero element.

Again, a much handier test is available, and given by the following theorem, which will not be proven here [F1].

Theorem. Linear system (3.79), (3.94) is observable if and only if the observability matrix **N** has rank n, where

$$\mathbf{N} = [\mathbf{C}^{\mathrm{T}} \quad \mathbf{A}^{\mathrm{T}}\mathbf{C}^{\mathrm{T}} \quad \cdots \quad \mathbf{A}^{Tn-1}\mathbf{C}^{\mathrm{T}}]. \tag{3.95}$$

It can be shown that the rank of **N** is equal to the dimension of the observable subspace, $Q \subseteq \mathbb{R}^n$ (see [P1]).

Example 3.18

Consider Example 3.8 with an observation

$$y = x_1.$$

Consequently,

$$\mathbf{A} = \begin{bmatrix} 0 & 1 & 0 \\ 0 & 0 & 1 \\ 0 & -\sigma_1\sigma_2 & -\sigma_1 - \sigma_2 \end{bmatrix}, \qquad \mathbf{C} = \mathbf{c} = [1 \quad 0 \quad 0].$$

Then

$$\mathbf{N} = \begin{bmatrix} 1 & 0 & 0 \\ 0 & 1 & 0 \\ 0 & 0 & 1 \end{bmatrix} \qquad \text{with} \quad \text{rank}\ \rho(N) = n = 3,$$

and (3.95) is observable. Similarly, (3.94) becomes

$$y = [1 \quad 0 \quad 0 \quad]\mathbf{Pz} = [1 \quad 1 \quad 1]\mathbf{z}$$

and shows observability of all canonical normal modes. Note that in general the observable subspace is of dimension equal to the number of observed normal modes.

The significance of controllability and observability is further shown by transfer-function analysis. That is, when the input-output of a LTIS is represented by its transfer function $G(s) = y(s)/r(s)$, it is assumed that there is no pole-zero cancellation—in other words, that all the normal modes are affected by the controls (controllability) and that they are observed (observability). If the transfer function is of lower order than the system (number of state variables, n), the system is either uncontrollable or unobservable, or both. Although this can lead to poor performance even for the stable case, it is dangerous, particularly when unstable poles (poles in the right-hand complex plane) are canceled by right-hand-plane zeros. In the latter case, any random disturbance will cause a divergence with a nonzero initial state.

If (3.79), (3.94) is observable with $\mathbf{x} \in \mathbb{R}^n$, $\mathbf{y} \in \mathbb{R}^p$ there exists a $(n-p)$th-order observer with $(2n-p)$ of the eigenvalues for the composite plant–observer system chosen as desired. Here the observer is designed by appropriate selection of **L**, so that the observer $\hat{\mathbf{x}}(t)$ of $\mathbf{x}(t)$ has the form

$$\dot{\hat{\mathbf{x}}} = \mathbf{A}\hat{\mathbf{x}} + \mathbf{B}\mathbf{u} + \mathbf{L}(\mathbf{y} - \mathbf{C}\hat{\mathbf{x}}), \tag{3.96}$$

and the linear feedback control may be designed so that

$$\mathbf{u} = \mathbf{K}\hat{\mathbf{x}}. \tag{3.97}$$

If all the unstable modes of the system are controllable and observable, (3.96) generally will provide a satisfactory observer and the feedback control (3.97) can be appropriately selected for stable operation with usually satisfactory performance. In short, system (3.79), (3.85) controllable and observable is often written as (**A**, **B**) controllable and (**A**, **C**) observable, respectively.

For linear time-invariant systems, several further comments are worth noting in regard to controllability and observability.

1. Controllability and observability are independent of the length of time of control or observation.
2. Since the outputs are different for two different initial conditions for any input, only one input need be considered to distinguish between two initial states.
3. Controllability with respect to the origin and observability with respect to the initial state are sufficient for all states and all times, respectively.

Unfortunately, none of these are true for nonlinear systems in general. From the analysis above, it is apparent that the order n of a linear system can be obtained from input–output observations if and only if the system is controllable and observable. The most basic identification problem, that of estimating system order, can be quite complicated for an inaccessible physiological system, as well as others. Even with tracers (which are introduced in Vol. 2), it should be ascertained that there is adequate experimental access before data are extracted for identification purposes.

Another way of stating the desired theorem for application here is that the lowest order of the linear system that will "realize" the observed input–output data is $n = n_0$ if and only if both M and N have rank n. Otherwise, $n \geq n_0$. Such lowest-order systems are called *minimal realizations*. While a system based on input–output data may generally be identified by any number of realizations, we are concerned primarily with minimal realizations in this book.

Now it is desired to apply this result to a biological tracer experiment. A tracer can be inserted by special routes, such as by injection, directly into a particular compartment, or a tracer can be ingested by normally existing pathways. Here, suppose that a tracer is inserted directly into q system compartments and that its behavior can be monitored in p compartments. Define an input matrix **Q** with unity elements in each row corresponding to a compartment accessible for tracer insertion and with zero elements elsewhere. Each column of **Q** has exactly one unity element and the rank of **Q** is q. Also, define a $p \times n$ output matrix **P** with zero elements except for unity in each column that corresponds to an observable compartment. Then it is shown in Vol. 2 that experimental access is described by

$$\frac{d\mathbf{a}}{dt} = \mathbf{S}\mathbf{a} + \mathbf{X}_d^{-1}\mathbf{Q}\mathbf{f}_i, \tag{3.98}$$

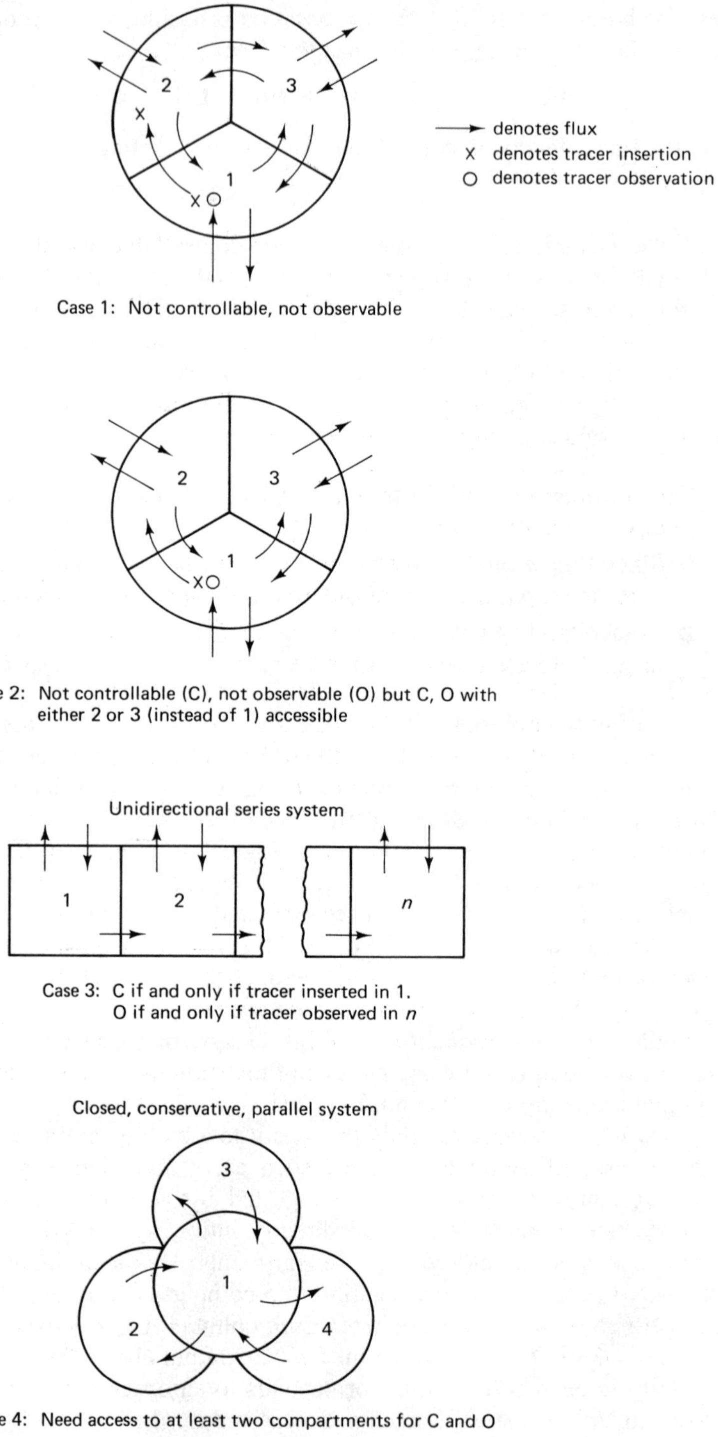

Figure 3.20 Tracer accessibility tests.

and

$$\mathbf{w} = \mathbf{P}\mathbf{a}, \tag{3.99}$$

where $\mathbf{X}_d$ is the diagonal matrix of elements $x_1, \ldots, x_n$ (assumed to be at steady state); $\mathbf{f}_i$ is the q-dimensional vector of inserted total activity fluxes, and therefore $\mathbf{X}_d^{-1}\mathbf{Q}\mathbf{f}_i$ is the vector of inserted specific activity fluxes; and $\mathbf{w}$ is the p-dimensional output vector of observed compartmental specific activities.

From (3.93) and (3.95), it is apparent that (3.98) and (3.99) are controllable and observable, respectively, if and only if

$$[\mathbf{Q} \quad \mathbf{S}\mathbf{Q} \quad \cdots \quad \mathbf{S}^{n-1}\mathbf{Q}]$$

and

$$[\mathbf{P}^{\mathrm{T}} \quad \mathbf{S}^{\mathrm{T}}\mathbf{P}^{\mathrm{T}} \quad \cdots \quad \mathbf{S}^{\mathrm{T}n-1}\mathbf{P}^{\mathrm{T}}]$$

are of rank n, respectively. Examples showing application of the rank test to several compartmental systems are given in Fig. 3.20. For case 4, the conservative system, it is assumed that the fluxes divided by the amount of substance from the originating compartment are all equal (i.e., $\phi_{12}/x_1 = \phi_{21}/x_2 = \cdots$).

Linear system controllability and observability are basic assumptions for nonlinear regulator design in Chapter 7 by the Popov method [A2].

EXERCISES

3.1. For Exercise 2.1 and Fig. 2.13:

(a) Define the equilibrium states.

(b) Derive the linearized perturbation equations that describe the motion for small perturbations from equilibrium. Assume that spring stiffness

$$k_i(x_i) = c_i(a_i x_i^2 + 1),$$

where a_i, c_i, $i = 1, 2$, are positive constants.

(c) Determine the solution to the linearized model, that is, $\delta x_1(t)$, $\delta x_3(t)$ for arbitrary initial condition δx_{10}, δx_{30} and $a_1 = 8$, $a_2 = 1$, $c_1 = 2$, $c_2 = 1$, $m_2 = 4m_1$, and $m_1 = 1/2g$.

(d) Sketch the state-plane portrait x_1 versus x_3 for various initial states. Use the *method of isoclines* to sketch the trajectories from lines of local tangency with slope $dx_3/dx_1 = f_3(x_1, x_3)/f_1(x_1, x_3)$.

(e) Discuss stability of the equilibrium states.

3.2. Analyze the stability of the singular point $x_1 = x_2 = 0$ by Liapunov's first method for the following systems.

(a) $\dot{x}_1 = -x_1^3 - 2x_2$,
$\dot{x}_2 = 2x_1 - 5x_2^3$.

(b) $\dot{x}_1 = x_1^2 + x_2$,
$\dot{x}_2 = x_1 + x_2^2$.

(c) $\dot{x}_1 = x_1^2 - x_2^2$,
$\dot{x}_2 = -x_1 x_2$.

(d) $\dot{x}_1 = x_2$,
$\dot{x}_2 = a_1 x_1 + a_2 x_2 + (b_1 x_1 + b_2 x_2)^2 x_2$,
where a_i are negative constants and b_i are constants, $i = 1, 2$.

3.3. Consider the two-link robot arm shown in Fig. 3.21. Here it is assumed that the link masses are negligible relative to the joint masses $2m$ and m as shown. Assume a spring and damping restraint torque at each link as shown with

$$T_1 = r\psi_1 + m_1\dot{\psi}_1,$$
$$T_2 = r(\psi_2 - \psi_1) + m_2(\dot{\psi}_2 - \dot{\psi}_1).$$

Also, a follower force F acts along the axis of the second link.
(a) Derive the dynamic equations of motion.
(b) Derive the linear perturbation equations at the zero equilibrium solution.
(c) Analyze the stability of the zero solution by Liapunov's first method.
(d) Analyze the controllability of this robot arm relative to T_1 and/or T_2.

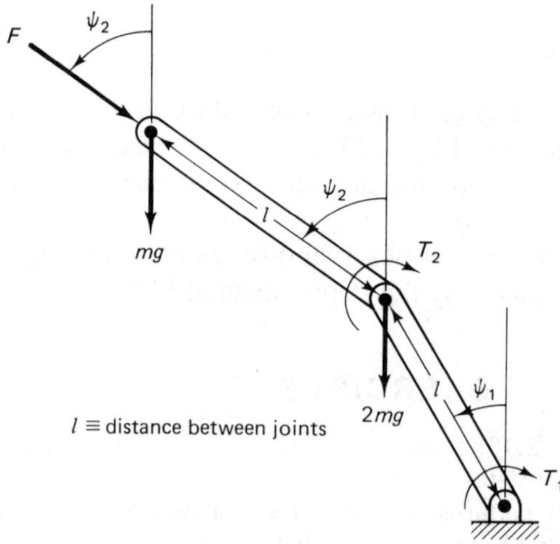

Figure 3.21 Robot arm for Exercise 3.3.

3.4. **(a)** Derive the closed-loop dynamic equations for Fig. 3.22. Write in canonical phase-variable form, $x_1 = x$, $x_2 = \dot{x}$, and in state form with $x_1 = x$, $x_2 = u$.
(b) Compute and classify all system equilibrium points.
(c) Sketch the parameter plane (k versus τ) to show the corresponding portrait types.

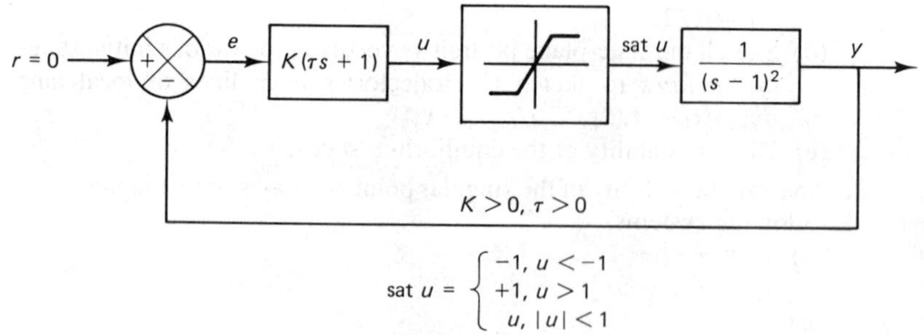

Figure 3.22 Feedback control Exercise 3.4.

3.5. Analyze linear control of the robot arm given by Exercise 3.3 for the small-perturbation case.

3.6. Show the stabilization of the linearized neutron kinetics (2.122) by traditional proportional-plus-integral control.

3.7. Suggest a classical feedback control for good performance of the system given by Exercises 2.1 and 3.1.

3.8. Compute heat-exchange transfer functions, $\delta T_u/\delta q(s)$, $\delta T_u/\delta w(s)$ according to small perturbations of (2.124)–(2.126) with $T_i = O$.

3.9. **(a)** Is the linearized neutron kinetics (2.122) controllable?
(b) Is the linearized state observable from measurements $\delta n(t)$?

3.10. **(a)** Analyze stability of the closed-loop control system for the nuclear reactor dynamics which is given by the perturbation equations $\delta T_u/\delta q(s)$ in Exercise 3.8 with $\delta w = 0$, (2.132) with $\delta k_p = 0$. Suggest a form of control compensation $\delta u/\delta T_u$ for zero δT_{ud}, desired. Here it is assumed that $\delta u = \delta k_c$ and $\delta q = \delta n$.
(b) Repeat part (a) with $\delta T_u/\delta w(s)$ and $\delta q = 0$ from Exercise 3.8, and δw replacing $\delta \dot{m}$ in (3.32). That is, part (a) refers to power (or temperature) control with coolant changes assumed to be zero, and part (b) refers to temperature control with power changes neglected. Actually, the assumptions are not too bad due to the system stiffness with rapid $\delta n(t)$ response and relative slow $\delta T_u(t)$.
(c) Discuss the overall closed-loop control characteristics.

3.11. Analyze controllability of the simple oscillator

$$\dot{x}_i = x_2, \qquad \dot{x}_2 = -w_n^2 x_1 + u$$

for **(a)** u unconstrained and **(b)** u constrained in magnitude such that $|u| \le 1$.

3.12. For the one-link telescopic robot of Exercise 2.7:
(a) Linearize the equations of motion about point $p(x_0, y_0)$ and $\dot{u} = 0$, $w^* = 0$.
(b) Linearize the equations of motion for constant $v = \alpha$ and $w^* = w(t)$ and for a constant w and $v^* = v(t)$.
(c) Study stability and controllability of both linearized models in part (b).
(d) Linearize the equations of motion about the reference trajectory $u(t) = (\alpha_2 - \alpha_1)e^{-t}$ and $w = (\theta_1 - \theta_0)e^{-t}$ with $\tan\theta_0 = y_0/x_0$ and $x_0^2 + y_0^2 = \alpha_1^2$.

3.13. For the system

$$\dot{x}_1 = x_3,$$
$$\dot{x}_2 = x_1 + x_2,$$
$$\dot{x}_3 = -2x_1 + 2x_3,$$

compute the modal matrix **P**. Use the eigenvector transformation to solve the system in the eigenvector coordinates and then in the original x_1, x_2, x_3 space.

CHAPTER

4

TIME-VARIANT SYSTEMS

In Chapter 3, linear perturbation analysis was introduced. It was shown that linearized models are convenient in many cases for ascertaining the stability of an equilibrium state, and indeed it is quite common for control engineers to design system controllers that are based on such analysis. A classic example of this is the design of an aircraft or missile control system which may be intended to hold the trajectory deviations near a trim condition. Another such example, which was introduced previously, is the dynamics for a nuclear reactor at some operating design power. Again, a task of a linear controller may be to correct the system deviations that result from plant disturbances to desired operational state.

In these cases time-invariant linear models were derived for small perturbations about equilibrium conditions. It was also shown, however, that time-variant linear models evolve when the perturbations are made about some time-variant reference trajectory as shown by (3.8). The purpose of this chapter is to summarize some important aspects of time-variant systems—in particular, linear time-variant systems (LTVS)—that can be useful to nonlinear systems with small perturbations.

On the other hand, linear time-variant processes are a significant class in their own right. Unfortunately, even for this class of linear systems, convenient analytical and design methodologies are not very prevalent. Linear systems with periodically time-varying parameters, however, are perhaps the most conveniently analyzed LTVS. They are summarized in Section 4.3 and applied in Section 4.4 to the stability analysis of limit cycles that occur for nonlinear systems.

4.1 FUNDAMENTAL PROPERTIES OF STABILITY

In Chapter 3, nonlinear systems were introduced by (3.1), that is, the nonautonomous (forced) system

$$\frac{d\mathbf{x}}{dt} = \mathbf{f}(\mathbf{x}, \mathbf{u}), \tag{4.1}$$

which for a specified feedback control $\mathbf{u}(\mathbf{x})$ becomes an autonomous time-invariant nonlinear system

$$\frac{d\mathbf{x}}{dt} = \mathbf{f}(\mathbf{x}, \mathbf{u}(x))$$

or

$$\frac{d\mathbf{x}}{dt} = \mathbf{f}_c(\mathbf{x}). \tag{4.2}$$

Here $\mathbf{x} \in \mathbb{R}^n$, $\mathbf{u} \in \mathbb{R}^m$, and $\mathbf{f}(\cdot)$, $\mathbf{f}_c(\cdot)$ are appropriate vector functions. Equilibrium states and their stability for autonomous systems were studied in Chapter 3.

4.1.1 Equilibrium States for Time-Variant Systems

If the control is a specified function of time $u(t)$, however, the nonlinear system (4.1) is also time variant or nonautonomous, such that

$$\frac{d\mathbf{x}}{dt} = \mathbf{f}(\mathbf{x}, t) \tag{4.3}$$

for $t_0 \leq t$ with $\mathbf{x}(t_0) = \mathbf{x}_0$. Even so, the basic concept of equilibrium is a bit more complicated for time-variant systems than for time-invariant ones. In general, the condition $\mathbf{f}(\mathbf{x}, t) = \mathbf{0}$ will require that some $\mathbf{x}_e$ and t_0, initial time, be specified. Then $\mathbf{x}_e$ is defined as an *equilibrium state* for (4.3) at $t \geq t_0$ if $\mathbf{f}(\mathbf{x}_e, t_0) = \mathbf{0}$.

Since equilibrium implies that the trajectory remains at $\mathbf{x}_e$ once it is reached, t_0 commonly is dropped from the definition. Consequently,

$$\mathbf{f}(\mathbf{x}_e, t) = \mathbf{0}$$

specifies $\mathbf{x}_e$ for all $t \geq t_0$. Notice the significance of t_0, however, for time-variant systems (4.3); that is, the same $\mathbf{x}(t; \mathbf{x}_0, t_0)$ generally may be attained by more than one path or trajectory with different t_0 and subsequent finite t. This is similar to a possible situation with non-Lipschitzian $\mathbf{f}(\mathbf{x})$ whereby an equilibrium point $\mathbf{x}_e$ may be reached or departed from in finite time for autonomous time-invariant (3.37).

For linear time-variant systems,

$$\frac{d\mathbf{x}}{dt} = \mathbf{A}(t)\mathbf{x}, \tag{4.4}$$

where $\mathbf{A}(t)$ is an appropriate $n \times n$ matrix function of time, $t \geq t_0$. Then

$$\mathbf{x}_e = \mathbf{0}$$

for all $t \geq t_0$. Furthermore, it is seen (as for LTIS) that if $\mathbf{A}(t)$ is nonsingular [i.e., $\mathbf{A}(t)\mathbf{x} = \mathbf{0}$ implies that $\mathbf{x} = \mathbf{0}$], then $\mathbf{x}_e = \mathbf{0}$ is the only and therefore an isolated equilibrium point. For (4.4) it is noted that a continuous $\mathbf{A}(t)$ is sufficient for a unique solution.

The perturbation linearized model for (4.3) evaluated at an equilibrium point $\mathbf{x}_e$ becomes

$$\delta\dot{\mathbf{x}} = \frac{\partial \mathbf{f}}{\partial \mathbf{x}}(\mathbf{x}_e, t)\, \delta\mathbf{x} \tag{4.5}$$

or

$$\delta\dot{\mathbf{x}} = \mathbf{A}(\mathbf{x}_e, t)\, \delta\mathbf{x}. \tag{4.6}$$

Then it can be shown that $\mathbf{x}_e$ is an isolated equilibrium state of the nonlinear time-variant system (4.3) if $\mathbf{A}(\mathbf{x}_e, t)$ is nonsingular and continuous [V1].

4.1.2 Equilibrium-Point Stability

Again, consider the nonlinear time-variant system (4.3) with equilibrium state $\mathbf{x}_e$ shown in Fig. 4.1. Here $\mathbf{x}_e$ is a *stable solution* to (4.3) if for any given t_0 and every real $\epsilon > 0$, there exists a real $\delta(\epsilon, t_0)$ such that

(a) $\|\mathbf{x}_0 - \mathbf{x}_e\| < \delta$

implies that

(b) $\|\mathbf{x}(t) - \mathbf{x}_e\| < \epsilon$,

where $\|\cdot\|$ are appropriate vector norms. Notice that this definition is similar to that for autonomous systems in Chapter 3, but here $\delta(\cdot)$ is a function of time t_0 as well as ϵ.

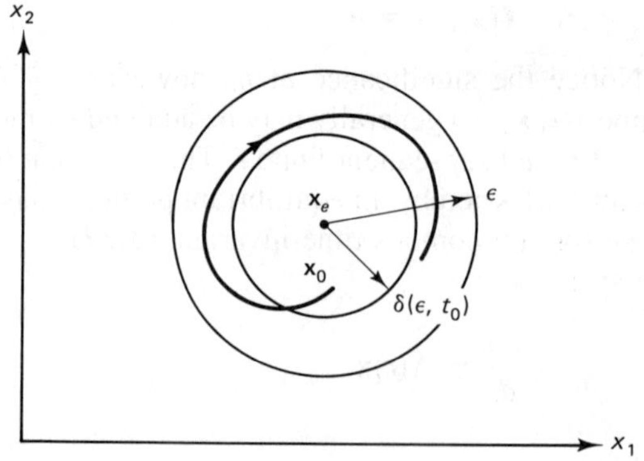

Figure 4.1 Equilibrium-point stability.

Again, *asymptotic stability* of $\mathbf{x}_e$ can be defined similar to Section 3.2.1, roughly stable and convergent (see Fig. 4.2), except again the dependence of $\delta_1(t_0)$ and $T(\epsilon, t_0)$ on t_0. Similarly, *global asymptotic stability* means that $\mathbf{x}_e$ is a unique equilibrium point which is stable such that every trajectory for (4.3) converges to it. If $\delta(\epsilon, t_0) = \delta(\epsilon)$, that is, the starting region is independent of t_0, and conditions (a) and (b) are satisfied, then $\mathbf{x}_e$ is *uniformly stable*. Similarly, with this added relaxation, we can define uniform asymptotic and global uniform asymptotic stability. More precisely, *uniform asymptotic stability* of $\mathbf{x}_e$ means that it is uniformly stable and the corresponding trajectories are uniformly convergent (i.e., δ_1 and T are independent of t_0).

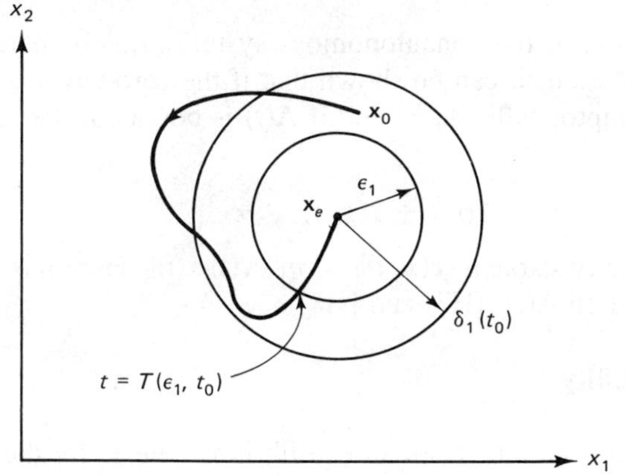

Figure 4.2 Equilibrium-point asymptotic stability.

Global uniform asymptotic stability or uniform asymptotic stability in the large for $\mathbf{x}_e$, however, also requires that $\mathbf{x}_e$ of (4.3) is bounded or equivalently Lagrange stable. Also, $\mathbf{x}_e$ is said to be *bounded* or *Lagrange stable* if

$$\|\mathbf{x}(t)\| < B(t_0)$$

for all $t \geq t_0$, and where $\mathbf{x}(t) = \mathbf{x}(t; \mathbf{x}_0, t_0)$. If B is independent of t_0, such $\mathbf{x}(t)$ is *uniformly bounded*. Then $\mathbf{x}_e$ is *globally uniformly asymptotically stable* if it is (1) uniformly stable, (2) uniformly bounded, and (3) all trajectories are uniformly convergent.

It is apparent from the definitions that all cases of stability for autonomous systems (3.37) are also cases of uniform stability. Also, since linear system solutions, $\mathbf{x}(t) = \mathbf{x}(t; \mathbf{x}_0, t_0)$, are proportional to $\mathbf{x}_0$, Lagrange stability or boundedness, which refer to a trajectory, is identical to system stability for linear systems.

Example 4.1 [M11]

Consider the first-order LTVS

$$\dot{x} = (4t \sin t - 2t)x$$

with solution

$$x(t; x_0, t_0) = x_0 \exp(4 \sin t - 4t \cos t - t^2 - 4 \sin t_0 + 4t_0 \cos t_0 + t_0^2).$$

Then, as shown in Fig. 4.3, $x_e = 0$ is globally asymptotically stable, but not uniformly stable since T depends on t_0.

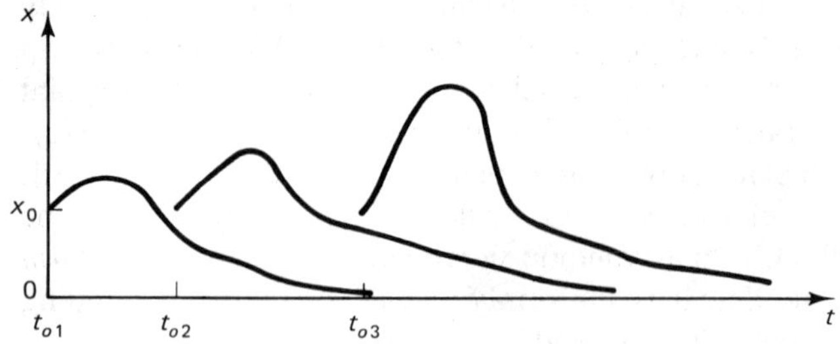

Figure 4.3 Different t_0 approximate trajectories for Example 4.1.

In regard to linearization of the nonautonomous system (4.3) by perturbations about a zero equilibrium solution, it can be shown that if the zero trivial solution of $\dot{\mathbf{x}} = \mathbf{A}(t)\mathbf{x}$ is uniformly asymptotically stable and if $\mathbf{A}(t)$ is bounded, the zero trivial solution (if there is one) of

$$\dot{\mathbf{x}} = \mathbf{A}(t)\mathbf{x} + \mathbf{r}(\mathbf{x}, t)$$

is also uniformly asymptotically stable if $\|\mathbf{r}(\mathbf{x}, t)\| < \eta\|(\mathbf{A}(t)\,\mathbf{x}(t)\|$. Here η is a constant which generally may depend on $\mathbf{A}(t)$ ([H7] and [H8]).

4.1.3 Free Trajectory Stability

Similar to the motion for perturbations from an equilibrium state $\mathbf{x}_e$ for the unforced system (4.3), stability of some reference trajectory $\mathbf{x}^*(t)$ for perturbations from this reference may be of importance. The satellite Example 3.12 is a case of interest, where $\mathbf{x}^*(t)$ may refer to a 24-hour orbit of a communication satellite, and a thrust pulse controller may be used to stabilize the motion $\delta\mathbf{x}(t)$ relative to $\mathbf{x}^*(t)$.

It is convenient to consider Figs. 4.1 and 4.2 and replace $\mathbf{x}_e$ by $\mathbf{x}^*(t)$. Consequently, all the previous definitions of stability, asymptotic stability, and uniform stability can be made relative to $\mathbf{x}^*(t)$. To do this, consider $\boldsymbol{\xi}(t) = \mathbf{x}(t) - \mathbf{x}^*(t)$, where $\dot{\mathbf{x}}^* = \mathbf{f}(\mathbf{x}^*, t)$ [i.e., $\mathbf{x}^*(t)$ satisfies (4.3)], and $\dot{\mathbf{x}}^*(t) + \dot{\boldsymbol{\xi}}(t) = \mathbf{f}(\boldsymbol{\xi} + \mathbf{x}^*, t)$. Then

$$\dot{\boldsymbol{\xi}} = \mathbf{f}_1(\xi, t),$$

where

$$\mathbf{f}_1(\xi, t) = \mathbf{f}(\xi + \mathbf{x}^*, t) - \mathbf{f}(\mathbf{x}^*, t)$$

and

$$\mathbf{f}_1(\mathbf{0}, t) = \mathbf{0}.$$

It should be noted here that $\boldsymbol{\xi}(t)$ is not restricted to small perturbations [as $\delta\mathbf{x}(t)$ was previously when linearizing the system].

This concept of stability is shown, more clearly perhaps, in Fig. 4.4. In summary, $\mathbf{x}^*(t; \mathbf{x}_0^*, \mathbf{t}_0)$ is *stable* if for all t_0 and $\epsilon > 0$, there is a $\delta(\epsilon, t_0) > 0$ such that

(a) $\|\mathbf{x}(t; \mathbf{x}_0, t_0) - \mathbf{x}^*(t; \mathbf{x}_0^*, t_0)\| < \epsilon$

for all $t \geq t_0$ if

(b) $\|\mathbf{x}_0 - \mathbf{x}_0^*\| < \delta$.

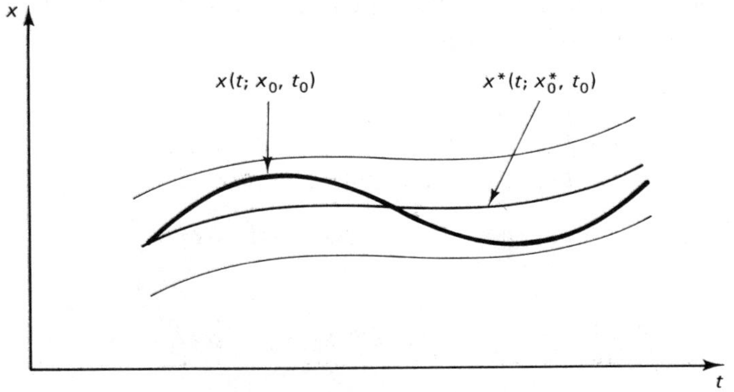

Figure 4.4 Trajectory–stability illustration.

Furthermore, $\mathbf{x}^*(t;\ \mathbf{x}_0^*,\ t_0)$ is *asymptotically stable* if it is stable and convergent where convergence roughly requires that for any t_0 there is a $\delta_1(t_0)$ such that

$$\|\mathbf{x}_0 - \mathbf{x}_0^*\| < \delta_1$$

implies that

$$\lim_{t\to\infty}\|\mathbf{x}(t;\ \mathbf{x}_0,\ t_0) - \mathbf{x}^*(t;\ \mathbf{x}_0^*,\ t_0)\| = 0.$$

It should be noted here that examination of state space Fig. 4.1 or 4.2 rather than Fig. 4.4 can be misleading if the time reference is not considered carefully. For example, consider the closed-solution path C in Fig. 4.5. It can be seen that a $\delta(\epsilon,\ t_0)$ cannot be found for every $\epsilon > 0$ such that conditions (a) and (b) of the stability definition are satisfied. Hence $\mathbf{x}^*(t)$, trajectory C is unstable. Intuitively, however, C does satisfy our sense of stability since a disregard of the time scale (i.e., ignoring "slippage" between trajectory time reference points in Fig. 4.5) does show convergence. Indeed, such stability is a very useful concept for periodic closed-solution paths and it is called *orbital stability*. Formally, a periodic motion $\mathbf{x}^*(t;\ \mathbf{x}_0;\ t_0)$ with closed state-space trajectory C is orbitally stable if any trajectory starting from a state very close to C remains very close to C. Furthermore, it is *asymptotically orbitally stable* if it is orbitally stable and any motion approaches C as $t \to \infty$.

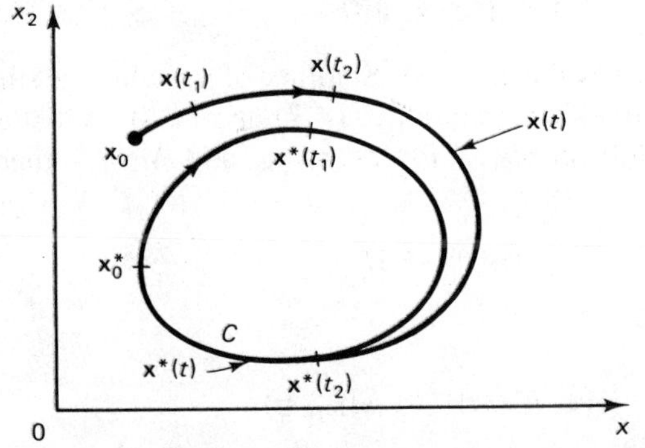

Figure 4.5 Asymptotically orbitally stable trajectory.

More precisely, periodic closed trajectory C is *orbitally stable* if there exists a moving point $\mathbf{p}(t)$ on C such that

$$\|\mathbf{x}(t;\, \mathbf{x}_0,\, t_0) - \mathbf{p}(t)\| < \epsilon$$

for all $t \geq t_0$. Furthermore, C is *asymptotically orbitally stable* if it is orbitally stable and for any $\mathbf{x}_0^*$ on C there exists a $\delta_1 > 0$ such that

$$\|\mathbf{x}_0 - \mathbf{x}_0^*\| < \delta_1$$

implies that

$$\lim_{t\to\infty}\|\mathbf{x}(t;\, \mathbf{x}_0,\, t_0) - \mathbf{p}(t)\| = 0.$$

Hence trajectory C shown in Fig. 4.5 in fact is asymptotically orbitally stable. Such orbital stability will be studied in more detail for nonlinear systems at the end of this chapter and early in Chapter 5. First, it is necessary to recall some fundamental concepts of linear time-variant systems (4.4) in Section 4.2 after the implications of trajectory stability to system stability are summarized.

4.1.4 System Stability

Certain conclusions of stability for system (4.3), as opposed to individual solution stability as studied above, may be drawn from these stability definitions. For example, if an isolated equilibrium point is *globally asymptotically stable*, it is also a unique equilibrium point, and the system sometimes is called completely stable.

Although Lagrange stability or boundedness referred to an equilibrium point above, they may be generalized so that a system is called *Lagrange stable* if all its solutions are bounded. Of course, for linear systems, $\dot{\mathbf{x}} = \mathbf{A}(t)\mathbf{x}$, it is readily seen that solutions are bounded if the zero equilibrium point (null solution) is stable. In general, it is seen that linear systems (LTVS as well as LTIS) are stable, asymptotically stable, or unstable (in the Lagrange sense) if the null solution is stable, asymptotically stable, or unstable (Lagrange), respectively.

In general, a forced time-variant system with input $\mathbf{u}(t) \in \mathbb{R}^m$, output $\mathbf{y}(t) \in \mathbb{R}^p$, and state $\mathbf{x}(t) \in \mathbb{R}^n$ is given by

$$\dot{\mathbf{x}} = \mathbf{f}(\mathbf{x},\, t;\, \mathbf{u}(t)), \tag{4.7}$$

$$\mathbf{y} = \mathbf{g}(\mathbf{x},\, t;\, \mathbf{u}(t)), \tag{4.8}$$

where $\mathbf{u}(t)$ is not given a priori as it is in (4.3). Stability of $\mathbf{x}_e$, which is studied above, refers to (4.7) with zero input. More generally, (4.7) and (4.8) are called *bounded-input, bounded-output (BIBO) stable* if for every $\mathbf{x}_0$ and $M > 0$ there exists an $N(\mathbf{x}_0,\, \mathbf{u}) > 0$ such that

$$\|\mathbf{u}(t)\| < M$$

for all $t \geq t_0$ implies that

$$\|\mathbf{g}(\mathbf{x},\, t;\, \mathbf{u}(t))\| < N(\mathbf{x}_0,\, \mathbf{u})$$

for all $t \geq t_0$. Here $\mathbf{x} = \mathbf{x}(t;\, \mathbf{x}_0,\, t_0)$ for a specified $\mathbf{u}(t)$. Such systems are said to be

BIBO stable. Unfortunately, useful BIBO stability results for nonlinear systems are rare except for very specific cases.

4.2 LINEAR TIME-VARIANT SYSTEMS

Unforced linear time-variant systems are given by (4.4), and as noted above, they have an equilibrium point at the origin which is isolated if $\mathbf{A}(t)$ is nonsingular. Then certain conclusions about stability were noted, and more will be investigated by Liapunov's direct (or second) method in Chapter 6. The forced linear time-variant system is given by

$$\frac{d\mathbf{x}}{dt} = \mathbf{A}(t)\mathbf{x} + \mathbf{B}(t)\mathbf{u}, \tag{4.9}$$

where $\mathbf{x}(t_0) = \mathbf{x}_0$, $\mathbf{x} \in \mathbb{R}^n$, $\mathbf{u} \in \mathbb{R}^m$, and $\mathbf{A}(t)$, $\mathbf{B}(t)$ are appropriate $n \times n$ and $n \times m$ matrix functions. Again, such LTVS were derived for the approximation of a nonlinear system under small perturbations about a specified time-variant reference trajectory as given by (3.8). In other cases, such models may arise naturally when a linear system has parameters that may change with time [e.g., a trimmed aircraft with rapid consumption of fuel so that the mass m in (3.46) is not constant]. Also, the nuclear fission process with a specified reactivity $u(t)$ in (3.15) becomes an LTVS, as would an electrical circuit with time-changing admittance.

4.2.1 Form of Solution

In Chapter 3 the state-transition matrix, $\mathbf{e}^{\mathbf{A}(t-t_0)}$, was introduced as the solution to the matrix equation $\dot{\mathbf{X}} = \mathbf{A}\mathbf{x}$ with $\mathbf{X}(t_0) = \mathbf{I}$. Then the general solution to the forced linear time-invariant system (3.79) $\dot{\mathbf{x}} = \mathbf{A}\mathbf{x} + \mathbf{B}\mathbf{u}$ is given by superposition or

$$\mathbf{x}(t) = \mathbf{\Phi}(t - t_0)\mathbf{x}_0 + \int_{t_0}^{t} \mathbf{\Phi}(t - \sigma)\mathbf{B}\mathbf{u}(\sigma)\, d\sigma,$$

where $\mathbf{\Phi}(t - t_0) = \exp[\mathbf{A}(t - t_0)]$.

Again, superposition may be applied to (4.9) to obtain the general solution as a summation of the zero-input response and the zero-state response. Again, we can obtain linearly independent solutions to the homogeneous matrix equation

$$\dot{\mathbf{X}} = \mathbf{A}(t)\mathbf{X}, \tag{4.10}$$

with $\mathbf{X}(t_0) = \mathbf{I}$, the identity matrix. Unfortunately, $\mathbf{\Phi}(\cdot)$ is not an exponential matrix in general. Fortunately, however, it can be used as a fundamental matrix solution for the general solution. Again, this fundamental matrix solution, called a state-transition matrix, depends on two arguments (i.e., when the initial state is applied and when it is observed). Consequently, the state-transition matrix, $\mathbf{\Phi}(t, t_0)$, is the solution to (4.10) with $\mathbf{\Phi}(t, t) = \mathbf{\Phi}(t_0, t_0) = \mathbf{I}$. Then the general solution to (4.9) becomes

$$\mathbf{x}(t) = \mathbf{\Phi}(t, t_0)\mathbf{x}_0 + \int_{t_0}^{t} \mathbf{\Phi}(t, \sigma)\mathbf{B}(\sigma)\mathbf{u}(\sigma)\, d\sigma, \tag{4.11}$$

which is readily verified by substitution into (4.9).

Several properties of $\boldsymbol{\Phi}(t, t_0)$ that may be derived include the following:

1. $\boldsymbol{\Phi}(t, t) = \mathbf{I}$.
2. $\boldsymbol{\Phi}(t, t_0)$ is unique.
3. $\boldsymbol{\Phi}(t_2, t_0) = \boldsymbol{\Phi}(t_2, t_1)\boldsymbol{\Phi}(t_1, t_0)$.
4. $\det[\boldsymbol{\Phi}(t, t_0)] = \exp \int_{t_0}^{t} \operatorname{Tr} \mathbf{A}(\sigma)\, d\sigma \neq 0$.
5. $\boldsymbol{\Phi}(t, t_0) = \exp \int_{t_0}^{t} \mathbf{A}(\sigma)\, d\sigma$ if and only if $\mathbf{A}(t_1)\mathbf{A}(t_2) = \mathbf{A}(t_2)\mathbf{A}(t_1)$ for all t_1, t_2 [includes diagonal $\mathbf{A}(\cdot)$ and constant $\mathbf{A}$].

Example 4.2

Consider the scalar LTVS

$$\dot{x} = [\operatorname{sgn}(t - 1)]x + b,$$

where $x(0) = x_0$, and

$$\operatorname{sgn} t = \begin{cases} 1, & t \geq 0, \\ -1, & t < 0. \end{cases}$$

(Here, sgn t is defined this way for convenience, but generally it is not defined for $t = 0$.) Then, solving for $0 \leq t < 1$ and then for $1 \leq t$, we obtain

$$x(t) = \begin{cases} (x_0 - b)e^{-t} + b, & 0 \leq t < 1, \\ (x_1 + b)e^{(t-1)} - b, & 1 \leq t, \end{cases}$$

where $x_1 = x(1) = (x_0 - b)e^{-1} + b$. If $x_1 = -b$, $x_e = -b$, which requires that $x_0 = (1 - 2e)b$. Obviously, x_e depends on x_0 and b, but equilibrium occurs only for very specific conditions, and it is never stable. If, on the other hand,

$$\dot{x} = [\operatorname{sgn}(1-t)]x + b,$$

$$x(t) = \begin{cases} (x_0 + b)e^{t} - b, & 0 \leq t < 1, \\ (x_1 - b)e^{(1-t)} + b, & 1 \leq t, \end{cases}$$

where $x_1 = ex_0 + (e - 1)b$. The new system has a stable equilibrium at $x_e = b$ which is attained under general x_0 and b. Unless $x_0 = b(2e^{-1} - 1)$, it takes infinite time to reach $x_e = b$.

Example 4.3

Consider LTVS,

$$t^2\frac{d^2x}{dt^2} + 4t\frac{dx}{dt} + 2x = u(t). \tag{4.12}$$

A change of variables, $t = e^{\tau}$, with

$$\frac{dx}{dt} = \frac{dx}{d\tau}\frac{d\tau}{dt} = \frac{1}{t}\frac{dx}{d\tau}$$

and

$$\frac{d^2x}{dt^2} = \frac{d}{dt}\left[\frac{1}{t}\frac{dx}{d\tau}\right] = -\frac{1}{t^2}\frac{dx}{d\tau} + \frac{1}{t}\frac{d}{d\tau}\left[\frac{dx}{d\tau}\right]\frac{d\tau}{dt}$$

$$= \frac{1}{t^2}\left[\frac{d^2x}{d\tau^2} - \frac{dx}{d\tau}\right],$$

leads to an LTIS. Now, by substitution, (4.12) becomes

$$\frac{d^2x}{d\tau^2} + 3\frac{dx}{d\tau} + 2x = u(\tau),$$

where $\tau = \ln t$. Let $x_1 = x$, $x_2 = dx/d\tau$. Then the modal matrix is

$$\mathbf{P} = \mathbf{V} = \begin{bmatrix} 1 & 1 \\ -1 & -2 \end{bmatrix}$$

and

$$\mathbf{P}^{-1} = \mathbf{V}^{-1} = \begin{bmatrix} 2 & 1 \\ -1 & -1 \end{bmatrix},$$

so that

$$\mathbf{e}^{\mathbf{A}\tau} = \mathbf{P}\mathbf{e}^{\Lambda\tau}\mathbf{P}^{-1}$$

$$= \begin{bmatrix} 2e^{-\tau} - e^{-2\tau} & e^{-\tau} - e^{-2\tau} \\ -2e^{-\tau} + 2e^{-\tau} & -e^{-\tau} + 2e^{-2\tau} \end{bmatrix}.$$

Then, with $\tau = \ln t$, it is seen that

$$\begin{bmatrix} x \\ t\dfrac{dx}{dt} \end{bmatrix} = \begin{bmatrix} 2\dfrac{t_0}{t} - \left(\dfrac{t_0}{t}\right)^2 & \dfrac{t_0}{t} - \left(\dfrac{t_0}{t}\right)^2 \\ -2\dfrac{t_0}{t} + 2\left(\dfrac{t_0}{t}\right)^2 & -\dfrac{t_0}{t} + 2\left(\dfrac{t_0}{t}\right)^2 \end{bmatrix} \begin{bmatrix} x(t_0) \\ t_0\dfrac{dx}{dt}(t_0) \end{bmatrix}.$$

Therefore, the state-transition matrix in terms of $x_1 = x$ and $x_2 = \dot{x}$ becomes

$$\mathbf{\Phi}(t, t_o) = \begin{bmatrix} 2\dfrac{t_0}{t} - \left(\dfrac{t_0}{t}\right)^2 & \dfrac{t_0^2}{t} - \dfrac{t_0^3}{t^2} \\ -2\dfrac{t_0}{t^2} + 2\dfrac{t_0^2}{t^3} & -\left(\dfrac{t_0}{t}\right)^2 + 2\left(\dfrac{t_0}{t}\right)^3 \end{bmatrix}. \tag{4.13}$$

Subsequently, it can be seen that the Euler linear differential equation

$$t^n\frac{d^nx}{dt^n} + a_1t^{n-1}\frac{dx^{n-1}}{dt^{n-1}} + \cdots + a_{n-1}t\frac{dx}{dt} + a_nx = u(t), \tag{4.14}$$

where a_k, $k = 1, \ldots, n$, can be transformed to a linear constant-coefficient state equation by $t = e^\tau$ as shown above for $n = 2$. In general,

$$\frac{d^nx}{dt^n} = \frac{1}{t^n}\frac{d}{d\tau}\left(\frac{d}{d\tau} - 1\right)\cdots\left[\frac{d}{d\tau} - (n-1)\right]x. \tag{4.15}$$

4.2.2 Adjoint System

The homogeneous adjoint system of the linear system (4.4) is commonly defined by

$$\frac{d\mathbf{p}}{dt} = -\mathbf{A}^{\mathrm{T}}(t)\mathbf{p}, \tag{4.16}$$

where $\mathbf{p} \in \mathbb{R}^n$, and again T refers to matrix transpose. Since transposition does not change the eigenvalues, but the negative sign reverses their sign, the eigenvalues of the adjoint system are merely the negative of those of the original $\mathbf{A}$ for the constant-matrix case. Or it is expected that the solution of (4.16) forward in time resembles the solution to $\dot{\mathbf{x}} = \mathbf{A}(t)\mathbf{x}$ with time reversed. This can be seen more precisely by examination of the state-transition matrix.

Let $\boldsymbol{\psi}(t, t_0)$ denote the state-transition matrix of (4.16). Then, by definition,

$$\frac{d}{dt}\boldsymbol{\psi}(t, t_0) = -\mathbf{A}^{\mathrm{T}}(t)\boldsymbol{\psi}(t, t_0), \tag{4.17}$$

where $\boldsymbol{\psi}(t_0, t_0) = \mathbf{I}$. Then

$$\begin{aligned}\frac{d}{dt}[\boldsymbol{\psi}^{\mathrm{T}}(t, t_0)\boldsymbol{\Phi}(t, t_0)] &= \left[\frac{d}{dt}\boldsymbol{\psi}^{\mathrm{T}}(t, t_0)\right]\boldsymbol{\Phi}(t, t_0) + \boldsymbol{\psi}^{\mathrm{T}}(t, t_0)\frac{d}{dt}\boldsymbol{\Phi}(t, t_0)\\ &= [-\mathbf{A}(t)\boldsymbol{\psi}(t, t_0)]^{\mathrm{T}}\boldsymbol{\Phi}(t, t_0) + \boldsymbol{\psi}^{\mathrm{T}}(t, t_0)\mathbf{A}(t)\boldsymbol{\Phi}(t, t_0)\\ &= \mathbf{0}\end{aligned}$$

or

$$\boldsymbol{\psi}^{\mathrm{T}}(t, t_0)\boldsymbol{\Phi}(t, t_0) = \mathbf{I}, \tag{4.18}$$

since $\boldsymbol{\psi}(t_0, t_0) = \boldsymbol{\Phi}(t_0, t_0) = \mathbf{I}$. Therefore,

$$\boldsymbol{\psi}^{\mathrm{T}}(t, t_0) = \boldsymbol{\Phi}^{-1}(t, t_0) = \boldsymbol{\Phi}(t_0, t) \tag{4.19}$$

for all t_0 and t.

Example 4.4

Consider (4.12) with $u = 0$, $x_1 = x$, and $x_2 = \dot{x}$ so that

$$\dot{\mathbf{x}} = \mathbf{A}(t)\mathbf{x},$$

where

$$\mathbf{A} = \begin{bmatrix} 0 & 1 \\ -\dfrac{2}{t^2} & -\dfrac{4}{t} \end{bmatrix}.$$

The adjoint system becomes

$$\dot{\mathbf{p}} = -\mathbf{A}^{\mathrm{T}}(t)\mathbf{p},$$

where

$$-\mathbf{A}^{\mathrm{T}}(t) = \begin{bmatrix} 0 & \dfrac{2}{t^2} \\ -1 & \dfrac{4}{t} \end{bmatrix}.$$

Then from (4.13) and (4.19),

$$\boldsymbol{\psi}(t, t_0) = \begin{bmatrix} 2\dfrac{t}{t_0} - \left(\dfrac{t}{t_0}\right)^2 & -2\dfrac{t}{t_0^2} + 2\dfrac{t^2}{t_0^3} \\ \dfrac{t^2}{t_0} - \dfrac{t^3}{t_0^2} & -\left(\dfrac{t}{t_0}\right)^2 + 2\left(\dfrac{t}{t_0}\right)^3 \end{bmatrix}. \tag{4.20}$$

The linear system (4.9), with output, a linear function of state, is given by

$$\dot{\mathbf{x}} = \mathbf{A}(t)\mathbf{x} + \mathbf{B}(t)\mathbf{u}, \tag{4.21}$$

$$\mathbf{y} = \mathbf{C}(t)\mathbf{x}, \tag{4.22}$$

where $\mathbf{y} \in \mathbb{R}^p$, and $\mathbf{C}(t)$ is an appropriate $p \times n$ matrix function of time. The *adjoint* or *dual system* (4.21)–(4.22) is given by

$$\dot{\mathbf{p}} = -\mathbf{A}^{\mathrm{T}}(t)\mathbf{p} + \mathbf{C}^{\mathrm{T}}(t)\mathbf{v}, \tag{4.23}$$

$$\mathbf{q} = \mathbf{B}^{\mathrm{T}}(t)\mathbf{p}. \tag{4.24}$$

In Vol. 2 adjoint systems will be utilized in the computation of optimal controls. Also, they are useful in the computation of mean-square error for time-variant linear systems. Adjoint systems are found to form a sometimes-convenient duality with linear systems regarding controllability and observability which establishes the relationship between optimal, quadratic-performance regulator design and filter or observer design [P1].

4.2.3 Controllability and Observability

As given in Chapter 3, the linear time-variant system (4.21) is *controllable* if and only if there exists a control $\mathbf{u}(t)$ which transfers the zero state to any finite final state in finite time. Then the following theorem may be derived [R4].

Theorem. System (4.21) is controllable on $[t_0, t_1]$ if and only if the controllability Grammian

$$\mathbf{Z}(t_0, t_1) \equiv \int_{t_0}^{t_1} \boldsymbol{\Phi}(t_1, t)\mathbf{B}(t)\mathbf{B}^{\mathrm{T}}(t)\boldsymbol{\Phi}^{\mathrm{T}}(t_1, t)\, dt \tag{4.25}$$

is positive definite. And if (4.25) is positive definite, it is easily shown that the following control will drive (4.21) from $\mathbf{0}$ to $\mathbf{x}$ during $[t_0, t_1]$:

$$\mathbf{u}(t) = \mathbf{B}^{\mathrm{T}}(t)\boldsymbol{\Phi}^{\mathrm{T}}(t_1, t)\mathbf{Z}^{-1}(t_0, t_1)\mathbf{x}_1. \tag{4.26}$$

Linear system (4.21)–(4.22) is *observable* if and only if an observation $\mathbf{y}(t)$, $t \in [t_0, t_1]$, yields $\mathbf{x}(t_0) = \mathbf{x}_0$ (as given in Chapter 3). Then the following equivalent theorems may be derived [R1].

Theorem. Linear system (4.21)–(4.22) is observable if and only if the observability Grammian

$$\mathbf{W}(t_0, t_1) \equiv \int_{t_0}^{t_1} \boldsymbol{\Phi}^{\mathrm{T}}(t, t_0)\mathbf{C}^{\mathrm{T}}(t)\mathbf{C}(t)\boldsymbol{\Phi}(t, t_0)\, dt \tag{4.27}$$

is positive definite.

Theorem. System (4.21)–(4.22) is observable if and only if there exists a sequence of time points $s_1, \ldots, s_q$ on $[t_0, t_1]$ such that the rank of the following matrix is n:

$$[\mathbf{\Phi}^T(s_1, t_0)\mathbf{C}^T(s_1) \cdots \mathbf{\Phi}^T(s_q, t_0)\mathbf{C}^T(s_q)]. \tag{4.28}$$

Theorem. System (4.21)–(4.22) is controllable (observable) if and only if the adjoint system (4.23)–(4.24) is observable (controllable).

Example 4.5

Consider Example 4.4 with an output so that

$$\dot{\mathbf{x}} = \mathbf{A}(t)\mathbf{x}, \qquad \mathbf{y} = \mathbf{c}^T \mathbf{x},$$

where

$$\mathbf{A}(t) = \begin{bmatrix} 0 & 1 \\ -\dfrac{2}{t^2} & -\dfrac{4}{t} \end{bmatrix}, \qquad \mathbf{c}^T = [0 \quad 1].$$

Then, from (4.13) with $t_0 = 1$,

$$\mathbf{\Phi}(t, 1) = \begin{bmatrix} \dfrac{2}{t} - \dfrac{1}{t^2} & \dfrac{1}{t} - \dfrac{1}{t^2} \\[2ex] -\dfrac{2}{t^2} + \dfrac{2}{t^3} & -\dfrac{1}{t^2} + \dfrac{2}{t^3} \end{bmatrix},$$

and substitution into (4.28) for $s_1 = 1$, $s_2 = 2$ yields

$$[\mathbf{\Phi}^T(1, 1)\mathbf{c} \quad \mathbf{\Phi}^T(2, 1)\mathbf{c}] = \begin{bmatrix} 0 & -\frac{1}{4} \\ 1 & 0 \end{bmatrix},$$

which has rank $n = 2$. Thus the system is observable on $[1, 2]$. Furthermore, the adjoint system

$$\dot{\mathbf{p}} = -\mathbf{A}^T(t)\mathbf{p} + \mathbf{c}v$$

is controllable. Here

$$-\mathbf{A}^T(t) = \begin{bmatrix} 0 & \dfrac{2}{t^2} \\[2ex] -1 & \dfrac{4}{t} \end{bmatrix}, \qquad \mathbf{c} = \begin{bmatrix} 0 \\ 1 \end{bmatrix}.$$

Consider the linearized perturbation system at equilibrium $\mathbf{x}_e$, $\mathbf{u}_e$:

$$\delta\dot{\mathbf{x}} = \mathbf{A}(t)\,\delta\mathbf{x} + \mathbf{B}(t)\,\delta\mathbf{u} \tag{4.29}$$

for

$$\dot{\mathbf{x}} = \mathbf{f}(\mathbf{x}, \mathbf{u}, t), \tag{4.30}$$

where

$$\mathbf{A}(t) = \frac{\partial \mathbf{f}}{\partial \mathbf{x}}(\mathbf{x}_e, \mathbf{u}_e, t),$$

$$\mathbf{B}(t) = \frac{\partial \mathbf{f}}{\partial \mathbf{u}}(\mathbf{x}_e, \mathbf{u}_e, t).$$

Then let local controllability mean that for any $\mathbf{x}_1$ sufficiently close to $\mathbf{x}_e$ there exists a control $\mathbf{u}(t) = \mathbf{u}_e + \delta\mathbf{u}(t)$, with $\delta\mathbf{u}(t)$ small on $[t_0, t_1]$, which steers (4.30) from $\mathbf{x}_0$ to $\mathbf{x}_e$ on $[t_0, t_1]$. In this sense, controllability of (4.29) as defined previously implies local controllability of (4.30) [R1].

4.3 LINEAR PERIODIC SYSTEMS

Again, consider perturbations to the autonomous nonlinear system, $\dot{\mathbf{x}} = \mathbf{f}(\mathbf{x})$, given by (3.37), but with $\mathbf{x}^*(t) = \mathbf{x}^*(t + T)$. That is, $\delta\mathbf{x}(t) = \mathbf{x}(t) - \mathbf{x}^*(t)$ represents small deviations about a periodic trajectory in time or a closed-solution path in state space with period $T > 0$. Then, as shown previously, the linear approximation becomes

$$\delta\dot{\mathbf{x}} = \mathbf{A}(t)\,\delta\mathbf{x}, \tag{4.31}$$

$$\mathbf{A}(t) = \frac{\partial \mathbf{f}}{\partial \mathbf{x}}(\mathbf{x}^*(t)), \tag{4.32}$$

or

$$\mathbf{A}(t) = \mathbf{A}(t + T) \qquad \text{for all } t.$$

Here it is assumed in (4.31) that the higher-order terms $\mathbf{r}(\delta\mathbf{x})$ converge to zero faster than $\delta\mathbf{x}$. Since

$$\dot{\mathbf{x}}^* = \mathbf{f}(\mathbf{x}^*(t)),$$

it is seen that

$$\frac{d}{dt}(\dot{\mathbf{x}}^*) = \left[\frac{\partial \mathbf{f}}{\partial \mathbf{x}}(\mathbf{x}^*(t))\right]\dot{\mathbf{x}}^*.$$

Therefore, $\dot{\mathbf{x}}^*(t)$ satisfies the variational Equation (4.31). Also, $\mathbf{x}^*(t)$ is periodic and $\dot{\mathbf{x}}^*(t) = \dot{\mathbf{x}}^*(t + T)$ is periodic. Consequently, we see that $\mathbf{x}^*(t)$ cannot be an asymptotically stable solution in the ordinary sense as noted in Section 4.1.3 and shown by Fig. 4.5. In fact, it can be shown from the conventional trajectory-stability definition that no nonconstant solution of an autonomous periodic system can be asymptotically stable. The concept of orbital stability was introduced in Section 4.1.3 and will be studied in more detail in Section 4.4.

It should be noted that if $\mathbf{x}^*(t)$ is a periodic solution to the time-variant system (4.3), $\dot{\mathbf{x}} = \mathbf{f}(\mathbf{x}, t)$, a perturbation analysis still can be made. Unfortunately, $\mathbf{A}(t)$ from the variational equation

$$\delta\dot{\mathbf{x}} = \frac{\partial \mathbf{f}}{\partial \mathbf{x}}(\mathbf{x}^*(t), t)\,\delta\mathbf{x}$$

or

$$\delta\dot{\mathbf{x}} = \mathbf{A}(t)\,\delta\mathbf{x}$$

will not be periodic in general. If $\mathbf{f}(\cdot)$ is assumed periodic in t, however, then $\mathbf{A}(t)$ is periodic. Periodic $\mathbf{f}(\cdot)$ may also result from a forcing function $\mathbf{u}(t)$ with a multiple of that frequency for the periodic initial-condition solution. With this as motivation let us summarize some convenient results for periodic linear systems.

4.3.1 Form of Solution and Stability

Now consider this interesting class of linear time-variant systems such that

$$\frac{d\mathbf{x}}{dt} = \mathbf{A}(t)\mathbf{x}(t), \tag{4.33}$$

where $\mathbf{A}(t) = \mathbf{A}(t + T)$ for some $T > 0$ and all t. Again, the state-transition matrix $\mathbf{\Phi}(t, t_0)$ is given by

$$\frac{d\mathbf{\Phi}}{dt} = \mathbf{A}(t)\mathbf{\Phi}, \tag{4.34}$$

where $\mathbf{\Phi}(t, t) = \mathbf{I}$ for all t. The theory developed here is referred to as Floquet theory, after its founder.

It is readily seen that

$$\mathbf{\Phi}(t + T, t_0) = \mathbf{\Phi}(t, t_0)\mathbf{C} \tag{4.35}$$

is a state-transition matrix with $\mathbf{C}$ a constant matrix. To check this, merely substitute (4.35) into (4.34) so that

$$\dot{\mathbf{\Phi}}(t + T, t_0) = \dot{\mathbf{\Phi}}(t, t_0)\mathbf{C} = \mathbf{A}(t)\mathbf{\Phi}(t, t_0)\mathbf{C}$$

or

$$\dot{\mathbf{\Phi}}(t + T, t_0) = \mathbf{A}(t)\mathbf{\Phi}(t + T, t_0).$$

Then there exists a constant matrix $\mathbf{R}T$ called the $\ln\mathbf{C}$ such that

$$\mathbf{C} \equiv e^{\mathbf{R}T}. \tag{4.36}$$

If $[\mathbf{C} - \mathbf{IA}]$ has all its eigenvalues inside the unit circle (i.e., its spectral radius is less than unity), then

$$\ln \mathbf{C} = [\mathbf{C} - \mathbf{I}] - \frac{1}{2!}[\mathbf{C} - \mathbf{I}]^2 + \frac{1}{3!}[\mathbf{C} - \mathbf{I}]^3 - \cdots$$

provides the logarithm by a convergent series.

Let

$$\mathbf{P}(t, t_0) = \mathbf{\Phi}(t, t_0)\mathbf{e}^{-\mathbf{R}(t-t_0)}$$

or

$$\mathbf{\Phi}(t, t_0) = \mathbf{P}(t, t_0)\mathbf{e}^{\mathbf{R}(t-t_0)}. \tag{4.37}$$

Since $\mathbf{\Phi}(\cdot)$ and $\mathbf{e}^{\mathbf{r}(\cdot)}$ are nonsingular, $\mathbf{P}(\cdot)$ is a nonsingular matrix. Also, $\mathbf{P}(\cdot)$ is periodic with

$$\mathbf{P}(t, t_0) = \mathbf{P}(t + T, t_0), \tag{4.38}$$

which is seen as follows:

$$\mathbf{P}(t + T, t_0) = \mathbf{\Phi}(t + T, t_0)\mathbf{e}^{-\mathbf{R}(t+T-t_0)}$$

or since $\mathbf{e}^{-\mathbf{R}(t+T-t_0)} = \mathbf{e}^{-\mathbf{R}T}\mathbf{e}^{-\mathbf{R}(t-t_0)}$,

$$\mathbf{P}(t + T, t_0) = \mathbf{\Phi}(t, t_0)\mathbf{e}^{-\mathbf{R}(t-t_0)} = \mathbf{P}(t, t_0).$$

Here $\mathbf{P}(t_0, t_0) = \mathbf{I}$.

From (4.37), with $\mathbf{\Phi}(t_0, t_0) = \mathbf{P}(t_0, t_0) = \mathbf{e}^{\mathbf{R}(t_0-t_0)} = \mathbf{I}$, it is seen that at $t = t_0 + T$,

$$\mathbf{\Phi}(t_0 + T, t_0) = \mathbf{P}(t_0 + T, t_0)\mathbf{e}^{\mathbf{R}T} = \mathbf{e}^{\mathbf{R}T}.$$

Consequently, the state-transition matrix, and solution to (4.33), consists of a periodically modulated exponential matrix function. Therefore, (4.33) is *asymptotically stable* if the eigenvalues of $\mathbf{R}$ all have negative real parts. That is,

$$\det[\mathbf{I}\rho - \mathbf{R}] = 0 \tag{4.39}$$

implies that

$$\mathrm{Re}\{\rho_i\} < 0, \qquad i = 1, \ldots, n, \tag{4.40}$$

or that

$$\det[\mathbf{I}\lambda - \mathbf{e}^{\mathbf{R}T}] = 0 \tag{4.41}$$

implies that

$$|\lambda_i| < 1. \tag{4.42}$$

Here ρ_i are defined as *characteristic exponents* of $\mathbf{A}(t)$, and λ_i as *characteristic multipliers* of $\mathbf{A}(t)$ for $i = 1, \ldots, n$.

Consider the transformation to periodic coordinates $\mathbf{y}$ by $\mathbf{x} = \mathbf{P}(t, t_0)\mathbf{y}$ so that

$$\mathbf{\Phi}(t, t_0)\mathbf{x}_0 = \mathbf{P}(t, t_0)\mathbf{e}^{\mathbf{R}(t-t_0)}\mathbf{x}_0$$

or

$$\mathbf{y} = \mathbf{P}^{-1}(t, t_0)\mathbf{x}. \tag{4.43}$$

Then a matrix set of independent solutions yields the fundamental matrix $\mathbf{Y}(t, t_0)$ such that

$$\mathbf{Y}(t, t_0) = \mathbf{P}^{-1}(t, t_0)\mathbf{\Phi}(t, t_0) = \mathbf{e}^{\mathbf{R}(t-t_0)},$$

which is then a solution to the LTIS

$$\dot{\mathbf{Y}} = \mathbf{R}\mathbf{Y},$$

with $\mathbf{Y}(t_0, t_0) = \mathbf{I}$. Consequently, the periodic linear system (4.33) can be transformed to the linear autonomous system

$$\dot{\mathbf{y}} = \mathbf{R}\mathbf{y}. \tag{4.44}$$

Obviously, this checks with the stability conditions which are given by (4.40).

Next we will attempt to relate the characteristic multipliers and the characteristic exponents more precisely. First assume that the λ_i are distinct and let

$$\mathbf{Q}^{-1}\mathbf{e}^{\mathbf{R}T}\mathbf{Q} = \mathbf{Q}^{-1}\left[\mathbf{I} + \mathbf{R}T + \frac{(\mathbf{R}T)^2}{2!} + \cdots\right]\mathbf{Q}.$$

But

$$\mathbf{Q}^{-1}(\mathbf{R}T)^m\mathbf{Q} = (\mathbf{Q}^{-1}\mathbf{R}T\mathbf{Q})^m$$

or

$$\begin{aligned}\mathbf{Q}^{-1}\mathbf{e}^{\mathbf{R}T}\mathbf{Q} &= \mathbf{I} + \mathbf{Q}^{-1}\mathbf{R}T\mathbf{Q} + \frac{1}{2!}(\mathbf{Q}^{-1}\mathbf{R}T\mathbf{Q})^2 + \cdots \\ &= \mathbf{e}^{\mathbf{Q}^{-1}\mathbf{R}T\mathbf{Q}}.\end{aligned}$$

Then select $\mathbf{Q}$ such that

$$\mathbf{e}^{\mathbf{Q}^{-1}\mathbf{R}T\mathbf{Q}} = \mathbf{Q}^{-1}\mathbf{e}^{\mathbf{R}T}\mathbf{Q} = \mathbf{\Lambda},$$

$$\mathbf{Q}^{-1}\mathbf{R}T\mathbf{Q} = \ln \mathbf{\Lambda} = \begin{bmatrix} \ln \lambda_1 & & & 0 \\ & \cdot & & \\ & & \cdot & \\ & & & \cdot \\ 0 & & & \ln \lambda_n \end{bmatrix}$$

or

$$\mathbf{R} = \mathbf{Q}\begin{bmatrix} \dfrac{\ln \lambda_1}{T} & & & 0 \\ & \cdot & & \\ & & \cdot & \\ & & & \cdot \\ 0 & & & \dfrac{\ln \lambda_n}{T} \end{bmatrix}\mathbf{Q}^{-1}.$$

Consequently, ρ_i, the eigenvalues of $\mathbf{R}$, are related to λ_i, the eigenvalues of $\mathbf{e}^{\mathbf{R}T}$, by

$$\rho_i = \frac{\ln \lambda_i}{T}, \tag{4.45}$$

where $i = 1, \ldots, n$. Again, ρ_i are the characteristic exponents and λ_i the characteristic multipliers of $\mathbf{A}$.

Figure 4.6 shows the mapping of the stable region for ρ (similar to the conventional s-plane) to λ (similar to the conventional z-plane). Again,

$$\text{Re}\{\rho_i\} \le 0 \Leftrightarrow |\lambda_i| = |e^{\rho_i T}| \le 1,$$

$i = 1, \ldots, n$, define a stable system (4.33).

Since

$$\det[\mathbf{\Phi}(t, t_0)] = \exp\int_{t_0}^{t} \text{Tr}\, \mathbf{A}(\sigma)\, d\sigma$$

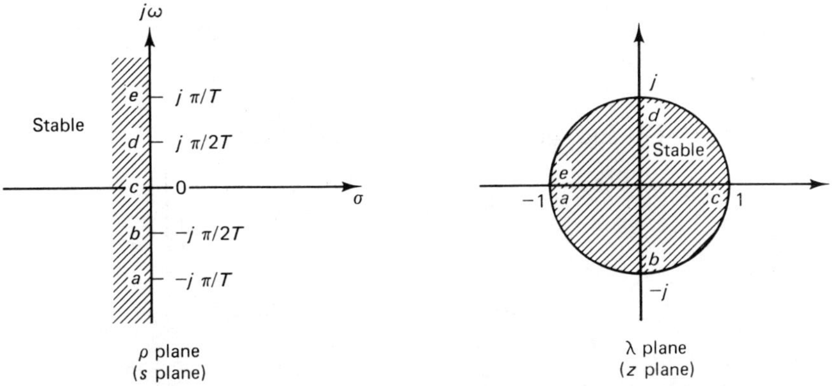

Figure 4.6 Stability regions in complex planes.

and

$$\mathbf{e}^{\mathbf{R}T} = \mathbf{\Phi}^{-1}(t, t_0)\mathbf{\Phi}(t + T, t_0)$$

with

$$\det[\mathbf{e}^{\mathbf{R}T}] = \prod_{j=1}^{n} \lambda_j$$

and

$$\mathbf{\Phi}(t + T, t_0) = \mathbf{\Phi}(t, t_0)\mathbf{e}^{\mathbf{R}T}$$

it is seen that

$$\prod_{j=1}^{n} \lambda_j = \exp\int_t^{t+T} \operatorname{Tr} \mathbf{A}(\sigma)\, d\sigma, \tag{4.46}$$

and with $\lambda_j = e^{\rho_j T}$,

$$\sum_{j=1}^{n} \rho_j = \frac{1}{T}\int_t^{t+T} \operatorname{Tr} \mathbf{A}(\sigma)\, d\sigma. \tag{4.47}$$

As will be shown, (4.46) and (4.47) can be useful for second-order systems in particular.

The similarity of linear periodic systems to linear sampled-data systems is worth noting, with T equal to the period and sample period, respectively. This is logical since the sampled signal can only be used to reconstruct the continuous signal up to $1/2T$ hertz. Also, note the similarity of constant-coefficient continuous-time system eigenvalues to ρ_i, and of such sampled-data system eigenvalues to λ_i.

Next, consider the *Hill equation*,

$$\ddot{x} + a(t)x = 0, \tag{4.48}$$

where $a(t) = a(t + T)$, $T > 0$ for all t. Then, with $x_1 = x$, $x_2 = \dot{x}$,

$$\dot{\mathbf{x}} = \mathbf{A}(t)\mathbf{x}, \tag{4.49}$$

where

$$\mathbf{A}(t) = \begin{bmatrix} 0 & 1 \\ -a(t) & 0 \end{bmatrix}.$$

Then let $t_0 = 0$ and $\mathbf{\Phi}(T, 0) = \mathbf{\Phi}(T) = \mathbf{e}^{\mathbf{R}T}$, so that

$$\mathbf{C} = \mathbf{e}^{\mathbf{R}T} = \begin{bmatrix} \phi_1(T) & \phi_2(T) \\ \dot{\phi}_1(T) & \dot{\phi}_2(T) \end{bmatrix}. \tag{4.50}$$

Then

$$\det[\mathbf{I}\lambda - \mathbf{C}] = 0$$

or

$$\det\begin{bmatrix} \lambda - \phi_1(T) & -\phi_2(T) \\ -\dot{\phi}_1(T) & \lambda - \dot{\phi}_2(T) \end{bmatrix} = 0. \tag{4.51}$$

Let

$$F \equiv F(a(T)) \equiv \text{Tr } \mathbf{C} = \phi_1(T) + \dot{\phi}_2(T).$$

Also, from (4.46),

$$\lambda_1\lambda_2 = \exp \int_t^{t+T} \text{Tr } \mathbf{A}(\sigma)\, d\sigma = 1.$$

Then (4.51) yields the characteristic equation for λ,

$$\lambda^2 - F\lambda + 1 = 0 \tag{4.52}$$

with roots plotted in Fig. 4.7. The following conclusions can be drawn for the Hill equation (4.49):

1. $|F| > 2$ implies that λ_i real, system unstable.
2. $|F| < 2$ implies that λ_i complex conjugates with $|\lambda_i| = 1$, system stable (not asymptotically stable).
3. $|F| = 2$ implies that $\lambda_i = \lambda_2 = \pm 1$. If $\mathbf{C}$ has two linearly independent eigenvectors, the system is stable; otherwise, the system is unstable (with off-diagonal Jordon form).

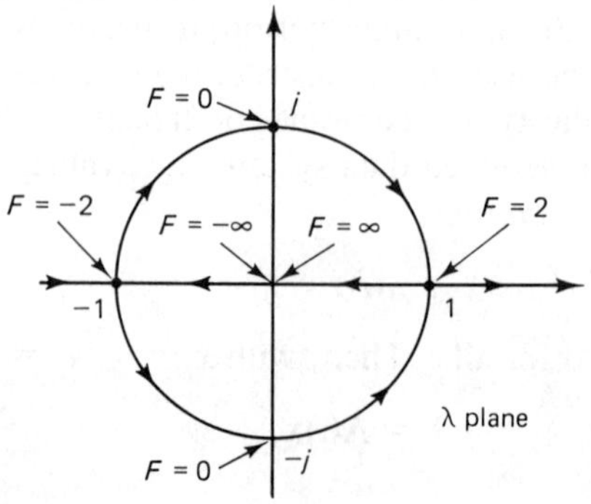

Figure 4.7 Characteristic multipliers for Hill equation.

The *Mathieu equation* is defined as a particular Hill equation such that $a(t) = -(\delta + \epsilon \cos t)$, or

$$\dot{x}_1 = x_2,$$
$$\dot{x}_2 = -(\delta + \epsilon \cos t)x_1, \qquad (4.53)$$

where δ and ϵ are constants (Fig. 4.8). Then, in (4.50), $\phi_1(T)$ is even (odd) if $\phi_2(T)$ is odd (even) since $\phi_1(0) = [1 \quad 0]^T$, $\phi_2(0) = [1 \quad 0]^T$ with $\mathbf{\Phi}(T) = \mathbf{\Phi}(T, 0)$ composed of linearly independent solutions. It is noted that if $\mathbf{x}(t)$ is neither an odd nor an even function which satisfies (4.53), then $\mathbf{x}(t)$, $\mathbf{x}(-t)$ are linearly independent solutions. If $|F| = 2$, as shown above, $\lambda_1 = \lambda_2 = \pm 1$, which yields a periodic solution if the eigenvectors are independent. This condition corresponds to marginal stability or borderline values of δ, ϵ between stability and instability. The derivation of such values is given in [S1] and [S2].

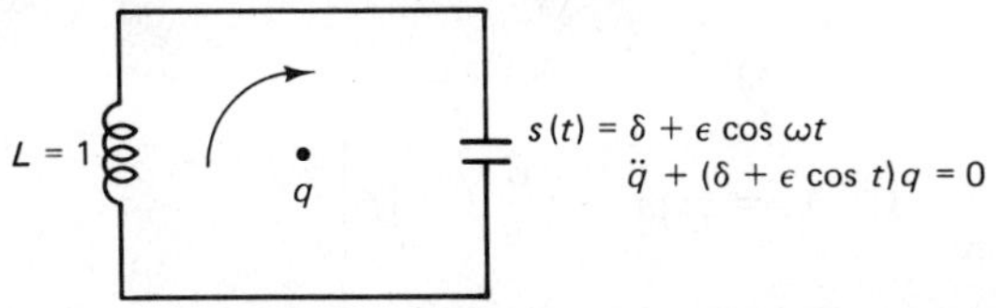

Figure 4.8 Parametrically excited circuit.

The transitional values of δ, ϵ can be computed with $F(\delta, \epsilon) = 2$ for one case and $F(\delta, \epsilon) = -2$ in the other. From the Fourier series for the two cases, it can be shown that either only sines for one case or only cosines for the other will appear [S1]. Then substitution of the Fourier series into (4.53) yields the transitional values of δ, ϵ shown in Fig. 4.9. For example, it shows that $\epsilon = 0$, $\delta > 0$ yields a stable system, but

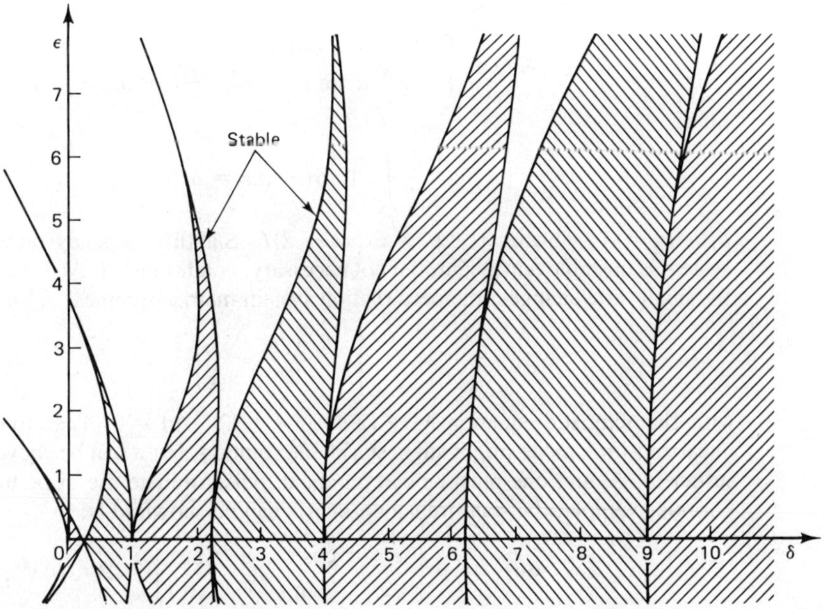

Figure 4.9 Stability regions for the Mathieu equation. (After Stoker, J.J., *Nonlinear Vibrations, Interscience,* John Wiley & Sons, Inc., 1950 [S1].)

$\delta \leq 0$ implies an unstable system. If $\epsilon = 0$, $\delta = k^2/4$, where k is an integer, it is seen that $T = 4\pi$ for k odd and $T = 2\pi$ for k even. Note that the system becomes a classic linear oscillator for $\epsilon = 0$, and only stable transition points occur for this case.

Example 4.6

Consider the single-precursor nuclear fission process again with

$$\dot{n} = \frac{u - \beta}{l} n + \lambda c,$$

$$\dot{c} = \frac{\beta}{l} n - \lambda c,$$

where reactivity, $u(t) = a \sin \omega t$, $T = 2\pi/\omega$. Then from (4.47),

$$\rho_1 + \rho_2 = -\left(\frac{\beta}{l} + \lambda\right),$$

and from (4.46),

$$\lambda_1 \cdot \lambda_2 = \exp\left[-\left(\frac{\beta}{l} + \lambda\right)T\right].$$

This alone yields inconclusive conclusions about stability. Indeed, since the Example 3.4 yielded the $\lambda_1 = 0$, $\lambda_2 = -\beta/l + \lambda$ for the linearized fission model about equilibrium conditions, stability might very well be anticipated. But recall that it was not conclusive from Liapunov's first method since the system is structurally unstable. Further analysis or computer simulation indeed shows that the system is unstable for $u(t) = a \sin \omega t$.

Example 4.7

Consider $\dot{\mathbf{x}} = \mathbf{A}(t)\mathbf{x}$, where

$$\mathbf{A}(t) = \begin{bmatrix} -1 + a\cos^2 t & 1 - a \sin t \cos t \\ -1 - a \sin t \cos t & -1 + a \sin^2 t \end{bmatrix}.$$

Since

$$\frac{1}{T}\int_0^T \operatorname{Tr} \mathbf{A}(\sigma)\, d\sigma = a - 2,$$

$\rho_1 + \rho_2 = a - 2$ and $\lambda_1 \cdot \lambda_2 = \exp(a - 2)T$. Stability depends on a. If $a > 2$, the system is unstable (sufficient, but not necessary, conditions). If $\mathbf{A}(t)$ is evaluated at $t = 0$ for small time changes, the perturbation system matrix becomes

$$\begin{bmatrix} -1 + a & 1 \\ -1 & -1 \end{bmatrix}$$

with characteristic equation $\lambda^2 + (2 - a)\lambda + (2 - a) = 0$. Therefore, if $a < 2$, the linearized-time mode might suggest a stable solution, but it can be shown that this is not sufficient. Again, one must be careful about indiscriminate use of perturbation models. For this example, the state-transition matrix can be shown to be

$$\mathbf{\Phi}(t, 0) = \begin{bmatrix} e^{(a-1)t} \cos t & e^{-t} \sin t \\ -e^{(a-1)t} \sin t & e^{-t} \cos t \end{bmatrix} = \mathbf{P}(t, 0)\mathbf{e}^{\mathbf{R}T},$$

where $\rho_1 = -1$, $\rho_2 = a - 1$, which is unstable for $a > 1$.

Example 4.8

Here the kinetic energy of the pendulum bob (Fig. 4.10) is given by

$$T = \tfrac{1}{2}\, m(l\dot{\theta})^2,$$

where the $\dot{l}^2$ term is neglected. Its potential energy (relative to its rest position) is given by

$$V = mgl(1 - \cos\theta).$$

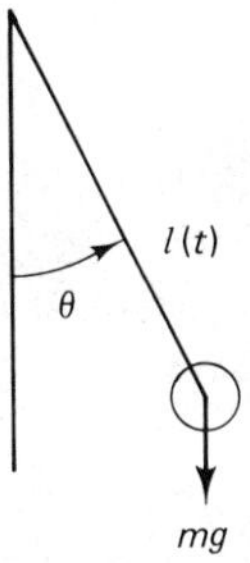

Figure 4.10 Pendulum configuration for Example 4.8.

Then the Euler–Lagrange equation (2.12) for this conservative system becomes

$$\frac{d}{dt}\left(\frac{\partial L}{\partial \dot{\theta}}\right) - \frac{\partial L}{\partial \theta} = 0,$$

or with

$$\frac{\partial L}{\partial \dot{\theta}} = ml^2\dot{\theta}, \qquad \frac{\partial L}{\partial \theta} = mgl\sin\theta,$$

$$\ddot{\theta}(t) + \frac{2\dot{l}(t)}{l(t)}\dot{\theta}(t) + \frac{g}{l(t)}\sin\theta(t) = 0.$$

Further assume that $\sin\theta(t) \approx \theta(t)$, and let $p(t) = \dot{l}(t)/l(t)$ and $q(t) = g/l(t)$. Then $\ddot{\theta} + 2p(t)\dot{\theta}(t) + q(t)\theta(t) = 0$ is a linear time-variant system which is a periodic linear system if $l(t) = l(t + T)$ for $T > 0$ and all t. By a transformation

$$y(t) = \theta(t)\exp\int_0^t p(\sigma)\,d\sigma,$$

the system takes the form of $\ddot{y} + [q(t) - \dot{p}(t) - p^2(t)]y = 0$, or

$$\ddot{y}(t) = \frac{g}{l(t)}\left[1 - \frac{\ddot{l}(t)}{g}\right]y(t) = 0.$$

Now we have the Hill equation with

$$f(t) = f(t + T) = \frac{g}{l(t)}\left[1 - \frac{\ddot{l}(t)}{g}\right]$$

or the Mathieu equation if

$$f(t) = \delta + \epsilon\cos t.$$

4.4 CLOSED-SOLUTION PATHS

Given again the nonlinear autonomous system

$$\dot{\mathbf{x}} = \mathbf{f}(\mathbf{x})$$

with nonconstant periodic solution $\mathbf{x}^*(t) = \mathbf{x}^*(t + T)$ for $T > 0$ and all t. It was shown in Section 4.3 that such $\dot{\mathbf{x}}^*(t)$ satisfies the variational equation (4.31) given by

$$\delta\dot{\mathbf{x}} = \mathbf{A}(t)\,\delta\mathbf{x},$$

where

$$\mathbf{A}(t) = \mathbf{A}(t + T) = \frac{\partial \mathbf{f}}{\partial \mathbf{x}}(\mathbf{x}^*(t)).$$

Then, since $\mathbf{x}^*(t)$ is periodic, at least one characteristic multiplier, λ_1, has

$$|\lambda_1| = 1.$$

That is, the closed-solution path is not asymptotically stable according to the definition given in Section 4.1.3. However, a somewhat weaker concept of stability, called orbital stability, was introduced in Section 4.1.3 for analyzing the stability of a closed-solution path C relative to a moving point $\mathbf{p}(t)$ on C in state space.

Figure 4.11 shows five different examples of closed-solution paths with different stability properties. Here Fig. 4.11(a) shows typical center trajectories for a simple linear harmonic oscillator. The motion is isochronous (equal periods T for each closed trajectory and equal δt for equal angles θ). Obviously, C is orbitally stable and even Liapunov stable since there is no slippage between corresponding time points on C_1 and C. That is, the $\delta(\epsilon, t_0)$ and ϵ conditions for Liapunov stability are satisfied, which is stronger than that required for orbital stability. But certainly C is in no way asymptotically stable.

Figure 4.11(b) shows nonisochronous motion, which typifies a nonlinear conservative system. Here the motion is orbitally stable relative to C, but obviously not asymptotically orbitally stable. It is a little more difficult to see that it is not Liapunov stable. However, with the slippage between corresponding time points on adjacent trajectories, it is readily seen that it is not possible to find a $\delta(\epsilon, t_0) > 0$ for every $\epsilon > 0$ such that trajectories starting interior to the region defined by $\delta(\cdot)$ never leave the region defined by ϵ. Figure 4.11(c) shows a limit cycle (isolated closed-solution path) which is asymptotically orbitally stable. Again, C is not Liapunov stable for the same reason as given for Fig. 4.11(b). Figure 4.11(d) shows a typical unstable limit cycle. It is not stable in any sense defined in Section 4.1.3. A semistable (sometimes called semiasymptotically stable) limit cycle is shown by path C in Fig. 4.11(e). Here the trajectories are asymptotically stable from states interior to C, but unstable from states exterior to C. Notice here if $\mathbf{0}$ is an equilibrium state, C encloses an unstable equilibrium point. Of course, the semistable case can be further explained. The limit cycle in Fig. 4.11(e) may also be called orbitally semiasymptotically stable and semiunstable, and the reader may readily identify all possible combinations of such semistable cases.

As shown in Fig. 4.11, a *limit cycle* is defined as an *isolated* closed-solution path C. That is, there exists no other closed trajectory within an infinitesimal neighborhood of C; for example, Fig. 4.11 (c) to (e) depict C as limit cycles but not (a) and (b).

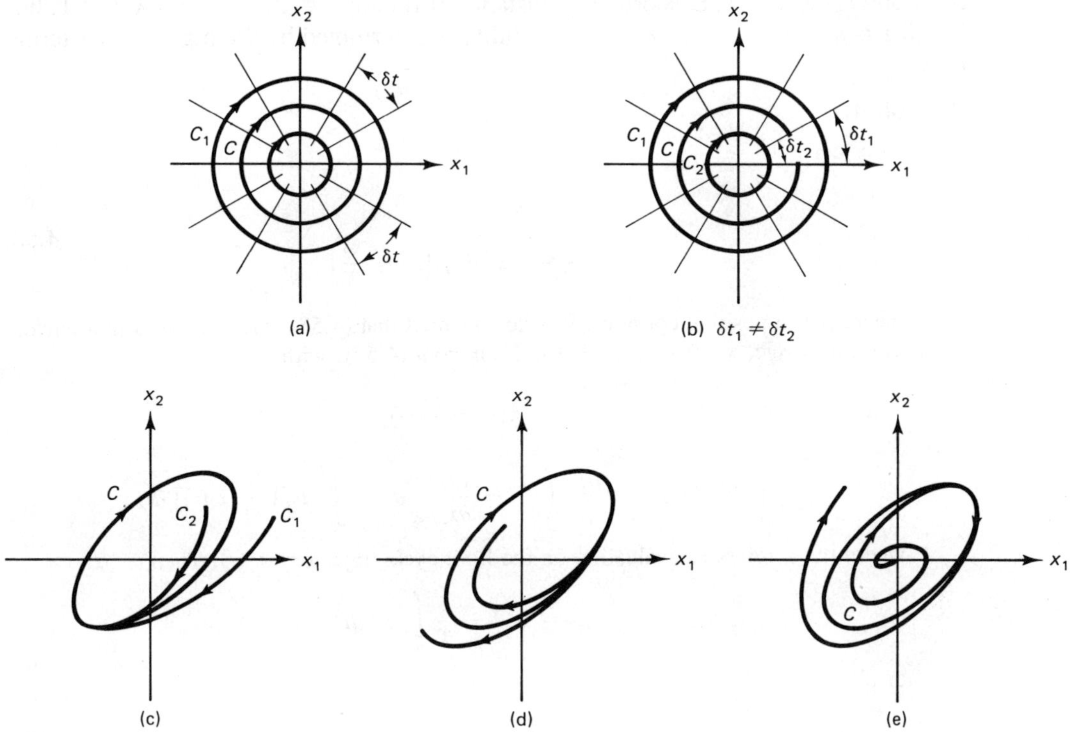

Figure 4.11 Closed-trajectory stability examples.

Now Floquet theory for periodic linear systems can be useful for studying stability of the closed-solution path C in state space. In particular, (4.46) is useful, so that

$$\prod_{i=1}^{n} \lambda_i = \exp \int_0^T \mathrm{Tr}\left[\frac{\partial \mathbf{f}}{\partial \mathbf{x}}(\mathbf{x}^*(\sigma))\right] d\sigma = \exp \int_0^T \mathrm{Tr}[\mathbf{A}(\mathbf{x}^*(\sigma))]\, d\sigma. \tag{4.54}$$

Then for $n = 2$, since $\lambda_1 = 1$,

$$|\lambda_2| = \exp \int_0^T \mathrm{Tr}[\mathbf{A}(\mathbf{x}^*(\sigma))]\, d\sigma < 1 \tag{4.55}$$

implies that C is *asymptotically orbitally stable*. Unfortunately, the general nth-order system is more difficult to analyze for stability since the following must hold for all $(n - 1)$ characteristic multipliers:

$$|\lambda_i| < 1, \tag{4.56}$$

$i = 2, \ldots, n$, for asymptotic orbital stability of C. Here

$$\prod_{i=2}^{n} \lambda_i = \exp \int_0^T \mathrm{Tr}[\mathbf{A}(\mathbf{x}^*(\sigma))]\, d\sigma. \tag{4.57}$$

If for any i, $|\lambda_i| > 1$, C is orbitally unstable. If for any $i = 2, \ldots, n$, $|\lambda_i| = 1$, but for all $i = 1, 2, \ldots, n$, $|\lambda_i| \leq 0$, stability is determined by the higher-order terms $\mathbf{r}(\delta\mathbf{x})$.

Example 4.9

Consider

$$\dot{x}_1 = x_2,$$
$$\dot{x}_2 = -x_1 + \mu\left(x_2 - \frac{x_2^3}{3}\right), \tag{4.58}$$

where μ is a positive constant. It is documented that (4.58) may have an origin-enclosing limit cycle, $\mathbf{x}^*(t) = \mathbf{x}^*(t + T)$. Then from (4.57), with

$$\mathbf{A}(\cdot) = \frac{\partial \mathbf{f}}{\partial \mathbf{x}}(\cdot),$$

$$\int_0^T \text{Tr } \mathbf{A}(\cdot)\, dt = \int_0^T \left(\frac{\partial f_1}{\partial x_1} + \frac{\partial f_2}{\partial x_2}\right)_* dt = \int_0^T \mu[1 - x_2^2(t)]\, dt,$$

where the $_*$ refers to evaluation on the limit-cycle trajectory $\mathbf{x}^*(t)$. Then

$$\rho_2 = \mu\left(1 - \frac{1}{T}\int_0^T x_2^2\, dt\right),$$

or if

$$\int_0^T x_2^2(t)\, dt > T,$$

so that for $\mu > 0$, the limit cycle would be asymptotically orbitally stable.

Example 4.10

Consider

$$\dot{x}_1 = x_2 - \mu(x_1^2 + x_2^2 - 1)x_1, \tag{4.59}$$
$$\dot{x}_2 = -x_1 - \mu(x_1^2 + x_2^2 - 1)x_2,$$

where μ is constant. There is an equilibrium state, $\mathbf{x}_e = \mathbf{0}$, with perturbation model

$$\delta\dot{x}_1 = \mu\, \delta x_1 + \delta x_2,$$
$$\delta\dot{x}_2 = -\delta x_1 + \mu\, \delta x_2.$$

Comparing this form with the complex-λ canonical form in Sections 3.3.2 and 3.3.3, the polar coordinate representation then is

$$\delta\dot{r} = \mu\, \delta r,$$
$$\delta\dot{\theta} = -1,$$

and the corresponding perturbation state-plane portrait corresponds to a focus with

$$\lambda_{1,2} = \mu \pm j1.$$

Consequently, $\mathbf{x}_e = \mathbf{0}$ is unstable for $\mu > 0$ and stable for $\mu < 0$.

Representation of (4.59) in polar form becomes

$$\dot{r} = \mu(1 - r^2)r,$$
$$\dot{\theta} = -1. \tag{4.60}$$

If $r^*(0) = 1$, $\dot{r}^* = 0$, $r^*(t) = 1$, and $\dot{\theta}^* = -1$; or $\theta^*(t) = \theta_0^* - t$, where $\theta_0^* = \tan^{-1}[x_2^*(0)/x_1^*(0)]$; or

$$x_1^*(t) = \cos(\theta_0^* - t),$$
$$x_2^*(t) = \sin(\theta_0^* - t)$$

is the periodic reference solution to (4.59).

Perturbations about this apparent limit cycle of radius unity yield in polar coordinates

$$\delta\dot{r} = -2\mu\ \delta r,$$
$$\delta\dot{\theta} = -1.$$

Or, $\delta\dot{\mathbf{x}} = \mathbf{A}(t)\ \delta\mathbf{x}$ such that

$$\delta\dot{x}_1 = (-\mu + \mu \cos 2t)\ \delta x_1 + (1 - \mu \sin 2t)\ \delta x_2,$$
$$\delta\dot{x}_2 = (-1 - \mu \sin 2t)\ \delta x_1 - (\mu + \mu \cos 2t)\ \delta x_2.$$

If $\mu = 0$,

$$\mathbf{A} = \begin{bmatrix} 0 & 1 \\ -1 & 0 \end{bmatrix}$$

and $\lambda_{1,2} = \pm j1$, which checks above.

Now $\mathbf{A}(t) = \mathbf{A}(t + T)$, $T = \pi$, and $\rho_1 = 0$, $\lambda_1 = 1$ for the limit-cycle motion. Then, from (4.47),

$$\rho_2 = \frac{1}{T}\int_0^T \mathrm{Tr}\ \mathbf{A}(\sigma)\ d\sigma = -2\mu,$$

and from (4.46),

$$\lambda_2 = e^{-2\pi\mu}.$$

Therefore, $\mu > 0$ yields an asymptotically orbitally stable limit cycle as shown in Fig. 4.12. If $\mu < 0$, the limit cycle is unstable and encloses a stable origin. If $\mu = 0$, again, there is no limit cycle, but there is a linear center portrait.

The perturbation solutions about the limit cycle $r^* = 1$ becomes

$$\delta r = \delta r_0 e^{-\mu t},$$
$$\delta\theta = \delta\theta_0 - t,$$

where $\delta r_0 = r(0) - r^*$, $\delta\theta_0 = \theta(0) - \theta^*$, or

$$\delta x_1 = \delta x_{10} e^{-2t} \cos t - \delta x_{20} \sin t,$$
$$\delta x_2 = \delta x_{10} e^{-2t} \sin t + \delta x_{20} \cos t,$$

where $\delta x_{10} = x_1(0) - x_1^*(t_0)$, $\delta x_{20} = x_2(0) - x_2^*(t_0)$.

The alert reader, however, will notice that (4.60) has its variables of integration

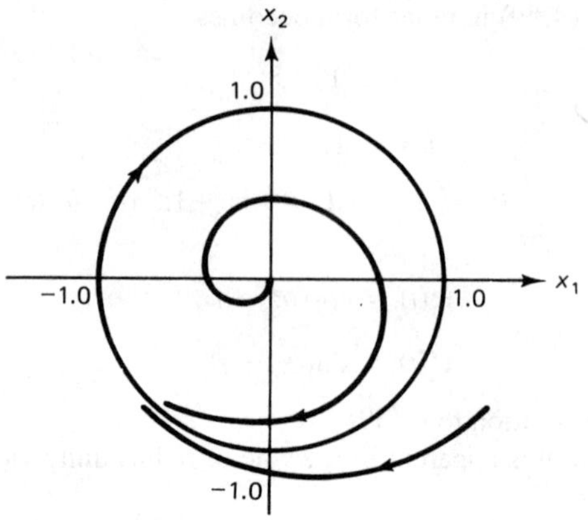

Figure 4.12 State-portrait sketch for Example 4.10.

separable in the convenient polar form. Direct integration of (4.60) with $r(0) = r_0$, $\theta(0) = \theta_0$ yields

$$r(t) = [1 - c\,\exp(-2\mu t)]^{1/2},$$
$$\theta(t) = \theta_0 + t,$$

where $c = (r_0^2 - 1)/r_0^2$. Furthermore, if $r_0 < 1$ and hence $c < 0$, and $\mu < 0$,

$$\lim_{t \to -\infty} r(t) = 1.$$

Also, $r_0 > 1$, $\mu < 0$ implies the same limit. Therefore, motion spirals away from the limit cycle on both sides and is orbitally unstable as indicated above. Similarly, if $\mu > 0$, motion is toward the limit cycle and it is orbitally stable. Again, $\mu = 0$ yields a simple linear oscillator.

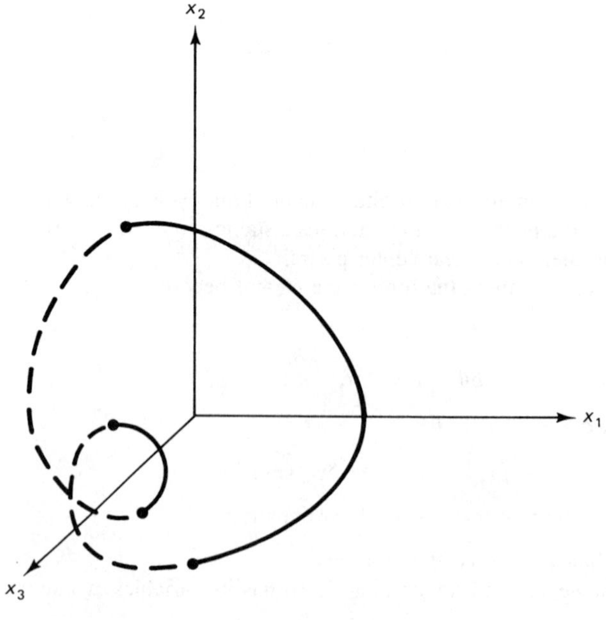

Figure 4.13 Limit cycle for a third-order system.

It has been shown that Floquet theory is useful particularly for the stability of limit cycles for second-order systems. Other techniques, which do not depend on computing the limit-cycle trajectory, for showing the existence or nonexistence of closed-solution paths in the state plane will be studied in Chapter 5. The complexity of such trajectories and the consequences of using state-plane theorems if a third (or higher) state variable is neglected is shown in Fig. 4.13.

EXERCISES

4.1. Show that the linear system (4.21)–(4.22) is controllable (observable) if the adjoint system is observable (controllable).

4.2. **(a)** Compute the solution to

$$\dot{x} = (\cos t - t \sin t - 2)x.$$

with $x(t_0) = x_0$.

(b) Is the origin asymptotically stable? Why?

(c) Is the origin uniformly stable? Why?

4.3. Consider the system $\dot{\mathbf{x}} = \mathbf{A}(t)\mathbf{x}$, where

$$\mathbf{A} = \begin{bmatrix} \cos \ln t & \sin \ln t \\ \sin \ln t & \cos \ln t \end{bmatrix},$$

and $t \geq 1$. One canonical normal solution can be shown to be $z_1 = \exp\,(t \sin \ln t)$. Compute another canonical normal solution $z_2(t)$.

4.4. It is sometimes suggested that the system $\dot{\mathbf{x}} = \mathbf{A}(t)\mathbf{x}$ is asymptotically stable if $\mathbf{A}(t)$ has eigenvalues $\lambda_i(t)$, $i = 1, \ldots, n$, with negative real parts for all $t \geq t_0$. Here, as for the constant-coefficient case, $\det\,[\mathbf{I}\lambda_i(t) - \mathbf{A}(t)] = 0$. In this regard, consider the system with

$$\mathbf{A}(t) = \begin{bmatrix} -1 - 9\cos^2 6t + 6 \sin 12t & 12 \cos^2 6t + \frac{9}{2} \sin 12t \\ -12 \sin^2 6t + \frac{9}{2} \sin 12t & -(1 + 9 \sin^2 6t + 6 \sin 12t) \end{bmatrix}.$$

One characteristic exponent can be shown to be $\rho_1 = -13$. Check if this example follows the suggestion, or is a counterexample.

4.5. In Exercise 4.4 it is seen that the periodic part of the solution for a given initial state consists of a sum of a sine component and a cosine component (both of the same angle). With this hint, calculate the solution $\mathbf{x}(t)$ for the system given by Exercise 4.4 if $\mathbf{x}^T(t_0) = [1 \quad -2]$.

4.6. Consider the nonlinear system

$$\ddot{y} + \omega_0^2(y + ay^2) = A \cos \omega t,$$

where a is a small positive constant.

(a) Transform the equation to normalized form so that

$$\frac{d^2x}{d\tau^2} + x = \cos \beta\tau - \mu x^2.$$

(b) Analyze stability of the perturbation dynamics $\delta x(\tau)$ for $\beta \neq 1$, where $\delta x(\tau) \approx x(\tau) - x^*(\tau)$ and $x^*(\tau)$ is the solution to the linear equation

$$\frac{d^2x^*}{d\tau^2} + x^* = \cos \beta\tau.$$

(c) Compute the periodic solutions

$$x^*(\tau), \quad \delta x(\tau).$$

4.7. An open-loop plant is given by

$$\dot{x}_1 = x_2,$$

$$\dot{x}_2 = - \quad x_1 - 2x_2 + u.$$

Suppose that the feedback consists of a time-variant gain so that

$$u = k(t)x_1,$$

where

$$k(t) = \begin{cases} -100, & j \le t \le j + \tau, \quad j = 1, 2, \ldots, \\ 0, & \text{elsewhere}. \end{cases}$$

Here τ is a nonnegative constant which is less than unity. Compute the solution $\mathbf{x}(t)$ for a given $\mathbf{x}(0) = \mathbf{x}_0$ with $\tau = 0.17$. Is the system stable?

4.8. Derive the general solution to the linear periodic system (4.33) if the characteristic exponents are repeated. Note stability requirement of repeated exponents.

4.9. Show for $n = 2$ that the antiperiodic solution [i.e., $x(t + T) = - \quad x(t)$ in (4.33)], has at least one characteristic multiplier which must equal -1.

4.10. Show for $n = 2$ that Tr $\mathbf{A}(t) \equiv 0$ in (4.33) is necessary and sufficient for (4.33) to be in canonical normal form $\dot{\mathbf{z}} = \mathbf{\Lambda}(t)\mathbf{z}$, where $\mathbf{\Lambda}(t)$ is a diagonal matrix of periodic functions.

4.11. Consider the second-order vector differential equation

$$\ddot{\mathbf{y}} + \mathbf{Q}(t)\dot{\mathbf{y}} + \mathbf{P}(t)\mathbf{y} = 0,$$

where $\mathbf{Q}(-t) = -\mathbf{Q}(t)$, $\mathbf{P}(-t) = \mathbf{P}(t)$ and both are periodic. Show that this equation in $\mathbf{y}(t)$ can be expressed in the form of (4.33) with $\mathbf{x}(t) = \mathbf{x}(-t)$, (i.e., the system is t-invariant).

CHAPTER

5

NONLINEAR STATE-PLANE ANALYSIS

As the system order increases, of course its analysis becomes more complex, and this is even more true for nonlinear systems than linear systems. However, it is possible to get a good appreciation of nonlinear system dynamics by first investigating the state-plane behavior. As a consequence of its relative simplicity, there also are certain analytical tools available for the second-order case which do not carry over to the general higher-order case.

5.1 CLOSED-TRAJECTORY ANALYSIS

Closed trajectories, including limit cycles, and their stability were introduced in Section 4.4.2, where it was shown that Floquet theory for linear periodic systems is most convenient for the stability analysis of closed trajectories for autonomous second-order nonlinear systems. This theory is reviewed briefly in Section 5.1.2. First recall the definition of a limit cycle. That is, a *limit cycle* is an isolated closed-solution path. "Isolated" means that there exist no other closed-solution paths in its infinitesimal neighborhood in $\mathbb{R}^n$. (It should be noted, however, that it is quite common to make no distinction between limit cycles and other closed-solution paths, that is, that they need not be isolated.) Consequently, in the strict sense, linear systems do not exhibit limit-cycle motion.

5.1.1 Perturbation Theory

Again, assume an autonomous nonlinear system that is given by

$$\frac{d\mathbf{x}}{dt} = \mathbf{f}(\mathbf{x}), \tag{5.1}$$

where $\mathbf{x} \in \mathbb{R}^2$, and which has a limit-cycle solution $\mathbf{x}^*(t) = \mathbf{x}^*(t + T)$ for all t, where $T > 0$ is the period. Also, it is assumed that $\mathbf{f}(\mathbf{x})$ is continuously differentiable. Then stability is determined from the perturbation equations

$$\delta\dot{\mathbf{x}} = \frac{\partial \mathbf{f}}{\partial \mathbf{x}}(\mathbf{x}^*(t))\,\delta\mathbf{x}. \tag{5.2}$$

Since $\dot{\mathbf{x}}^*(t)$ also satisfies the variational equation (5.2), it was shown that the motion cannot be asymptotically stable, and the corresponding characteristic multiplier has modulus equal to 1 (spectral radius of 1), so that

$$|\lambda_1| = 1. \tag{5.3}$$

Still, the limit cycle may be asymptotically stable in the orbital sense. As a consequence of the application of Floquet theory to the second-order perturbation equation (5.2), it was shown in Section 4.4.2 that the limit cycle $\mathbf{x}^*(t)$ is *asymptotically orbitally stable* if and only if

$$\int_0^T \operatorname{Tr}\left[\frac{\partial \mathbf{f}}{\partial \mathbf{x}}(\mathbf{x}^*(t))\right] dt < 0. \tag{5.4}$$

Similarly, if the integral that is given by (5.4) is positive, $\mathbf{x}^*(t)$ is unstable. If it is equal to zero, stability depends on the higher-order terms $\mathbf{r}(\delta\mathbf{x})$.

5.1.2 Existence of Closed Trajectories

Here conditions are presented for the existence or absence of limit cycles and other closed trajectories throughout certain regions of the state plane in $\mathbb{R}^2$. Again consider (5.1), which is written more explicitly as

$$\frac{dx_1}{dt} = f_1(x_1, x_2),$$

$$\frac{dx_2}{dt} = f_2(x_1, x_2), \tag{5.5}$$

where $f_1(\cdot), f_2(\cdot)$ are continuously differentiable functions of x_1, x_2.

To study a region of nonexistence of limit cycles, it is convenient first to define connected and simply connected regions in $\mathbb{R}^2$, the real plane. A *connected region* is one in which every two points in the region can be connected by a curve that is entirely within the region. Then a *simply connected region* is one that is connected and also its boundary is connected. Figure 5.1 shows an annular region $D_1 \subseteq \mathbb{R}^2$, which is connected but does not have a connected boundary. Hence it is not simply connected.

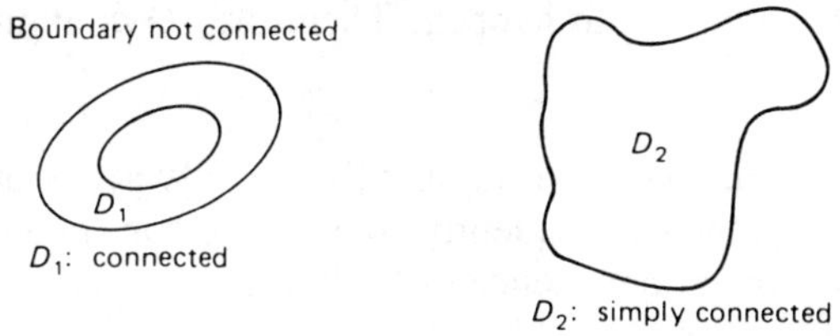

Figure 5.1 Connected regions in the real plane.

The region $D_2 \subseteq \mathbb{R}^2$, however, is a continuously deformed circular region that is simply connected. Now the Bendixson theorem for the nonexistence of closed trajectories in the state plane is stated and then proved [A3], [A4].

Theorem (Bendixson). A closed, bounded, simply connected domain D in $\mathbb{R}^2$ contains *no closed trajectories* if

$$\text{(a)}\ \nabla \mathbf{f}(\mathbf{x}) = \frac{\partial f_1}{\partial x_1} + \frac{\partial f_2}{\partial x_2} \neq 0. \qquad (5.6)$$

(b) $\nabla \mathbf{f}(\mathbf{x})$ does *not* change sign throughout D.

Proof: For the closed-solution curve C which encloses the region S in Fig. 5.2, the well-known Green's theorem is given by

$$\iint_S \nabla \mathbf{f}\, ds = \oint_C \mathbf{f} \cdot \mathbf{n}\, dC = 0,$$

which is equal to zero since the state velocity vector $\mathbf{f}(\cdot)$ is tangent to the solution curve C to which $\mathbf{n}$ is defined to be orthogonal. Or, more explicitly,

$$\iint_S \left(\frac{\partial f_1}{\partial x_1} + \frac{\partial f_2}{\partial x_2}\right) dx_1\, dx_2 = \oint_C f_2\, df_1 - f_1\, df_2 = 0.$$

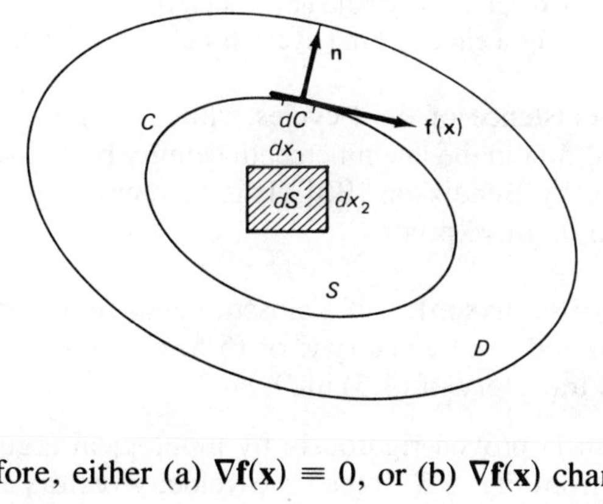

Figure 5.2 Illustration of the Bendixson theorem.

Therefore, either (a) $\nabla \mathbf{f}(\mathbf{x}) \equiv 0$, or (b) $\nabla \mathbf{f}(\mathbf{x})$ changes sign $\mathbf{x} \in S$, that is,

(a) $(\partial f_1/\partial x_1 + \partial f_2/\partial x_2)$ is identically zero interior to C, or

(b) $(\partial f_1/\partial x_1 + \partial f_2/\partial x_2)$ changes sign interior to C.

But if S is a subregion of D, neither can happen. Therefore, D contains no closed-solution paths.

It is important to realize that conditions (a) and (b) are only necessary conditions for C to be a closed-solution path. Consequently, the Bendixson theorem only provides sufficient conditions for the nonexistence of limit cycles.

Example 5.1

Consider the celebrated Van der Pol equation

$$\dot{x}_1 = x_2,$$

$$\dot{x}_2 = \mu(1 - x_1^2)x_2 - x_1, \tag{5.7}$$

where $\mu \neq 0$. Then

$$\frac{\partial f_1}{\partial x_1} + \frac{\partial f_2}{\partial x_2} = \mu(1 - x_1^2),$$

and the Bendixson theorem shows that there cannot exist a closed-solution path which lies entirely in the region defined by $|x_1| < 1$.

Example 5.2

For Example 4.10,

$$\dot{x}_1 = x_2 - \mu x_1(x_1^2 + x_2^2 - 1),$$

$$\dot{x}_2 = -x_1 - \mu x_2(x_1^2 + x_2^2 - 1),$$

where $\mu \neq 0$.

$$\frac{\partial f_1}{\partial x_1} + \frac{\partial f_2}{\partial x_2} = 2\mu(1 - 2x_1^2 - 2x_2^2), \tag{5.8}$$

and there cannot be a closed-solution path within the region about the origin enclosed by a circle of radius $r = 1/\sqrt{2}$. However, notice here that (5.8) satisfies the conditions for the exterior of this region except that D would not be simply connected. Indeed, Example 4.10 showed the existence of a circular limit cycle of radius 1 about the origin.

With respect to the existence of limit cycles within a region D, a sometimes useful theorem was derived first in the late nineteenth century by Poincaré and later in the early twentieth century by Bendixson [B3]. This theorem is given briefly here, along with an outline of an intuitive proof.

Theorem (Poincaré–Bendixson). If a closed, bounded region D contains no equilibrium points of (5.5) and if a trajectory C of (5.5) remains in D for all $t \geq t_0$, there is at least one closed trajectory of (5.5) in D.

Proof: The theorem is proved rigorously by topological arguments in [C1]. Roughly, it can be shown from Fig. 5.3. If the state-velocity vectors all point into the annular region D (as required for the solutions to remain in D), it is obvious that the solution will approach either an equilibrium point or a closed trajectory as $t \to \infty$. (It

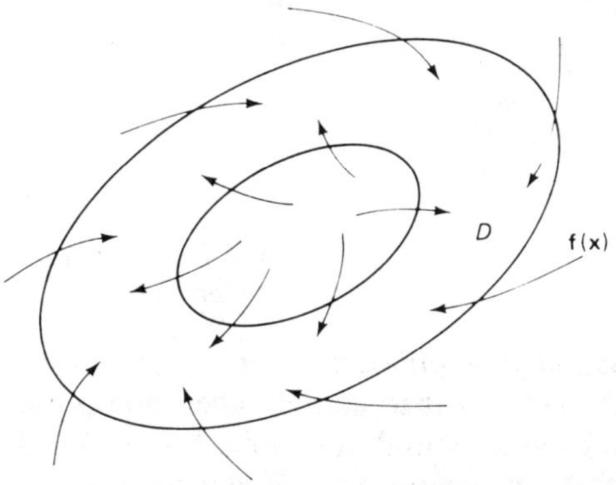

Figure 5.3 Illustration of the Poincaré–Bendixson theorem.

can be shown that no other positive limit set can exist here for this time-invariant system.) Since there are no equilibrium points in D, there must be at least one closed trajectory.

Of course, the difficulty with the Poincaré–Bendixson theorem is the necessity to find the region D such that $\mathbf{f}(\mathbf{x})$ always points into D. But the advent of inexpensive fast computing somewhat alleviates the problem. Also, it may be possible that the search/generation of Liapunov functions, which will be studied in Chapter 6, could lead to such regions. In principle at least, the concept might be extended to higher-order systems (the curse of dimensionality not withstanding).

Example 5.3

Continuation of Examples 4.10 and 5.2 shows trajectories never leave the region defined by

$$D = \{r : 1 - \epsilon \leq r^2 \leq 1 + \epsilon\},$$

where $r^2 = x_1^2 + x_2^2$ and ϵ is a small positive number.

5.1.3 Index Theory

The *index* of a closed curve C with respect to $\mathbf{f}(\mathbf{x})$ in (5.1) and on which $\mathbf{f}(\mathbf{x}) \neq \mathbf{0}$ is the number of counterclockwise revolutions made by $\mathbf{f}(\mathbf{x})$ as $\mathbf{x}$ traverses C once in the counterclockwise direction [C4]. Or, from Fig. 5.4, the index of C is given by

$$I = \frac{\Delta\theta}{2\pi} = \frac{1}{2\pi}\oint_C d\theta, \tag{5.9}$$

where

$$\theta = \tan^{-1}\frac{f_2}{f_1},$$

$$d\theta = \frac{f_1\, df_2 - f_2\, df_1}{f_1^2 + f_2^2}$$

and $f_1(\cdot), f_2(\cdot)$ are nonzero on C.

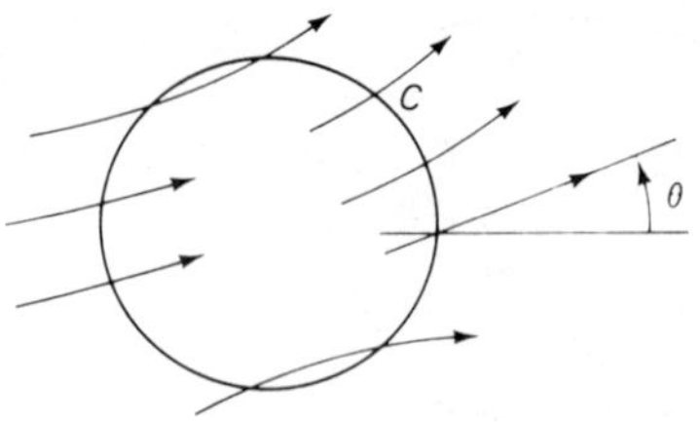

Figure 5.4 Illustration of the index of a closed curve with respect to $\mathbf{f}(\mathbf{x})$.

Next, the *index of an isolated equilibrium point* (or singular point) is defined by the index of the curve which encircles that, and no other, equilibrium point of (5.5). Suppose that the origin is an isolated equilibrium point for (5.5) and for its linearized perturbation equations $\delta\dot{\mathbf{x}} = \mathbf{A}\ \delta\mathbf{x}$, where $\delta\mathbf{x}(t) \sim \mathbf{x}(t)$ for small deviations from $\mathbf{0}$ equilibrium. Then it can be shown that the isolated equilibrium point for the nonlinear system (5.5) has the same index as its linearized representation at that equilibrium point. The index of various linear state-plane portraits is shown in Fig. 5.5. Consequently, it is seen that the *index for saddle points,* with $\Delta\theta = -2\pi$ in (5.9),

$$I_S = -1, \tag{5.10}$$

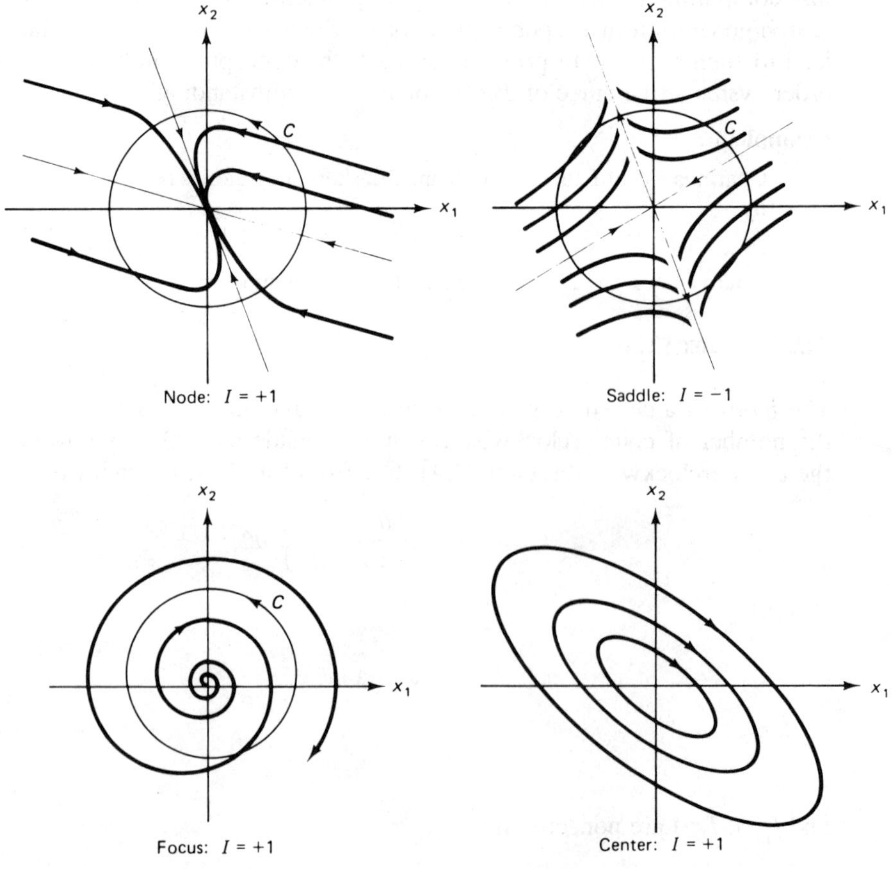

Figure 5.5 Indices for classical linear singular points.

and for *nonsaddles,* with $\Delta\theta = 2\pi$ in (5.9),

$$I_N = +1. \tag{5.11}$$

Then from Figs. 5.4 and 5.5, it can readily be shown that the index of C is zero if there is no equilibrium point inside the region bounded by C (nor on C itself as originally assumed). From Fig. 5.6 it is seen that the index of a closed curve C, which encloses N other closed curves as shown, each with index $I_1, \ldots, I_N$, is equal to the sum of the individual indices. That is,

$$I_C = \sum_{j=1}^{N} I_j. \tag{5.12}$$

The proof is obvious by substitution into the definition with cancellation of $\Delta\theta_j$ along curves with common segments as seen in Fig. 5.6.

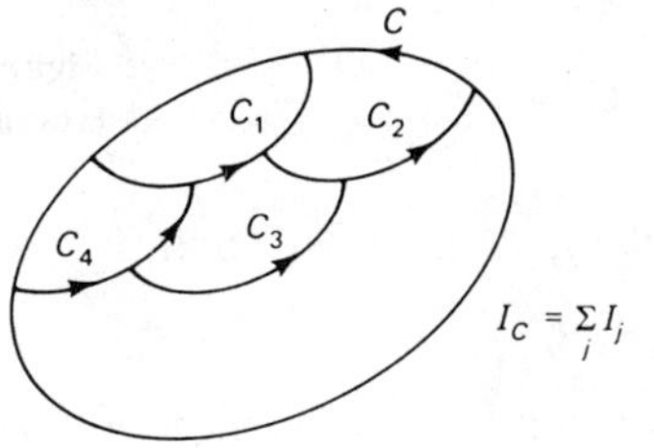

Figure 5.6 Combination of enclosed indices.

As exemplified for the center point in Fig. 5.5, the index of any closed trajectory, including a limit cycle, is equal to 1. In conclusion, a closed trajectory (solution path) for (5.5) may enclose a combination of singular points in state space [equilibrium points of (5.5)] such that

$$N = S + 1, \tag{5.13}$$

where N is the number of nonsaddle points and S is the number of saddle points. Obviously, the total number of such enclosed singular points must be odd. Now it is possible to examine the characteristic equation for the second-order perturbation equations at an isolated equilibrium state for (5.5) to examine the possibility of limit cycles (or other closed trajectories) about various combinations of such equilibrium states. For example, a limit cycle might enclose a combination of one saddle and two foci.

Example 5.4

Consider the following nonlinear system:

$$\begin{aligned} \dot{x}_1 &= x_2, \\ \dot{x}_2 &= -|x_1|. \end{aligned} \tag{5.14}$$

Obviously, there is a unique equilibrium $x_1 = x_2 = 0$. The trajectories for

$$\frac{dx_2}{dx_1} = -\frac{|x_1|}{x_2}$$

are sketched in Fig. 5.7. For $x_1 < 0$, the portrait is for a linear saddle point, and for $x_1 \geq 0$, it is a center. Apparently, the corresponding singular point at the origin is degenerate (no

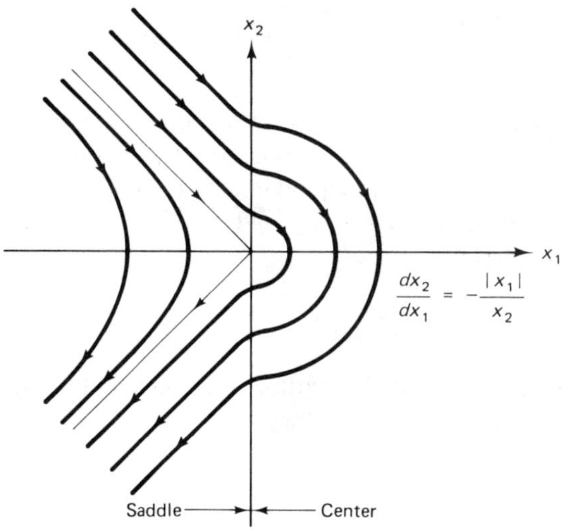

Figure 5.7 Phase-plane portrait for Example 5.4.

valid linear representation), so that $I = 0$. Still, it is seen that no closed trajectory can enclose the origin.

Suppose that (5.5) is a continuous function of a parameter α such that

$$\begin{aligned} \frac{dx_1}{dt} &= f_1(x_1,\ x_2,\ \alpha), \\ \frac{dx_2}{dt} &= f_2(x_1,\ x_2,\ \alpha). \end{aligned} \tag{5.15}$$

Then from the index definition (5.9), it is seen that $I_C(\alpha)$, the index of a closed curve relative to $\mathbf{f}(\mathbf{x})$ in (5.15) is a continuous function of α. Figure 5.8 shows how portraits may change characteristics as a consequence of continuous parametric changes and the index definition, $I_C(\alpha)$. Figure 5.8(a) shows the manner in which a linearized saddle point, $I = -1$, and a linearized center portrait, $I = +1$ (as for a conservative system to be studied in Section 5.2), may come together as α varies continuously over some range of values until at some critical value they meet to form a degenerate point with zero index. At this degenerate configuration it depicts Example 5.4. Note the existence of closed trajectories for α_1 with a center point in Fig. 5.7.

Figure 5.8(b) shows the case where the portrait near the origin changes from a stable focus to a center to an unstable focus for continuously increasing parameter α. The index of the origin is unity for all values. Finally, Fig. 5.8(c) shows the culmination of a stable and unstable limit cycle to a semistable limit cycle and finally no limit cycle as α increases continuously. Obviously, the complexities of nonlinear systems are apparent even for second-order autonomous systems.

It does become apparent, however, that certain combinations of limit cycles can exist while others cannot. For example, a stable or semistable limit cycle might encircle an unstable focus, while an unstable or semistable limit cycle may enclose a stable focus. Also, stable and unstable limit cycles may appear alternately about each other with semistable limit cycles possibly in between.

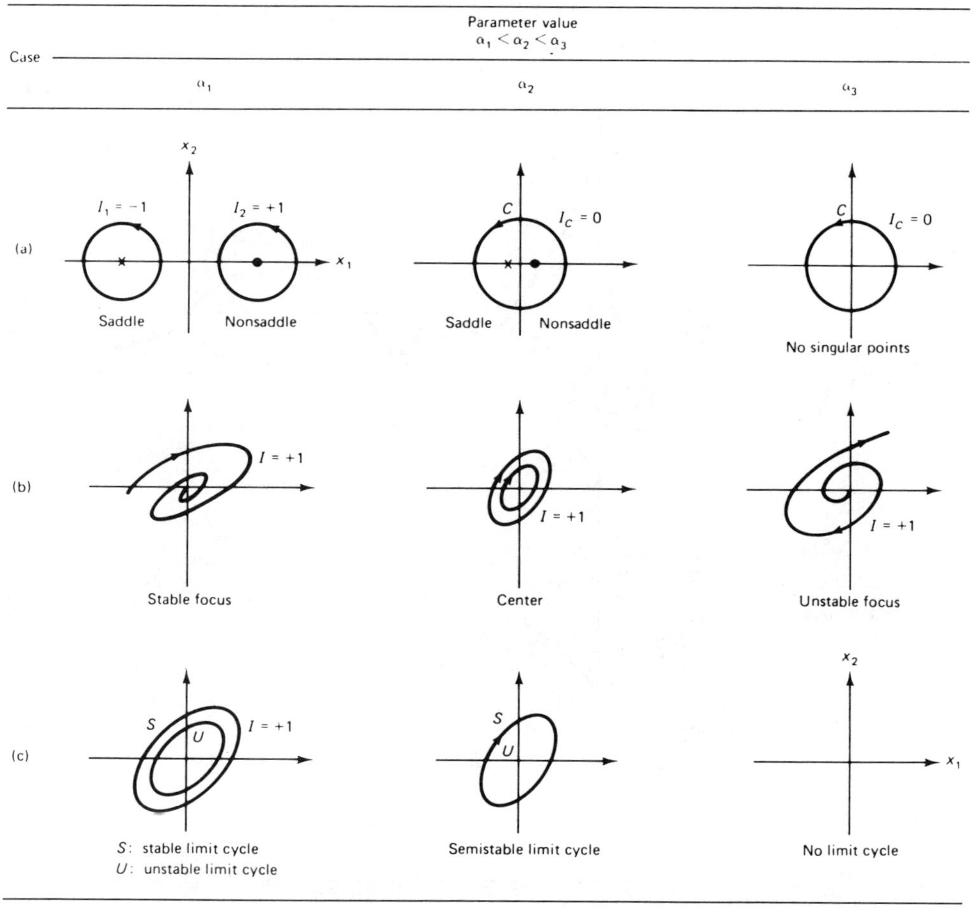

Figure 5.8 Index parametric effects.

5.2 CONSERVATIVE SYSTEMS

Recall from Chapter 2 that for a single degree of freedom, conservative systems are those which have total constant energy E, so that

$$T(\dot{x}) + V(x) = E, \tag{5.16}$$

where $T(\cdot)$ is kinetic energy, $V(\cdot)$ is potential energy, and x, $\dot{x}$ are position and velocity scalars, respectively. The Euler–Lagrange equation for such systems was shown to have the form

$$\frac{d}{dt}\left(\frac{\partial L}{\partial \dot{x}}\right) - \frac{\partial L}{\partial x} = 0, \tag{5.17}$$

where the Lagrangian, $L(x, \dot{x}) = T(\dot{x}) - V(x)$. Then we can represent (5.17) simply by

$$\frac{d^2x}{dt^2} + f(x) = 0. \tag{5.18}$$

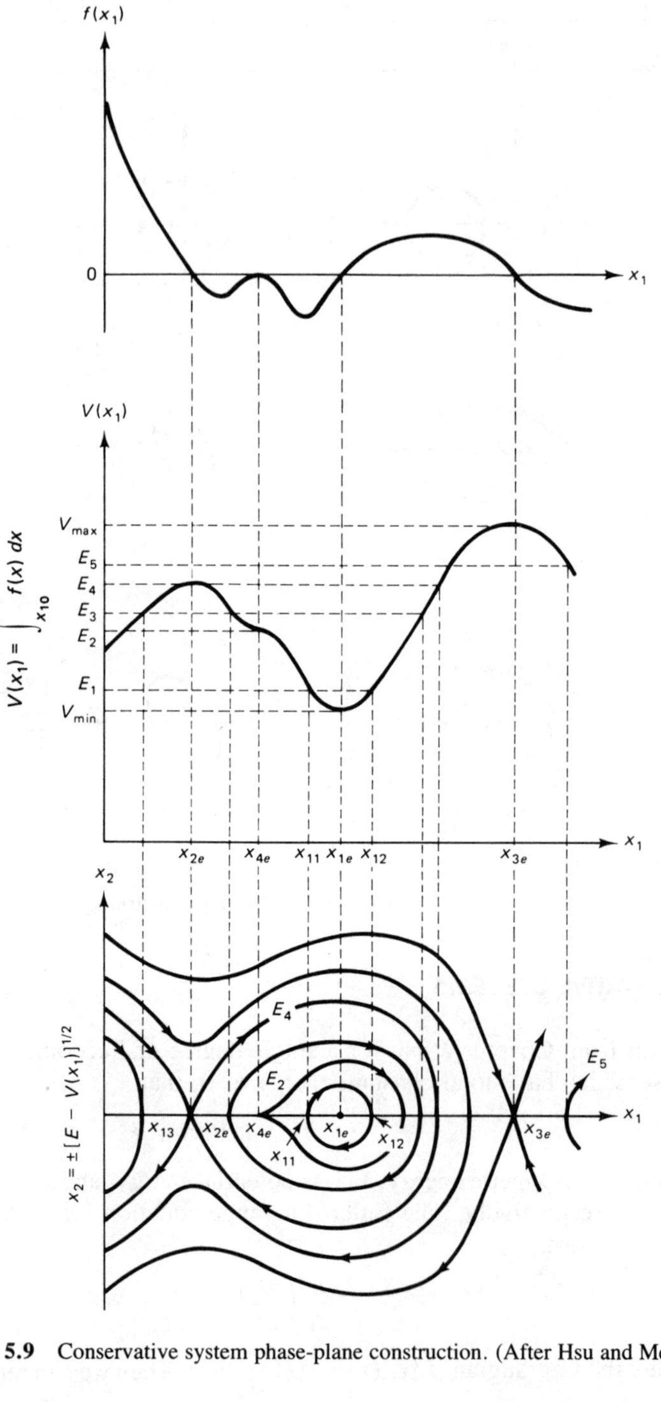

Figure 5.9 Conservative system phase-plane construction. (After Hsu and Meyer [H1].)

Then with $x_1 = x$, $x_2 = \dot{x} = \dot{x}_1$,

$$\dot{x}_1 = x_2,$$
$$\dot{x}_2 = -f(x_1). \tag{5.19}$$

Now the state-plane trajectories must satisfy

$$\frac{dx_2}{dx_1} = -\frac{f(x_1)}{x_2}. \tag{5.20}$$

Upon integration (5.20) becomes

$$\frac{x_2^2}{2} + \int_{x_{10}}^{x_1} f(\sigma)\ d\sigma = E \tag{5.21}$$

or

$$T(x_2) + V(x_1) = E,$$

where E is a constant of integration or total constant energy. Then, from (5.21),

$$x_2 = \pm 2[E - V(x_1)]^{1/2} \tag{5.22}$$

Figure 5.9 shows the construction for the corresponding conditions.

If the initial velocity $x_2(0) = 0$, $V(x_1(0)) = E$. Then Fig. 5.9 sketches the trajectories for different values of total energy or corresponding initial position. Several conclusions are apparent here. First, points of minimum potential energy with $\dot{x}_1 = 0 (T = 0)$ correspond to center singular points, and maximum potential energy with $\dot{x}_1 = 0$ corresponds to saddle points. Then saddle points and center points alternate on the x_1 axis with possible degenerate points in between. Obviously, conservative second-order systems of this form have a continuum of closed trajectories but no limit cycles.

Example 5.5

Consider the simple pendulum of Fig. 5.10 which is shown to be the direct analog of an approximation to the synchronous machine swing dynamics (see Example 3.4). In the former, the kinetic energy

$$T = \frac{1}{2} m v_\theta^2,$$

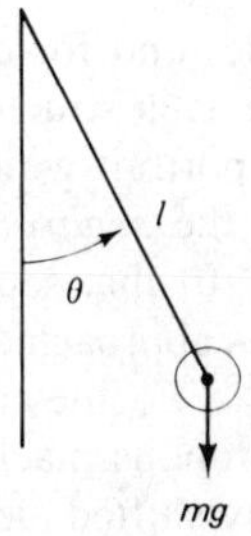

Figure 5.10 Simple pendulum analogy of electrical swing equation.

where $v_\theta = l\dot{\theta}$, and potential energy becomes

$$V = mgl(1 - \cos\theta).$$

Then, with $x = \theta$, (5.17) or (5.18) becomes

$$\ddot{\theta} + \omega^2 \sin\theta = 0, \tag{5.23}$$

where $\omega = g/l$.

For a synchronous machine, which is connected to an infinite bus with negligible damping torque and with zero power input, (5.23) describes the swing dynamics with θ, the load angle, and $\omega^2 = P_e/I_M$. Here P_e is the maximum electrical power output of the generator, and I_M is the machine inertia constant. Then the power input from the turbine is merely added to the right-hand side of (5.23).

With $x_1 = \theta$, $x_2 = \dot{\theta}$, (5.23) becomes

$$\dot{x}_1 = x_2,$$

$$\dot{x}_2 = -\omega^2 \sin x_1,$$

which has equilibrium states

$$x_{1e} = k\pi, \qquad k = 0, \pm 1, \ldots,$$

$$x_{2e} = 0.$$

Then the perturbation equations about $\mathbf{x}_e$ are

$$\delta\dot{x}_1 = \delta x_2,$$

$$\delta\dot{x}_2 = -\omega^2 \cos x_{1e}\, \delta x_1,$$

where

(a) $\cos x_{1e} = 1$, at $x_{1e} = 0, \pm 2\pi, \ldots$.
(b) $\cos x_{1e} = -1$, at $x_{1e} = \pm\pi, \pm 3\pi, \ldots$.

For case (a),

$$\mathbf{A} = \begin{bmatrix} 0 & 1 \\ -\omega^2 & 0 \end{bmatrix}, \qquad \lambda_{1,2} = \pm j\omega, \text{ center;}$$

for case (b),

$$\mathbf{A} = \begin{bmatrix} 0 & 1 \\ \omega^2 & 0 \end{bmatrix}, \qquad \lambda_{1,2} = \pm\omega, \text{ saddle point.}$$

Then case (b) is unstable, and for case (a) no conclusion can be drawn based on the perturbation model since it is structurally unstable. Figure 5.11 sketches the potential energy and phase-plane portrait which shows the alternating centers and saddle points along the x_1 axis. Note the *separatrices,* which are the specific trajectories (one for $x_2 > 0$ and one for $x_2 < 0$) that separate the motion characteristics. That is, only for $E = E_3$ do the trajectories approach equilibrium. For $E < E_3$, trajectories are closed and approach a center. For $E > E_3$, they remain outside the separatrices and never change the sign of $\dot{x}$. For the synchronous-machine swing equation with constant power input the singular points are merely shifted along the x_1 axis.

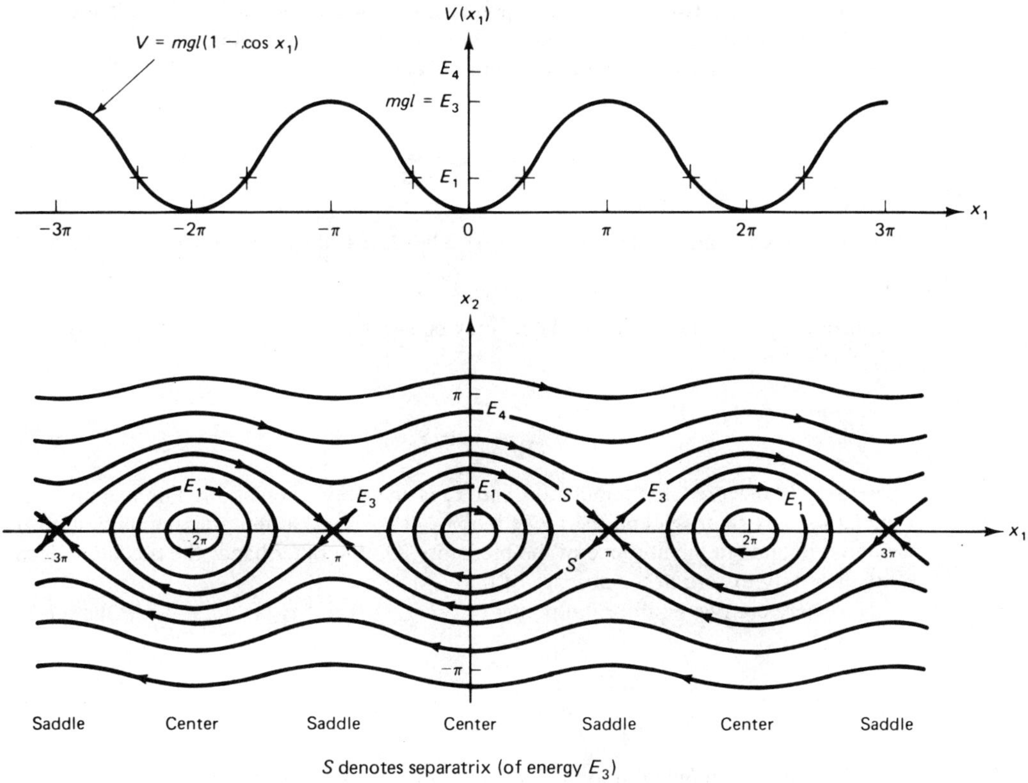

Figure 5.11 Phase-plane portrait for Example 5.5.

Example 5.6

Consider the motion $x(t)$ from rest state for m in Fig. 5.12 which is restrained by a nonlinear spring such that

$$\ddot{x} + \alpha x + \beta x^3 = 0,$$

where $\alpha = k_1/mg > 0$, $\beta = k_2/mg$. $\beta > 0$ refers to a hard spring, and $\beta < 0$ to a soft spring. It is apparent that the hard spring has one equilibrium state $\dot{x} = x = 0$ at minimum potential energy which corresponds to a center for small motion. The oscillations are periodic, but here the period depends on the initial state. Hence the motion is nonisochronous, which distinguishes it from the linear conservative case (simple harmonic motion).

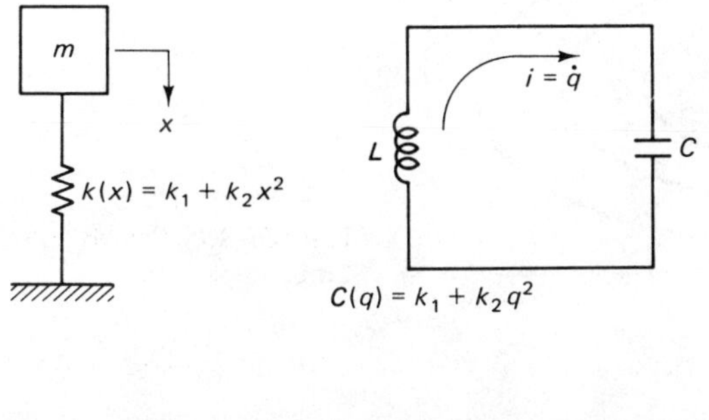

Figure 5.12 Nonlinear spring for Example 5.6.

For the soft spring, two other singular points also occur, located at $\dot{x} = 0$ and that x which corresponds to the maximum values of the potential energy. Hence in addition to the center at the origin, there are saddle points at $x_2 = \dot{x} = 0$ and

$$x_1 = x = \pm\sqrt{\frac{-\alpha}{\beta}}.$$

The solutions are analyzed in detail in [C4]. The electrical analogy is shown in Fig. 5.12.

Example 5.7

Consider the Volterra–Lotka predator–prey equations

$$\dot{x}_1 = -x_1 + x_1x_2,$$
$$\dot{x}_2 = x_2 - x_1x_2,$$

where x_1 is the predator population and x_2 is the prey population. For example, the predators may be foxes and the preys rabbits. It is seen that the prey population grows without bounds in the absence of the predators. On the other hand, the predators decay exponentially to zero in the absence of prey.

There are two equilibrium points $x_{1e} = x_{2e} = 0$ and $x_{1e} = x_{2e} = 1$. At the origin,

$$\delta\dot{x}_1 = -\delta x_1,$$
$$\delta\dot{x}_2 = \delta x_2,$$

which is in the canonical normal form with the eigenvectors along the x_1, x_2 axes. At (1, 1),

$$\delta\dot{x}_1 = \delta x_2,$$
$$\delta\dot{x}_2 = -\delta x_1,$$

which is a canonical form for a center with $\beta = 1$ [see (3.67)]. Index theory indicates the possibility of closed-solution paths about (1, 1). Figure 5.13 shows a sketch of the resulting phase-plane trajectories. In reality, of course, only the first quadrant, which includes a continuum of closed trajectories, is of physical interest.

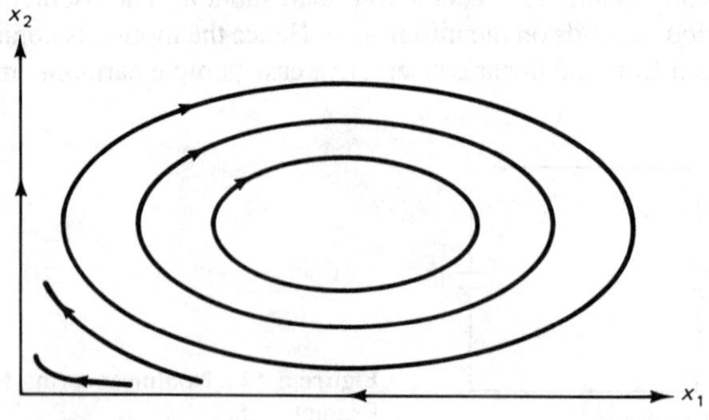

Figure 5.13 Predator–prey state plane for Example 5.7.

5.3 NONCONSERVATIVE SYSTEMS

If the conservative system (5.18) has an added nonlinear damping term $h(x)\dot{x}$, the nonconservative system becomes

$$\frac{d^2x}{dt^2} + h(x)\frac{dx}{dt} + f(x) = 0, \tag{5.24}$$

where $h(x) \geq 0$. For example, nonlinear damping of this form might be added to represent nonlinear friction on the pendulum or damping torque on the synchronous machine. In this case, the total energy $E = H(x, \dot{x})$ decreases monotonically toward an asymptotically stable equilibrium state where the potential energy becomes a minimum with zero kinetic energy. Here the linearized motion typically is a stable focus. In this case (5.24) becomes

$$\frac{dH}{dt} = -h(x)\dot{x}^2,$$

where $H(\cdot) = T(\cdot) + U(\cdot) = \dot{x}^2/2 + \int^x f(\sigma)\, d\sigma$. If $h(x)$ is allowed to become negative, the energy may increase. Also, a limit cycle may occur. Obviously, the addition of a nonlinear damping or self-excitation drastically alters the system structure. Whereas conservative systems were characterized by a continuum of closed trajectories, (5.24) is more likely to have limit cycles as shown in Section 4.3.1.

5.3.1 The Van der Pol Equation

One special case of (5.24) is the Van der Pol equation, where $f(x) = x$, $h(x) = -\mu(1 - x^2)$, and $\mu > 0$. Restated, it becomes

$$\frac{d^2x}{dt^2} - \mu(1 - x^2)\frac{dx}{dt} + x = 0. \tag{5.25}$$

This popular equation was analyzed by the Bendixson theorem in Example 5.1. It was shown that (5.25) cannot exhibit a limit cycle within $|x| < 1$. With $\ddot{x} = \dot{x} = 0$, (5.25) is at equilibrium with $x = 0$. Again, for canonical phase variables, (5.25) is given by the state formulation

$$\begin{aligned} \frac{dx_1}{dt} &= x_2, \\ \frac{dx_2}{dt} &= \mu(1 - x_1^2)x_2 - x_1; \end{aligned} \tag{5.26}$$

$x_{1e} = x_{2e} = 0$ is the equilibrium state.

Now, the perturbation equations at $\mathbf{x}_e$ are given by

$$\begin{aligned} \delta\dot{x}_1 &= \delta x_2, \\ \delta\dot{x}_2 &= \mu\delta x_2 - \delta x_1, \end{aligned} \tag{5.27}$$

which has the characteristic equation

$$\lambda^2 - \mu\lambda + 1 = 0.$$

Then for $\mu > 2$, λ_1 and λ_2 are distinct, real, and positive—an unstable node; for $\mu = 2$, $\lambda_1 = \lambda_2 = 1$ (a degenerate unstable node with one unique eigenvector). If $0 < \mu < 2$, λ_1 and λ_2 are complex with positive real parts. Obviously, the origin is unstable.

From index theory [see (5.13)] it is possible that at least one limit cycle may exist about the origin. It is shown that the Van der Pol equation (5.26) arises as a natural model for an oscillator circuit with nonlinear resistance [V3]. A limit cycle about the origin is well established. These periodic solutions are studied in some detail in the literature (e.g., [V3], [L3], [S2], and [H1]). The changing form of limit cycles for different values of μ are shown in Fig. 5.14.

The block diagram of a feedback control system that satisfies (5.25) is shown in Fig. 5.15. It is also of interest that (5.26) may be envisioned as a bilinear control system with the feedback control $u = -x_1^2$, where

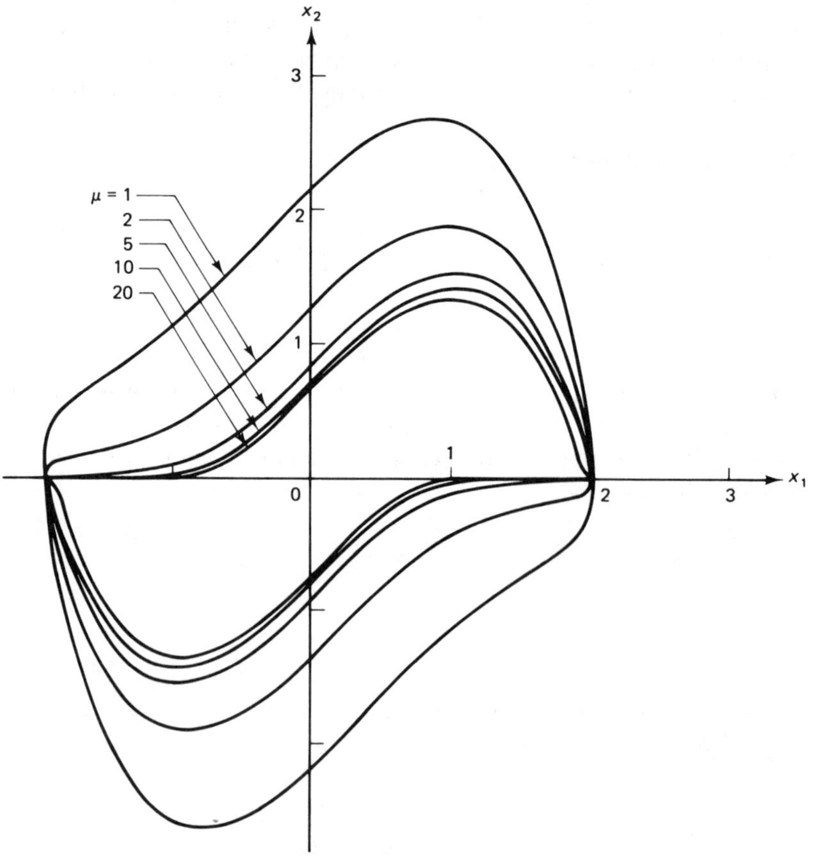

Figure 5.14 Limit cycles for the Van der Pol equation.

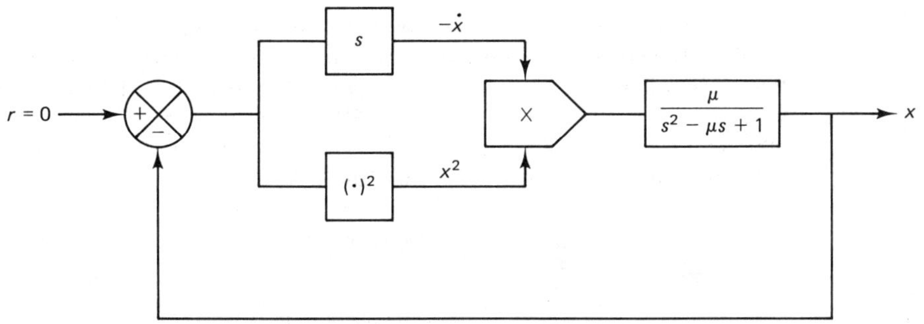

Figure 5.15 Feedback synthesis of the Van der Pol equation.

$$\frac{dx_1}{dt} = x_2,$$
$$\frac{dx_2}{dt} = -x_1 + \mu x_2(1 + u). \tag{5.28}$$

Or if x_1 is measured, (5.26) and (5.28) become

$$\frac{dx_1}{dt} = x_2,$$
$$\frac{dx_2}{dt} = -y + (1 - y^2)x_2, \tag{5.29}$$
$$y = x_1,$$

which is called a conditionally linear system. That is, with respect to given measurements, the process is linear. Bilinear systems (BLS) and conditionally linear systems (CLS) are studied in Vol. 2 and are shown to have some interesting properties, and in some cases, are conducive to analysis. Also, the Van der Pol equation undergoes a stability analysis by Liapunov's direct method in Chapter 6.

Equation (5.24) is a generalization of the Van der Pol equation which has been studied in some depth. With certain constraints on $f(x)$ and $h(x)$, (5.24) is sometimes referred to as a special case of Lienard's equation. Among the most common conditions, which have been shown to yield a unique limit cycle for (5.24), are that $f(x)$ is odd with $f(0) = 0$; $h(x)$ is even with $h(0) < 0$, and as $x \to \infty$, $\lim \int_0^x h(\sigma)\, d\sigma = \infty$. Although other conditions generally are required, the following example shows the existence of limit cycles for a special case.

Example 5.8

Consider the circuit in Fig. 5.16 for a dc generator with an RC load and iron saturation. From Kirchhoff's law it is seen that

$$LI\frac{dx_1}{dt} = -rIx_1 - Ex_2 + E \text{ sat } x_1,$$
$$CE\frac{dx_2}{dt} = Ix_1 - \frac{E}{R}x_2,$$

where $x_1 \equiv i/I$ and $x_2 \equiv e/E$ are normalized dimensionless state variables, and the other circuit parameters are defined in Fig. 5.16. For convenience, normalize time with $\tau \equiv rt/L$, where L/r is the generator time constant. Also, define the fractional load resistance, $\rho \equiv RI/E$, fractional generator resistance, $\alpha \equiv rI/E$ and ratio of load time constant to generator time constant, $\gamma = Rrc/L$. Here each of ρ, α, γ are positive constants. Then the normalized circuit equations are

$$\frac{dx_1}{d\tau} = -x_1 + \frac{1}{\alpha}(\text{sat } x_1 - x_2),$$

$$\frac{dx_2}{d\tau} = \frac{1}{\gamma}(\rho x_1 - x_2).$$

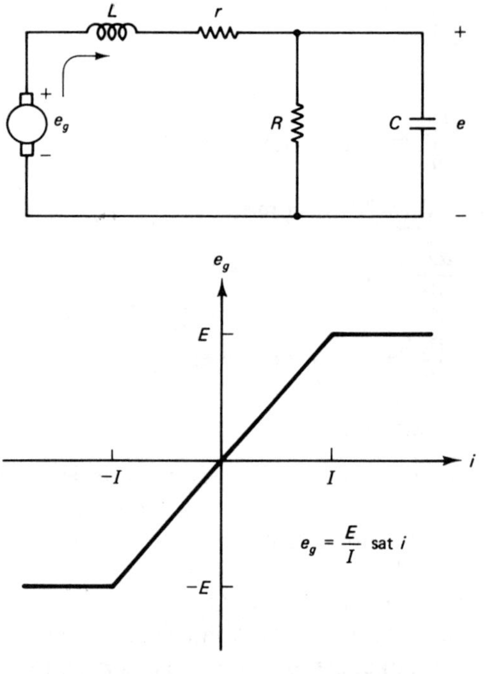

Figure 5.16 Generator circuit for Example 5.8.

Now the equilibrium states are found with $dx_1/d\tau = dx_2/d\tau = 0$. Therefore, there exists one equilibrium state for $\rho > (1 - \alpha)$ which yields $\mathbf{x}_e = \mathbf{0}$. For $\rho < (1 - \alpha)$, there are three $\mathbf{x}_e$: $\mathbf{x}_e = \mathbf{0}$ for $|x_1| < 1$, $\mathbf{x}_e^T = [1/(\alpha + \rho) \quad \rho/(\alpha + \rho)]$ for $x_1 > 1$, and $\mathbf{x}_e^T = [-1/(\alpha + \rho) \quad -\rho/(\alpha + \rho)]$ for $x_1 < -1$. If $\rho = 1 - \alpha$, there exists an infinite number of equilibrium states.

Obviously, the system is piecewise linear so that there are three different linear models: one for $|x_1| < 1$, another for $x_1 > 1$, and still another linear model for $x_1 < -1$. For $|x_1| < 1$, the equivalent linear system matrix is given by

$$\mathbf{A} = \begin{bmatrix} -1 + \dfrac{1}{\alpha} & -\dfrac{1}{\alpha} \\ \dfrac{\rho}{\gamma} & -\dfrac{1}{\gamma} \end{bmatrix},$$

with characteristic equations of the form

$$\lambda^2 + 2\zeta\omega_n\lambda + \omega_n^2 = 0$$

or

$$\lambda^2 + \frac{\alpha - \gamma + \alpha\gamma}{\alpha\gamma}\lambda + \frac{\alpha - 1 + \rho}{\alpha\gamma} = 0.$$

For $\omega_n^2 < 0$, the singular point of the origin in the x_1–x_2 plane is a saddle point, which also implies that $\rho < (1 - \alpha)$, so that two other singular points will exist, as noted above. For the nonsaddle point, $\omega_n^2 > 0$ with $\rho > (1 - \alpha)$, which only leads to a singular point at **0**. For a stable nonsaddle point at the origin, $\zeta > 0$ or $\alpha > \gamma/(1 + \gamma)$.

For $x_{1e} > 1$,

$$\mathbf{A} = \begin{bmatrix} -1 & -\dfrac{1}{\alpha} \\ \dfrac{\rho}{\gamma} & -\dfrac{1}{\gamma} \end{bmatrix},$$

so that $\det[\mathbf{I}\lambda - \mathbf{A}] = 0$ yields

$$\lambda^2 + \frac{\gamma + 1}{\gamma}\lambda + \frac{\alpha + \rho}{\alpha\gamma} = 0.$$

Here, this yields a stable equilibrium state since α, ρ, γ are positive. Similarly, $x_{1e} < -1$ is stable. For the above characteristic equation, complex eigenvalues exist if $\alpha(\gamma^2 - 2\gamma + 1)/4\gamma < \rho$. Otherwise, a stable-node portrait results. Figure 5.17 divides the α–ρ parameter plane into four basic characteristic regions: I—$\mathbf{x}_e = \mathbf{0}$, an unstable focus or unstable node; II—$\mathbf{x}_e = \mathbf{0}$, a stable focus or stable node; III and IV—saddle point at $\mathbf{x}_e = 0$ and two other stable nodes or stable foci as given above. For $(\alpha/4\gamma)(\gamma^2 - 2\gamma + 1) > \rho$, it is a stable focus.

From Bendixson's theorem [equation (5.6)],

$$\nabla\mathbf{f} = \frac{\partial f_1}{\partial x_1} + \frac{\partial f_2}{\partial x_2} = -1 + \frac{\text{pulse } x_1}{\alpha} - \frac{1}{\gamma},$$

where

$$\text{pulse } x_1 = \begin{cases} 0, & |x_1| > 1, \\ 1, & |x_1| < 1. \end{cases}$$

Then, since $\nabla\mathbf{f}$ does not change sign (and is not identically zero) for $\alpha > \gamma/(1 + \gamma)$, there is no limit cycle for parameter values in zones II and IV of Fig. 5.17. However, they are not ruled out in zones I and III.

It is seen that this system is a special case of Lienard's equation with $x = x_1$, $f(x) = (\rho/\alpha\gamma + \gamma - \text{sat}/\alpha\gamma)x$ and $h(x) = 1 - 1/\alpha$ pulse $x - 1/\gamma$. Then in zone I, the conditions that were stated above for Lienard's equation are satisfied, and indeed it can be shown that a unique limit cycle exists about the unstable focus at the origin. Index theory (5.13) also indicates the possibility of a limit cycle about either the unstable focus or the unstable node with $I = 1$. In zone III, there can be no limit cycle about the origin as above since (5.13) would be violated. However, a limit cycle might enclose all three

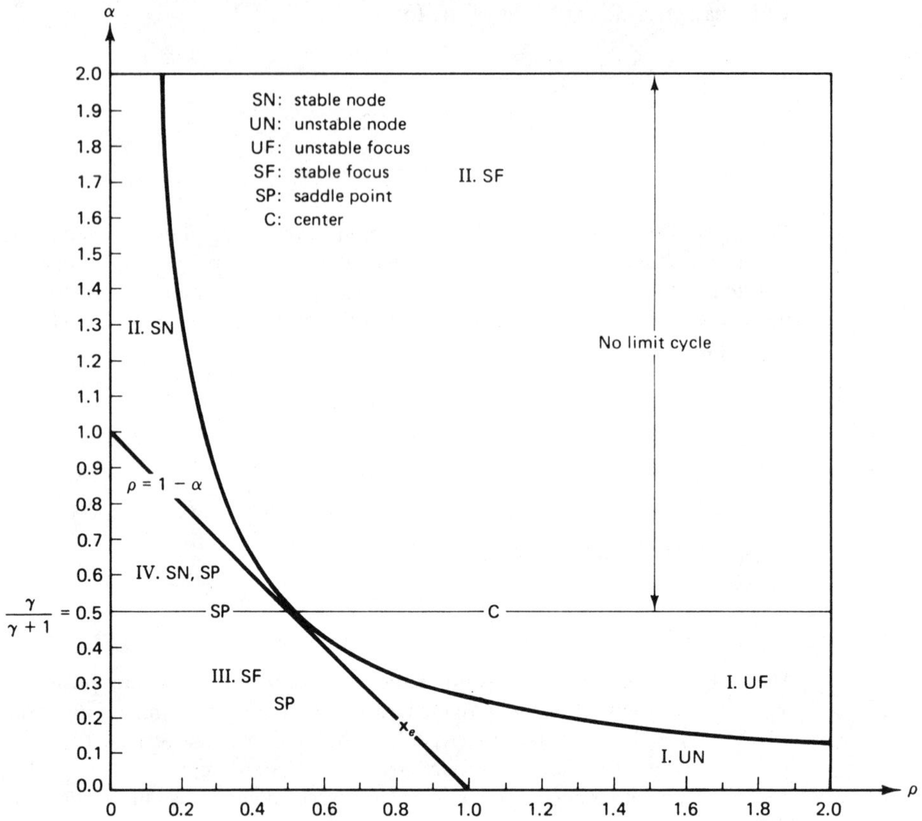

Figure 5.17 Parameter-plane characteristics for Example 5.8, with $k = 1$.

singular points or each of the nonorigin singular points alone. Figure 5.18 shows several state-plane trajectories for various parameter values. Apparently, zone IV represents a desired operating region for the dc generator system.

More generally, Lienard's equation is given by

$$\frac{d^2x}{dt^2} + h(x, dx/dt)\frac{dx}{dt} + f(x) = 0,$$

or in canonical phase-variable form,

$$\frac{dx_1}{dt} = x_2,$$

$$\frac{dx_2}{dt} = -f(x_1) - h(x_1, x_2)x_2,$$

where various appropriate assumptions are made on $f(\cdot)$ and $h(\cdot)$.

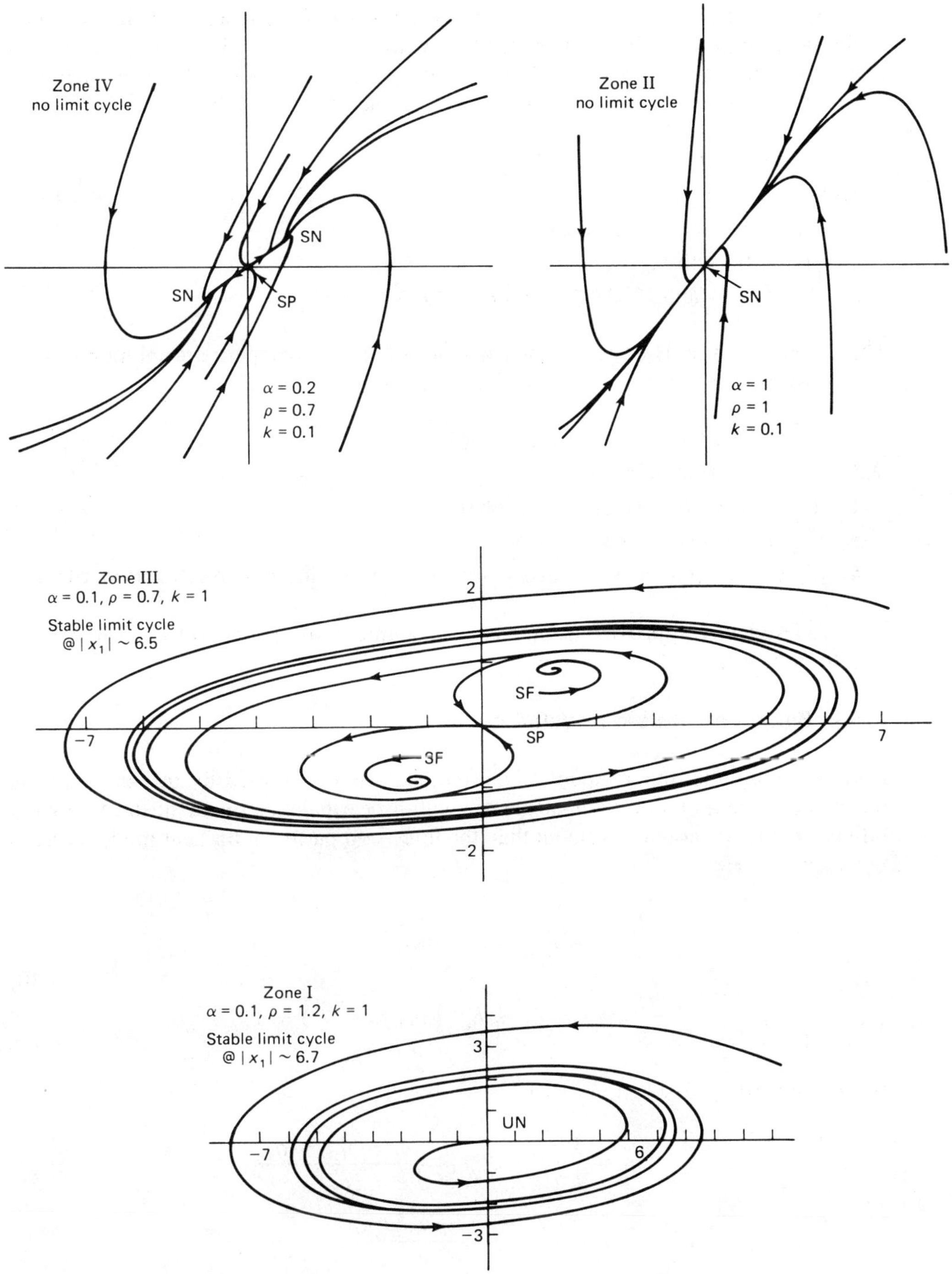

Figure 5.18 State-plane portraits for different parameter values for Example 5.8.

Several useful theorems have been established for Lienard's equation under different assumptions. These are stated here with the proofs available in the references as noted. The first theorem states that Lienard's equation has at least one periodic solution if there exist positive constants a, b, c such that

1. $h(x_1, x_2) > b$ if $|x_1| \geq a$ and $|x_2| \geq a$.
2. $h(0, 0) < 0$.
3. $h(x_1, x_2) > -c$ for all x_1, x_2.
4. $f(0) = 0$; with $x_1 f(x_1) > 0$, $x_1 \neq 0$.
5. $|f(x_1)| \to \infty$ and $f(x_1)/\int_0^{x_1} f(\sigma)\, d\sigma \to 0$ as $|x_1| \to \infty$.

The second theorem [L8] shows the existence of at least one periodic solution if there exist positive constants a, b, c, d with $a < d$ and

1. $h(x_1, x_2) \geq 0$ if $|x_1| \geq a$ for every x_2.

2,3. As for the first theorem.

4. $f(0) = 0$; with $x_1 f(x_1) > 0$, $x_1 \neq 0$.
5. $\int_0^{x_1} f(\sigma)\, d\sigma \to \infty$ as $|x_1| \to \infty$.
6. $\int_a^d h(x_1, x_2)\, dx_1 > bc$ for every arbitrary decreasing positive function $x_2(x_1)$.

Reference [L4] proves the existence of a unique limit cycle for the Van der Pol equation (5.26).

5.3.2 Body Temperature Regulation

Here a two-compartment model of thermoregulation is studied by means of a state plane. For core and skin compartments, with a heat balance on each, the following bilinear model is more convenient than the three-compartment bilinear model derived in Chapter 2. Here

$$\frac{dT_c}{dt} = u_1(T_s - T_c) + v_1,$$
$$\frac{dT_s}{dt} = bu_1(T_c - T_s) + u_2(T_s - T_a) + v_2, \tag{5.30}$$

where, according to Section 2.4.1,

$$u_1 = \frac{k_v A_{cs}}{l_{cs} C_{mc}}, \qquad v_1 = \frac{q_b - q_{re} + q_{ex}}{C_{mc}},$$

$$u_2(T_s - T_a) + v_2 \approx \frac{q_c + q_e + q_r}{C_{ms}}, \qquad b = \frac{C_{mc}}{C_{ms}},$$

and T_a is ambient temperature; u_2, v_1, v_2 may be manipulated to some extent, but most of the discussion here will study the effect of $u_1(t)$, the vasomotor control term.

From (5.30), the equilibrium state is defined by

$$T_{ce} = \frac{bv_1 + v_2}{u_2} + \frac{v_1}{u_1} + T_a,$$
$$T_{se} = \frac{bv_1 + v_2}{u_2} + T_a. \tag{5.31}$$

Figure 5.19 shows the region S of possible equilibrium points with constraints of the form

$$0 < u_{1a} \le u_1 \le u_{1b},$$
$$0 < u_{2a} \le u_2 \le u_{2b},$$
$$0 < v_{1a} \le v_1 \le v_{1b},$$
$$0 < v_{2a} \le v_2 \le v_{2b}. \tag{5.32}$$

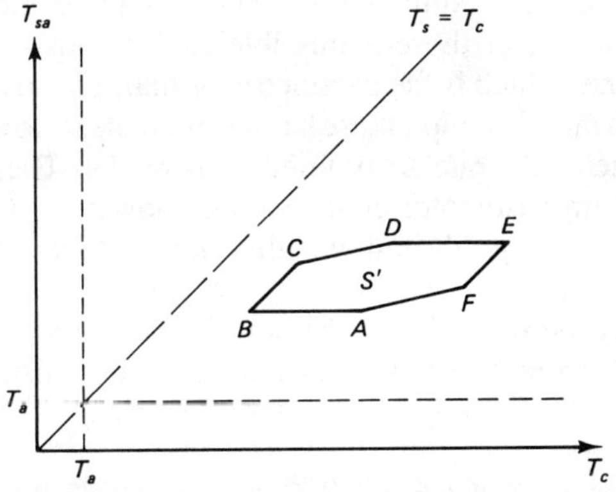

Figure 5.19 Thermal equilibrium region S for (5.30) and (5.31).

Point A is obtained from (5.31) with $u_1 = u_{1a}$, $u_2 = u_{2b}$, $v_1 = v_{1a}$, and $v_2 = v_{2a}$, and the other points are similarly located, as shown in Table 5.1. Line segments B–A and D–E locate two extreme equilibrium sets for $u_1(t)$, the main control of interest here.

TABLE 5.1 CONTROL VALUES FOR EQUILIBRIUM SET (FIG. 5.19)

Point	u_1	u_2	v_1	v_2
A	u_{1a}	u_{2b}	v_{1a}	v_{2a}
B	u_{1b}	u_{2b}	v_{1a}	v_{2a}
C	u_{1b}	u_{2a}	v_{1a}	v_{2b}
D	u_{1b}	u_{2a}	v_{1b}	v_{2b}
E	u_{1a}	u_{2a}	v_{1b}	v_{2b}
F	u_{1a}	u_{2b}	v_{1b}	v_{2a}

Now assume that u_2, v_1, v_2 are held constant, in order to study controllability aspects of $u_1(t)$. In Fig. 5.20 a typical viability region is superimposed on an equilibrium set corresponding to extreme values of ambient temperature (T_{aa}, T_{ab}) and to

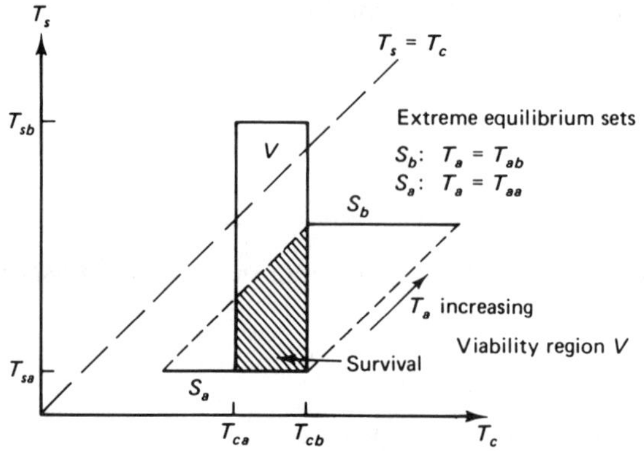

Figure 5.20 Thermal survival region with vasomotor control, $u(t)$.

extreme values of control (u_{1a}, u_{1b}). For example, $T_c > T_{cb}$ might cause irreversible damage to the nervous system; $T_c < T_{ca}$, core hypothermia, causes "cold narcosis"; $T_s > T_{sb}$ results in skin burn; and $T_s < T_{sa}$ may cause freezing. Certainly, larger tolerances on skin temperature T_s can be admitted than those on core temperature T_c. The cross-hatched region in Fig. 5.20 locates those admissible equilibrium states for survival with vasomotor control alone, which is the parametric or multiplicative control in this case. Note that altering the ambient temperature has a similar effect to adding clothing, exercising, or increasing metabolic rate according to this model. These take the form of additive control here. With vasomotor control alone, however, it is seen that the thermoregulator will maintain equilibrium viability for ambient temperatures on $[T_{aa}, T_{ab}]$.

It is of interest to study the state-plane behavior of the system (5.30) for extreme constant values of vasomotor control with the other control mechanisms constant. The eigenvalues of the system matrix in this case are described by

$$\lambda_{1,2} = -\tfrac{1}{2}[(b + 1)u_1 + u_2] \pm \tfrac{1}{2}\{[(b + 1)u_1 + u_2]^2 - 4u_1u_2\}^{1/2}.$$

From a Taylor series, the eigenvalues may be approximated by

$$\lambda_1 \approx \frac{-u_1u_2}{(b + 1)u_1 + u_2}$$

and

$$\lambda_2 \approx -[(b + 1)u_1 + u_2].$$

Therefore, equilibrium points for constant control are characterized as stable nodes with the characteristic behavior shown in Fig. 5.21 for two extreme equilibrium points, $\mathbf{x}_{ea}$ and $\mathbf{x}_{eb}$, corresponding to extreme controls u_{1a} and u_{1b}. The set of points S locates the equilibrium states for admissible constant controls.

From the solid trajectories for u_{1a} and the dashed trajectories for u_{1b}, the zone that can be reached from the admissible equilibrium S in finite time is defined by R, all points interior to the dotted trajectory segments. Similarly, it is apparent that the incident zone, that is, the set of points from which S is reached in finite time, includes

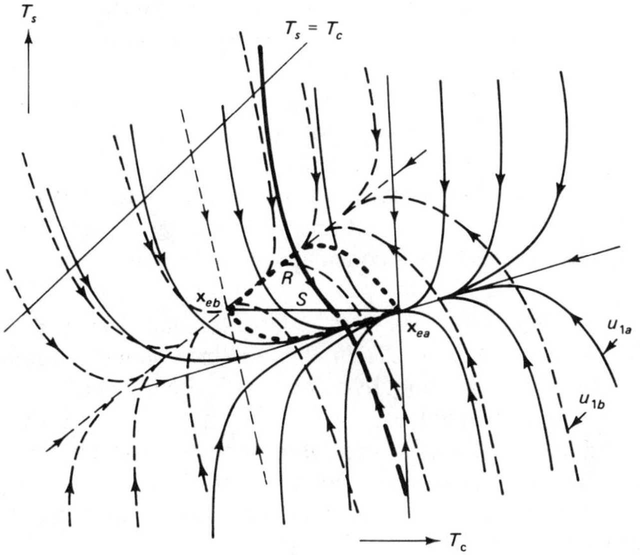

Figure 5.21 Bilinear state portrait for bang-bang control, u_{1a},u_{1b}.

the entire finite state plane. Consequently, it is apparent that the region of complete controllability is defined by R also. Apparently, the system cannot exhibit a limit cycle for constant u_1. However, it is seen from Fig. 5.21 that a periodic solution could result interior to R by a piecewise-constant periodic $u_1(t)$.

5.4 NONLINEAR CONTROLLERS

Typical examples of effective second-order nonlinear controllers are presented here, with more elegant analytical techniques to follow.

5.4.1 Nonlinear Feedback Control

Consider the second-order linear system

$$\dot{\mathbf{x}} = \mathbf{A}\mathbf{x} + \mathbf{B}u, \tag{5.33}$$

$$\mathbf{y} = \mathbf{c}\mathbf{x},$$

where

$$\mathbf{A} = \begin{bmatrix} \alpha & \beta \\ -\beta & \alpha \end{bmatrix}.$$

Since $\mathbf{A}$ is a convenient canonical matrix, the eigenvalues of $\mathbf{A}$ are given by

$$\lambda_1,\ \lambda_2 = \alpha \pm j\beta.$$

If $\beta > |\alpha|$, the configurations correspond to a lowly damped system.

Linear feedback control could be used to provide satisfactory response for many such cases. The feedback compensation might include a PID control. Here the integrator would provide zero steady-state error for regulation or for step-type inputs and the derivative would provide improved damping. If a fast rise time is desired, a somewhat low damping with considerable overshoot results. And for linear systems, the percentage of overshoot remains constant for a fixed second-order pole–zero configuration. This may provide perfectly satisfactory performance in many cases, but in certain cases (e.g., chemical process control) the overshoot may not be tolerable for large disturbances or step demands. Consequently, we consider a nonlinear feedback synthesis.

In polar coordinates, $\dot{r} = \alpha r$ and $\dot{\theta} = -\beta$. Then $r(t) = r_0 e^{\alpha t}$ and $\theta = \theta_0 - \beta t$ represents the homogeneous ($u = 0$) solution. Then the nonlinear control might achieve its aim of fast response but little overshoot by effective manipulation of α, the real part of the closed-loop system "eigenvalues," while keeping β at its already relatively large value to maintain bandwidth. Of course, it must be realized that this is merely an intuitive argument. Since with nonlinear feedback control, the closed-loop regulator is really nonlinear, any reference to eigenvalues, or poles, again refers to linearized perturbation dynamics.

Also, great care must be taken in adding nonlinear elements without a complete knowledge of its effects. For example, if $\beta = 1$, $\alpha = \mu$, $\mathbf{B} = \mathbf{I}$, $u_1 = -\mu(x_1^2 + x_2^2)x_1$, and $u_2 = -\mu(x_1^2 + x_2^2)x_2$, the system has a limit cycle at $r = 1$, as demonstrated by Example 4.10. Ordinarily, such a situation should be avoided, unless such an oscillator is desired. Then there is the problem of controllability. In the linear polar canonical form, a control in the radial direction has no effect on rotational motion, which, in many cases, would not be admissible.

Example 5.9

Consider (5.33) with scalar control u,

$$\mathbf{A} = \begin{bmatrix} -1 & 4 \\ -4 & -1 \end{bmatrix}, \qquad \mathbf{b} = \begin{bmatrix} 0 \\ \frac{1}{4} \end{bmatrix}, \tag{5.34}$$

so that the eigenvalues are

$$\lambda_\lambda, \ \lambda_2 = -1 \pm j4.$$

Obviously, the open-loop system is very lowly damped. It can be shown that a cubic control function of state provides fast response with little overshoot in x_1 or x_2. The following may be derived by "cut-and-try" simulation or by more elegant optimization [S3]:

$$u = -K(x_1^2 + x_2^2)x_2.$$

Here the feedback control can be shown to be stabilizing, as shown in Vol. 2, where such systems are studied as bilinear with quadratic state feedback. Also optimization of such a process is studied in Vol. 2.

Classical linear control reasoning may also be used to explain the effectiveness of this nonlinear control policy as demonstrated in [S3], with an earlier application to the variable-damping control of an electric motor demonstrated in [L5]. The closed-loop regulator system dynamics (for zero input), with $y = x_1$, becomes

$$\ddot{y} + [2 + M(\dot{y},\ y)]\dot{y} + N(y)y = 0, \tag{5.35}$$

where

$$M(\dot{y}, y) = \frac{K}{16}(\dot{y}^2 + 3y\dot{y} + 7y^2),$$

$$N(y) = 17 + \frac{5}{16}Ky^2.$$

Notice that since the coefficient of the $\dot{y}$ term is a function of $\dot{y}$, this nonlinear model is a generalization of (5.24). Then, assuming that $M(\cdot)$, $N(\cdot)$ are constant for small time intervals, the "damping coefficient" is given at $t = k$ by [S3]

$$\zeta_k^2 = \zeta_0^2 + \frac{K}{16N_k}[(\dot{y}_k^2 + 3y_k\dot{y}_k + 6.7y_k^2) + \frac{K}{64}(y_k^2 + 3y_k\dot{y}_k + 7y_k^2)^2],$$

where ζ_0 is the open-loop damping coefficient. Then, for large y_k,

$$\zeta_k^2 \approx \zeta_0^2 + \frac{K(1 + K/64)M_k}{16N_k}.$$

Since M_k is positive for all $(y_k, \dot{y}_k)$, the closed-loop damping is always greater than that of the open loop.

To estimate the region in the state plane (x_1 versus x_2) where no overshoot occurs, let $\zeta_k = 1$ above; then it can be shown that the following should be satisfied:

$$\frac{K}{16}(16x_1^2 + 4x_1x_2) + \left(\frac{K}{8}\right)^2(5x_1^2 + 4x_1x_2 + 16x_2^2)^2 \le 16.$$

The state trajectories with $x_1(0) = 1$, $x_2(0) = 0$ for the open-loop ($K = 0$) and closed-loop models ($K = 10, 20$) are shown in Fig. 5.22.

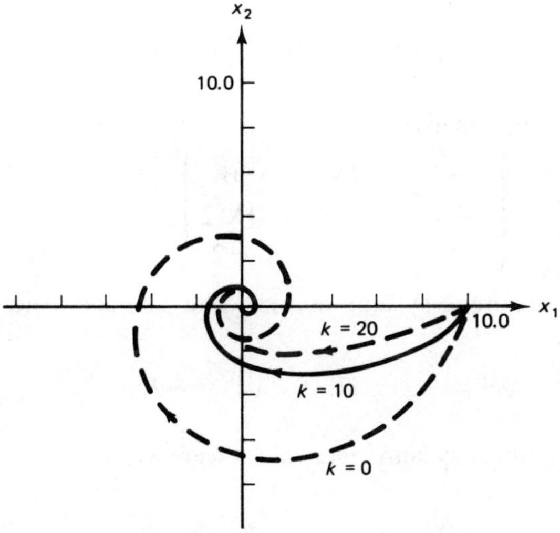

Figure 5.22 Nonlinear regulator Example 5.9. (After Sandor and Williamson [S6].)

5.4.2 Multilinear Second-Order Systems

The nonlinear feedback control, which is studied in Section 5.4.1, is an example of a multilinear system. Such systems do have some interesting dynamical properties relative to their linear system base. These properties can be useful for nonlinear feedback control design ([B1] and [S3]).

For the state vector $\mathbf{x}^T = [x_1 \quad x_2]$, it is possible to generate products such as x_1^2, x_1x_2, x_2^2, x_1^3, . . . , x_1^q, . . . , and the number of independent q-degree terms is equal to $(q + 1)$. If it is desired to preserve the Euclidean norm by such multilinear operations, then

$$\|\mathbf{x}\|^q = \|\mathbf{x}^{[q]}\|,$$
$$\|\mathbf{x}\| = (\mathbf{x}^T\mathbf{x})^{1/2} = \langle \mathbf{x}, \ \mathbf{x} \rangle^{1/2},$$

where $\mathbf{x}^{[q]} \in \mathbb{R}^2$ denotes the $(q + 1)$-tuple. For example,

$$\mathbf{x}^{[2]T} = [x_1^2 \quad \sqrt{2}\, x_1x_2 \quad x_2^2], \tag{5.36}$$
$$\mathbf{x}^{[3]T} = [x_1^3 \quad \sqrt{3}\, x_1^2x_2 \quad \sqrt{3}\, x_1x_2^2 \quad x_2^3]. \tag{5.37}$$

These forms are called *lexicographical listings*.

Consider the lexicographical form for $q = 2$ in Example 5.9:

$$\mathbf{x}^{[2]T} = [x_1^2 \quad \sqrt{2}\, x_1x_2 \quad x_2^2].$$

Then

$$\frac{d(x_1^2)}{dt} = 2x_1\dot{x}_1 = -2(x_1^2) + 4\sqrt{2}\,(\sqrt{2}\, x_1x_2),$$

and similarly,

$$\frac{d(\sqrt{2}\, x_1x_2)}{dt} = -4\sqrt{2}\,(x_1^2) - 2(\sqrt{2}\, x_1x_2) + 4\sqrt{2}\,(x_2^2) + \frac{\sqrt{2}}{4}\, ux_1,$$
$$\frac{d(x_2^2)}{dt} = -4\sqrt{2}\,(\sqrt{2}\, x_1x_2) - 2(x_2^2) + \frac{u}{2}x_2.$$

Then this multilinear system state matrix is

$$\mathbf{A}^{[2]} = \begin{bmatrix} -2 & 4\sqrt{2} & 0 \\ -4\sqrt{2} & -2 & 4\sqrt{2} \\ 0 & -4\sqrt{2} & -2 \end{bmatrix};$$

that is, $\mathbf{A}^{[2]}$ is the 3×3 quadratic-system matrix. The characteristic equation, $\det[\mathbf{I}\lambda^{[2]} - \mathbf{A}^{[2]}] = 0$, yields

$$\lambda_1^{[2]}, \qquad \lambda_3^{[2]} = -2 \pm j8, \qquad \lambda_2^{[2]} = -2.$$

Recall that the corresponding linear system eigenvalues were $\lambda_1, \lambda_2 = -1 \pm j4$. The relationship

$$\lambda_1^{[2]} = 2\lambda_1, \qquad \lambda_3^{[2]} = 2\lambda_2, \qquad \lambda_2^{[2]} = \lambda_1 + \lambda_2$$

is valid for all such second-order, second-degree forms. This can be seen from the transformation to canonical normal form $\mathbf{P}^{-1}\mathbf{A}\mathbf{P} = \mathbf{\Lambda}$ and $\mathbf{P}^{[q]-1}\mathbf{A}^{[q]}\mathbf{P}^{[q]} = \mathbf{\Lambda}^{[q]}$, where $\mathbf{J}$ replaces $\mathbf{\Lambda}$ for the nonunique eigenvector case. Such forms may be handy for studying approximate solutions to general nonlinear systems for optimal control and for bilinear systems [B1].

5.4.3 Variable-Structure Control

Variable-structure systems and adaptive control have been active areas of research on and off over the past 30 years or so. For a few examples, see [A5], [A6], and [M12]. These are systems which adapt their structure to an ever-changing environment. Biological control processes are good examples. For example, the immune system must control a multitude of ever-changing alien structures (in a chemical sense). Body temperature regulation (as studied in Section 5.3.2), similarly, must adapt to rapid environmental changes.

Variable-structure and adaptive-control processes are studied in more detail in Vol. 2. Here we introduce the concept of second-order piecewise-constant policies to implement parametric or variable structure control. The bilinear system (BLS) presents a natural and simple representation for such systems. For example, consider the following canonical phase-variable representation of a second-order system with damping coefficient ζ, natural frequency ω_n, scalar additive control $v(t)$, and parametric (multiplicative) control $u(t)$:

$$\begin{aligned} \dot{x}_1 &= x_2, \\ \dot{x}_2 &= -\omega_n^2 x_1 - 2\zeta\omega_n x_2 + u x_2 + v, \\ y &= x_1. \end{aligned} \tag{5.38}$$

Here $u(\cdot)$ manipulates the system damping coefficient, while $v(\cdot)$ provides the classical linear (additive) control. If ζ represents plant damping, the parametrically controlled system damping for any constant u is given by

$$\zeta_s = \zeta - \frac{u}{2\omega_n}. \tag{5.39}$$

It may be desired, for example, to use a small ζ_s to "minimize" rise time, which might be followed by a large ζ_s to "minimize" overshoot. Ideally, $\zeta_s = 0$, with $u = 2\zeta\omega_n$, could be used until $y(t)$ approaches unity for a unit-step input on $v(t)$. Then the controller might switch to some desired $\zeta_s = \zeta_d$, say 0.8 or 1.0, so that $u = 2\omega_n(\zeta - \zeta_d)$. Of course, the system may be further compensated, as may be desired, by classical linear feedback control $v(\mathbf{x})$. Actually, when the switching policy $u(\mathbf{x})$ is decided, it also provides feedback control of the form

$$u = \begin{cases} \eta_1, & s(\mathbf{x}) < 0, \\ \eta_2, & \text{elsewhere.} \end{cases}$$

For example, in the policy above, $\eta_1 = 2\omega_n(\zeta - \zeta_d)$ for $x_1 \geq 1 - \epsilon$, and $\eta_2 = 2\zeta\omega_n$ for $x_1 < 1 - \epsilon$.

In practice, however, further difficulties may arise in the implementation of such control policies. Measurement errors and switching time delay can result in multiple switchings and in stability problems. The resulting control may be called a sliding or chattering control. In some cases, as seen in Vol. 2 for optimal control of nonlinear systems, the switching function may go to zero for a finite time interval and yield a singular control policy. Both of these aspects of nonlinear optimal switching controls will be studied in Vol. 2.

EXERCISES

5.1. Consider all possible values of ω for Fig. 2.16 and Exercise 2.2.
- **(a)** Analyze the system for limit cycles.
- **(b)** Sketch the phase-space portrait of trajectories, which describes the motion of the head for various initial states.
- **(c)** Is the process conservative?

5.2. Given the system

$$\frac{dx_1}{dt} = x_2, \qquad \frac{dx_2}{dt} = -x_1 - 0.4x_2^3.$$

- **(a)** Analyze stability of the equilibrium from these equations.
- **(b)** Sketch the phase-plane portrait for various initial states.
- **(c)** Repeat part (a) if the second equation is driven by a step input $a\delta_{-1}(t)$ (i.e., added to the right-hand side).
- **(d)** Repeat part (a) if the second equation is driven by a ramp input $(a + bt)\delta_{-1}(t)$.

5.3. Analyze stability and the possible existence of limit cycles for the following cases.
- **(a)** $\dot{x}_1 = x_2,\ \dot{x}_2 = -|x_1|$.
- **(b)** $\dot{x}_1 = x_2,\ \dot{x}_2 = -x_1^{2n} \quad (n = 1, 2, \cdots)$.
- **(c)** $\dot{x}_1 = |x_1| - |x_2|$,
  $\dot{x}_2 = (|x_1 x_2|)\ \text{sgn}(x_1 x_2)$.
- **(d)** Evaluate the index of the origin for cases (a), (b), and (c).
- **(e)** Explain the not-so-common state-plane portraits of these systems (a), (b), and (c) for small perturbations about the origin.

5.4. Sketch the phase-plane portrait and the time response with $\mathbf{x}^T(t_0) = [x_{10} \quad x_{20}]$ for

$$\dot{x}_1 = x_2, \qquad \dot{x}_2 = -\omega_n^2 x_1 + bu,$$

where

$$u = \text{sgn}\ x_2 = \begin{cases} +1, & \text{for } x_2 > 0, \\ -1, & \text{for } x_2 < 0, \end{cases}$$

and u is undefined for $x_2 = 0$.

5.5. Consider the bilinear system

$$\dot{x}_1 = x_2, \qquad \dot{x}_2 = -x_1 + ux_2,$$

with linear feedback $u = -x_1$. Sketch the phase-plane portrait.

5.6. Given the nonlinear system

$$\dot{x}_1 = x_2,$$

$$\dot{x}_2 = (0.1 - \tfrac{10}{3}x_2^2)x_2 - x_1 + x_1^3.$$

- **(a)** Analyze the stability of the equilibrium states by Liapunov's first method.
- **(b)** Check for the possible existence of limit cycles.
- **(c)** If a limit cycle appears possible, approximate its solution by $x_1 = A \sin \omega t$ and compute A (method of equivalent linearization).

5.7. Consider the navigation for an "inertialess" zero-drag missile with respect to a target, as shown in Fig. 5.23. Let $I\ddot{\theta} = u$.

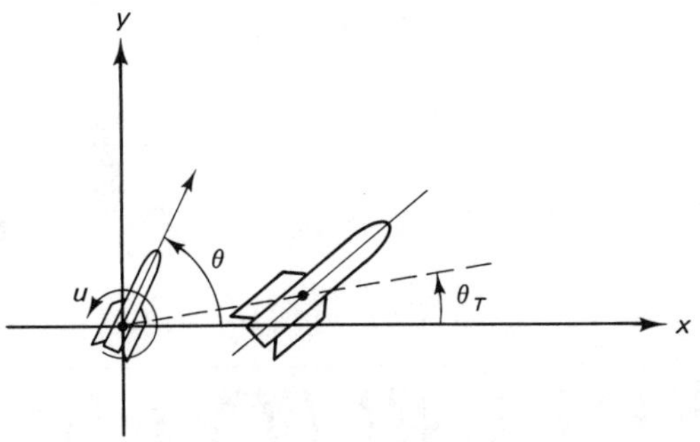

Figure 5.23 Missile navigation geometry for Exercise 5.7.

(a) Study the behavior of the missile by the phase plane (θ versus $\dot{\theta}$) portrait with heading control torque about the center of the mass given by the bang-bang policy

$$u = -k \ \text{sgn}(\theta - \theta_T),$$

where sgn(·) represents the sign of the function as defined before.

(b) Show the improved tracking performance in the phase plane if rate feedback is introduced so that

$$u = -k \ \text{sgn}(\theta - \theta_T + K_d\dot{\theta}).$$

(c) Compute the conditions that result in a singular solution with sgn(·) undefined. Note that the motion can become "trapped" on the switching line. Discuss the practical aspects of this case whereby the sliding mode in reality becomes a chattering process.

5.8. Analyze the systems of Exercise 3.2 for limit cycles by index theory, the Bendixson theorem, and the Poincaré–Bendixson theorem.

5.9. Study the dynamical behavior of the Van der Pol equation (5.25) for very small values of μ and very large values of μ.

5.10. Consider the displacement $x(t)$ of a unit mass supported by a nonlinear spring with an approximately resonant force so that a forced Duffing's equation results. Consequently,

$$\ddot{x} + (1 + \epsilon x^2)x = A \cos \omega t,$$

where $\omega \simeq 1$ and $\dot{x}(0) = 0$. For convenience, let $A = \epsilon F$. This allows one to identify effects of successive higher-order terms in the following expansion:

$$x = x_0 + \epsilon x_1 + \epsilon^2 x_2 + \cdots$$

and

$$\omega = 1 + \epsilon\omega_1 + \epsilon^2\omega_2 + \cdots$$

with $\tau = \omega t$ and $x_k' = dx_k/d\tau$, $k = 0, 1, 2, \cdots$, substitute these into the system equation, and equate like terms in ϵ^k to study the amplitude frequency response $|x(\omega)|$ for fixed $A = \epsilon F$ and ω in the neighborhood of unity. Consider the following three cases.

(a) Linear spring (no cubic term).

(b) Soft spring ($\epsilon < 0$).

(c) Hard spring ($\epsilon > 0$).

Note the multiple solutions in parts **(a)** and **(b)** over a range of frequencies.

5.11. For the telescopic robot arm of Exercise 2.13 with $w(t) = w_0$ and $v(t) = 0$, use the Bendixson theorem to analyze the system for limit cycles.

CHAPTER

6

NONLINEAR STABILITY: TIME-DOMAIN ANALYSIS

Basic definitions of stability for equilibrium points, for trajectories (e.g., limit cycles), and for systems have been introduced in Chapters 3 to 5. Consequently, if the necessary solutions are available or can be obtained, it is usually not too difficult to ascertain stability for particular cases. Unfortunately, nonlinear system solutions generally are not available except by approximations, such as may accurately be demonstrated by computer simulations. The objective here is to analyze system stability and to develop related design methods that do not require solution of the nonlinear differential equations. Although the methods presented do have limitations, they are very crucial to nonlinear system stability and design analysis.

The first method of Liapunov, presented in Chapter 3, is fundamental to the stability analysis of equilibrium points. Unfortunately, the method is applicable only to local stability at the point of interest. Now, Liapunov's second, or direct, method will provide somewhat stronger stability. While the method is powerful, it sometimes provides very conservative stability conditions which are sufficient but, except in very special cases, are not necessary. The method is demonstrated by several physically meaningful examples.

Then, in Chapter 7, two frequency-domain methods, which are similar in appearance, are studied. Both conveniently include nonlinear characteristics which are nondynamical in nature. The first method, due to Popov, is based on Liapunov's direct method and thereby provides sufficient, and sometimes conservative, stability criteria. Describing function methods, on the other hand, provides more relaxed and easy-to-apply stability criteria. This method is based on an equivalent fundamental-frequency representation by means of the Fourier series. The describing function method provides one of the simplest, yet most convenient apparatus for practical control-system design.

6.1 LIAPUNOV DIRECT METHOD

6.1.1 Stability and Energy

As a base for Liapunov's second or direct method (sometimes referred to as the direct method), consider system energy for a simple second-order nonlinear system such as might result from a force balance on a constant mass supported by a parallel nonlinear spring–damper combination. Or analogously, the model arises from Kirchhoff's law for a single-loop circuit combination of fixed inductance in series with nonlinear resistance and capacitance which are functions of charge, the dependent variable x in the differential equation. Hence, with the form given by (5.24) and repeated here for convenience,

$$\frac{d^2x}{dt^2} + h(x)\frac{dx}{dt} + f(x) = 0, \tag{6.1}$$

where the system is unforced or free. In Section 5.3 it was noted that this equation also results for the nonlinear damped pendulum or the synchronous machine with damping torque.

In canonical phase-variable form (6.1) is given by

$$\begin{aligned}\frac{dx_1}{dt} &= x_2,\\ \frac{dx_2}{dt} &= -f(x_1) - h(x_1)x_2,\end{aligned} \tag{6.2}$$

where $x_1 f(x_1) > 0$ and $f(0) = 0$ so that $x_{1e} = x_{2e} = 0$ is a unique equilibrium point. As shown previously, the total energy is equal to a sum of the kinetic energy of the mass, $m x_2^2$, and the potential energy, $\int_0^{x_1} f(\sigma)\, d\sigma$, which is stored in the spring. If we normalize the energy with respect to the mass, and let $V(x_1, x_2)$ designate the total energy, for purposes that will become apparent in Section 6.1.3,

$$V(x_1, x_2) = \frac{x_2^2}{2} + \int_0^{x_1} f(\sigma)\, d\sigma, \tag{6.3}$$

where

$$V(\mathbf{x}) > 0 \qquad \text{for } \mathbf{x} \neq \mathbf{0}$$

and

$$V(\mathbf{0}) = 0. \tag{6.4}$$

Equation (6.4) states that the energy goes to zero at equilibrium but is positive elsewhere.

The rate of change of the energy, $V(\mathbf{x})$, along a trajectory [solution to (6.2)] is given by

$$\frac{dV}{dt} = \frac{\partial V}{\partial x_1}\frac{dx_1}{dt} + \frac{\partial V}{\partial x_2}\frac{dx_2}{dt}, \tag{6.5}$$

or from (6.3) and (6.2),

$$\frac{dV}{dt} = -h(x_1)x_2^2. \tag{6.6}$$

Therefore, the total energy, $V(\mathbf{x})$ is dissipated along a solution path if the damping $h(x_1)$ is positive for all nonzero x_1, and x_2 is nonzero as well. At $\mathbf{x} = \mathbf{0}$, of course, the motion is at equilibrium. Consequently, it is seen that system energy, $V(\mathbf{x})$, cannot increase if the damping is positive. Furthermore, since $\dot{x}_2 = -f(x_1)$ is zero only when $\mathbf{x}_e = \mathbf{0}$, the motion cannot remain at $x_2 = 0$ unless it is at equilibrium with $x_1 = 0$ as well. In other words, $V(\mathbf{x})$ can neither increase nor remain constant for nonzero $\mathbf{x}$, and at $\mathbf{x} = \mathbf{0} = \mathbf{x}_e$, $V(\mathbf{x})$ remains constant. Hence $\mathbf{x}_e$ is a unique stable equilibrium or $\mathbf{x}_e = \mathbf{0}$ is, in an intuitive sense, "globally asymptotically stable."

If $h(x_1) < 0$, the system has negative damping, energy increases (as the dashpot adds kinetic energy to the mass), dV/dt is positive for all nonzero x_1, x_2, and hence the motion is divergent from $\mathbf{0}$, which is unstable. For $h(x_1) = 0$, of course, the system is conservative, so that $\dot{V}(\mathbf{x}) = 0$ and $V(\mathbf{x})$ is a constant. In the latter case, the origin is a stable equilibrium point. Figure 6.1 shows three cases, corresponding to asymptotically stable, stable, and unstable equilibrium points. Typical constant energy contours

$$V(x_1, x_2) = V_k,$$

$k = 1, 2, \ldots$, such that $V_{N+1} > V_N > \cdots > V_1 > 0$, are shown in Fig. 6.2. For a linear spring, $h(x_1) = kx_1$, the contours represent a family of ellipses with

$$V(x_1, x_2) = \frac{kx_1^2 + x_2^2}{2}. \tag{6.7}$$

As shown by Fig. 6.2, the motion corresponding to the positively damped case, $h(x_1) > 0$ with $x_1 \neq 0$, always moves from the exterior to the interior of constant $V(\cdot)$ contours as it increases. As the equilibrium point at the origin is approached, $V(\cdot)$ approaches zero. If $h(x_1) = 0$, the trajectory $\mathbf{x}(t)$ lies along a constant $V(\cdot)$ contour. And if $h(x_1) < 0$ with $x_1 \neq 0$, the motion moves from the interior to the exterior of the constant $V(\cdot)$ contours as t increases. Although this technique does not require the

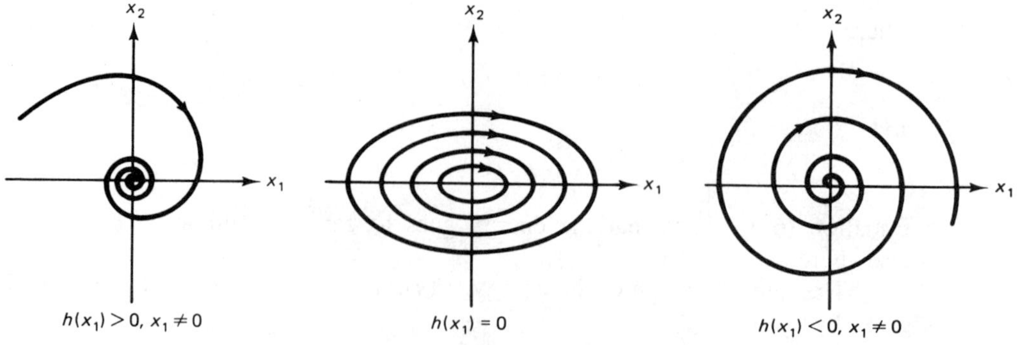

(a) Asymptotically stable origin (b) Stable origin (c) Unstable origin

Figure 6.1 Phase-plane sketches for characteristic cases of (6.2).

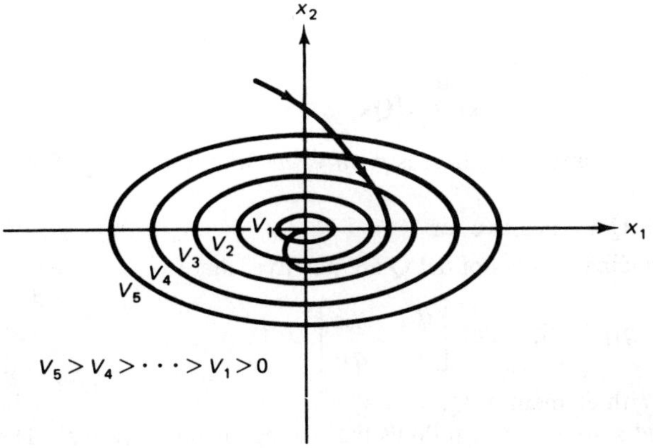

Figure 6.2 Typical constant energy contours.

solution to the original differential equations, it does require a knowledge of system energy and somewhat intuitive arguments. These are put on a more firm and general basis by the formulation of an energy-like function, $V(\mathbf{x})$, in the next two sections.

6.1.2 Positive-Definite Functions

The real scalar function $V(\mathbf{x})$ is *defined* to be *positive-definite* (p.d.) in some closed bounded region D of state space if for all $\mathbf{x}$ in D,

(a) $V(\mathbf{x})$ is continuously differentiable with respect to $\mathbf{x}$,
(b) $V(\mathbf{0}) = 0$,
(c) $V(\mathbf{x}) > 0$, for all $\mathbf{x} \neq \mathbf{0}$.

$V(\mathbf{x})$ by *definition* is *positive-semidefinite* (p.s.d.) on D if conditions (a) and (b) are satisfied and condition (c) is replaced by

(c) $V(\mathbf{x}) \geq 0$, for all $\mathbf{x}$.

$V(\mathbf{x})$ is *defined* to be *negative-definite* (n.d.) on D if $-V(\mathbf{x})$ is positive-definite on D. That is, $V(\mathbf{x})$ is negative-definite on D if and only if conditions (a) and (b) are satisfied, and (c) $V(\mathbf{x}) < 0$ for all $\mathbf{x} \neq 0$. Similarly, $V(\mathbf{x})$ is *negative-semidefinite* (n.s.d.) on D if and only if conditions (a) and (b) are satisfied, but (c) $V(\mathbf{x}) \leq 0$ for all $\mathbf{x}$.

Example 6.1

$$\mathbf{x}^T = [x_1, x_2]$$

Then

$$V(\mathbf{x}) = x_1^2 \text{ is positive-semidefinite (p.s.d.) on } R^2$$

and

$$V(\mathbf{x}) = x_1^2 + x_2^2 \text{ is positive-definite (p.d.) on } R^2.$$

Example 6.2

The *quadratic form*

$$V(\mathbf{x}) = \mathbf{x}^T\mathbf{Q}\mathbf{x}, \tag{6.8}$$

where $\mathbf{Q}$ is an $n \times n$ real symmetric matrix, is *positive-definite* if and only if

(a) The eigenvalues of $\mathbf{Q}$ are positive, or

(b) The successive principal minors of det $\mathbf{Q}$ are positive, that is,

$$q_{11} > 0, \quad \det \begin{bmatrix} q_{11} & q_{12} \\ q_{21} & q_{22} \end{bmatrix} > 0, \quad \ldots,$$

where q_{ij} designates the ijth element of $\mathbf{Q}$.

Since $\mathbf{Q}$ is real and symmetric, it follows that its eigenvalues are real. Then, by means of the transformation to normal canonical form $\mathbf{x} = \mathbf{P}\mathbf{z}$ with $V(\mathbf{z}) = \mathbf{z}^T\mathbf{\Lambda}\mathbf{z}$, and the diagonal eigenvalue matrix $\mathbf{\Lambda} = \mathbf{P}^T\mathbf{Q}\mathbf{P}$, it can be shown that conditions (a) and (b) follow ([B4] and [G5]). Here the matrix $\mathbf{Q}$ also is said to be positive-definite.

Example 6.3

Consider $V(\mathbf{x})$ as a sum of homogeneous polynomials in x_i, $i = 1, \ldots, n$, such that

$$V(\mathbf{x}) = V_1(\mathbf{x}) + V_2(\mathbf{x}) + \cdots,$$

where $V_1(\mathbf{x})$ is the lowest-degree polynomial. Then, for small $\|\mathbf{x}\|$, $V(\mathbf{x}) \sim V_1(\mathbf{x})$ and the sign $V(\cdot)$ is the same as the sign of $V_1(\cdot)$ in a sufficiently small neighborhood of the origin. Also $V(\mathbf{x})$ cannot be positive-definite if $V_1(\mathbf{x})$ is odd (necessary but not sufficient). For example, with $\mathbf{x} \in \mathbb{R}^2$,

$$V(\mathbf{x}) = x_2^2 - x_1^2 + x_1^3 + x_2^3$$

is not p.d.

For nonautonomous systems, time-variant positive-definite functions can be seen to play a role in stability analysis. By definition, a scalar $W(\mathbf{x}, t)$ is *positive-definite* in D if for all t,

(a) $W(\mathbf{x}, t)$ has continuous partial derivatives with respect to all its arguments,

(b) $W(0, t) = 0$,

(c) $W(\mathbf{x}, t) \geq V(\mathbf{x})$,

where $V(\mathbf{x})$ is a positive-definite (time-invariant) function.

Again, $W(\mathbf{x}, t)$ is *positive-semidefinite* if conditions (a) and (b) are satisfied but condition (c) is replaced by the weaker condition

(c) $W(\mathbf{x}, t) \geq 0$.

Another useful requirement of $W(\mathbf{x}, t)$ is that $W(\mathbf{x}, t)$ be decrescent. By definition, $W(\mathbf{x}, t)$ is *decrescent* if $W(\mathbf{x}, t) \to 0$ uniformly with respect to t as $\|\mathbf{x}\| \to 0$. Then any time-invariant $V(\mathbf{x})$ that is positive-definite is also decresent.

With regard to global stability analysis, it is sometimes convenient to introduce a property of radially unboundedness. $W(\mathbf{x}, t)$ is *radially unbounded* if $W(\mathbf{x}, t) \to \infty$ in t uniformly as $\|\mathbf{x}\| \to \infty$. That is, for any $M > 0$, there exists an $N > 0$ such that for all t, $W(\mathbf{x}, t) > M$ and $\|\mathbf{x}\| > N$. Then if $W(\mathbf{x}, t)$ is p.d. on any domain and also is radially unbounded, there exists a p.d. $V(\mathbf{x})$ such that

$$V(\mathbf{x}) \to \infty \qquad \text{as } \|\mathbf{x}\| \to \infty$$

and

$$W(\mathbf{x}, t) \geq V(\mathbf{x}) \qquad \text{for all } \mathbf{x} \text{ and } t.$$

Again, negative-definite and negative-semidefinite definitions follow as before.

Following [W2], it is seen that for $\mathbf{x} \in \mathbb{R}^2$,

(a) $(x_1^2 + x_2^2)e^{-t}$ is p.s.d. but not p.d.;

(b) $\dfrac{(x_1^2 + x_2^2)(t^2 + 1)}{(x_1^2 + 2)}$ is p.d., but neither decrescent nor radially unbounded;

(c) $(x_1^2 + x_2^2)(t^2 + 1)$ is p.d., radially unbounded, but not decrescent;

(d) $\dfrac{x_1^2 + 2x_2^2}{x_1^2 + 2}$ is p.d., decrescent, but not radially unbounded;

(e) $\dfrac{(x_1^2 + x_2^2)(t^2 + 1)}{(t^2 + 2)}$ is p.d., decrescent, and radially unbounded.

While time-variant functions can be useful in particular cases, their selection is generally difficult. In most cases we shall be concerned with time-invariant $V(\mathbf{x})$.

6.1.3 Liapunov Stability Criteria

In this section, stability of an equilibrium state is investigated. It is assumed that the equilibrium point of interest is transformed to the origin in order to simplify the use of Section 6.1.2 without any loss of generality. Following the analysis of Chapter 5, the equilibrium-point stability usually may be generalized to a time-variant reference trajectory.

Consider again the autonomous nonlinear system

$$\frac{d\mathbf{x}}{dt} = \mathbf{f}(\mathbf{x}), \tag{6.9}$$

where $\mathbf{f}(\mathbf{0}) = \mathbf{0}$. Then a Liapunov function $V(\mathbf{x})$ corresponding to (6.9) is *defined* as follows: $V(\mathbf{x})$ is a Liapunov function in a neighborhood D of the origin if

(a) $V(\mathbf{x})$ is positive-definite, and

(b) $dV/dt = (\partial V/\partial \mathbf{x})^{\mathrm{T}}\, \dot{\mathbf{x}} = \partial V/d\mathbf{x} \cdot \mathbf{f}(\mathbf{x})$ is negative-semidefinite for all $\mathbf{x} \in D$.

Sometimes $\partial V/\partial \mathbf{x}$ is written in the gradient form ∇V. Also, as shown here, the vector transpose or dot product notation may conveniently be utilized. Then the following theorems can be derived:

(1) Stability Theorem for (6.9). The origin is *stable* if there exists a Liapunov function $V(\mathbf{x})$ throughout D, a neighborhood of origin.

(2) Asymptotic Stability Theorem for (6.9). The origin is *asymptotically stable* if there exists a Liapunov function $V(\mathbf{x})$ throughout D, a neighborhood of the origin such that $\dot{V}(\mathbf{x})$ is negative-definite [i.e., $-\dot{V}(\mathbf{x})$ positive-definite].

(3) Instability Theorem for (6.9). The origin is *unstable* if there exists a $V(\mathbf{x})$ in the neighborhood D of the origin where $V(\mathbf{0}) = 0$ such that $\dot{V}(\mathbf{x})$ is positive definite on D, and $V(\mathbf{x}) > 0$ for $\|\mathbf{x}\|$ arbitrarily small.

(4) Global Asymptotic Stability Theorem for (6.9). The origin is asymptotically stable in the large (i.e., globally asymptotically stable) if it is asymptotically stable [Theorem (2)], and if, furthermore, $V(\mathbf{x})$ is radially unbounded [i.e., $V(\mathbf{x}) \to \infty$ as $\|\mathbf{x}\| \to \infty$].

(5) Region of Asymptotic Stability Theorem for (6.9). As Theorem (2) but with $V(\mathbf{x}) < \eta$ in D. Then the $\mathbf{0}$ is asymptotically stable and every solution with $\mathbf{x}(t_0)$ in D approaches $\mathbf{0}$ asymptotically as $t \to \infty$.

Relative to Theorem (4), if the origin is the only equilibrium state of (6.9), the system is said to be *completely stable*. Also, if $\dot{V}(\mathbf{x}) = 0$ for $\mathbf{x} \in S$, where S is a generalization of an isolated equilibrium point at $\mathbf{0}$ but contains no complete trajectories other than equilibrium states, all solutions approach an $\mathbf{x}_e \in S$ as $t \to \infty$. Furthermore, *any* $\mathbf{x}_e$, for which $V(\mathbf{x}_e)$ is an isolated local minimum, is asymptotically stable.

Detailed proofs of Theorems (1) to (4) are given in [L6], [L7], and [H9]. Intuitively, these criteria follow from the relationship of energy to stability that is introduced in Section 6.1.1. The requirement of a radially unbounded $V(\mathbf{x})$ in Theorem (4), however, is a bit subtle. It can be seen that by this condition, $V(\mathbf{x}) \to \infty$ as $\|\mathbf{x}\| \to \infty$; together with $\dot{V}(\mathbf{x}) \leq 0$, it ensures that no solution can grow without bound. That is, the system is Lagrange stable. Also, it is seen that the application to multiple equilibrium states can be made by *not* requiring $\dot{V}(\mathbf{x}) < 0$. The absence of complete trajectories in S precludes the presence of limit cycles there.

A sketch of the proof for Theorems (1) and (2) is given here to provide some insight into the basic concept of stability beyond that provided above. Here the reasoning follows that of [L3] and [S2], along with Fig. 6.3. From the definition of stability of $\mathbf{x}_e = \mathbf{0}$ in Chapter 3, it must be shown that the conditions of the theorem imply that for any $\epsilon > 0$ there exists a $\delta(\epsilon) > 0$ such that if $\|\mathbf{x}(\mathbf{0})\| < \delta$, then $\|\mathbf{x}(t)\| < \epsilon$ for all $t \geq 0$, and furthermore, that $\mathbf{x} \to \mathbf{0}$ as $t \to \infty$. Figure 6.3 illustrates the appropriate regions for $\mathbb{R}^2$ where $\|\mathbf{x}\|$ is taken as the Euclidean norm $(x_1^2 + x_2^2)^{1/2}$. Here $V(\mathbf{x}) = \beta$ is the minimum value of $V(\cdot)$ on the circle C which is defined by $\|\mathbf{x}\| = \epsilon$. Then, since $V(\cdot)$ is p.d., select $\delta > 0$ such that for all $\mathbf{x}$ with $\|\mathbf{x}\| < \delta$, $V(\mathbf{x}) < \beta$ as shown in Fig. 6.3. It is seen that all trajectories with $\|\mathbf{x}(t_0)\| < \delta$ have $\|\mathbf{x}(t)\| < \epsilon$. Therefore, the origin is *stable* [Theorem (1)]. Then for any $\mathbf{x}(0)$ with $\|\mathbf{x}(0)\| < \delta$ and $V(\mathbf{x}(0)) = V_0$, it can be shown by Fig. 6.3 that for any $\epsilon_1 > 0$, there exists a T such that for $t > T$, $\|\mathbf{x}(t)\| < \epsilon_1$ so that $\mathbf{x} \to 0$ as $t \to \infty$. Select a positive $\epsilon_1 < \delta$ so that $V(\mathbf{x})$ has a minimum α on the circle which is given by $\|\mathbf{x}\| = \epsilon_1$. Then, as before, an η can be found such that with $\|\mathbf{x}\| < \eta$, $V(\mathbf{x}) < \alpha$. Since $\dot{V}(\mathbf{x})$ is negative-definite, $\dot{V}(\cdot)$ has an

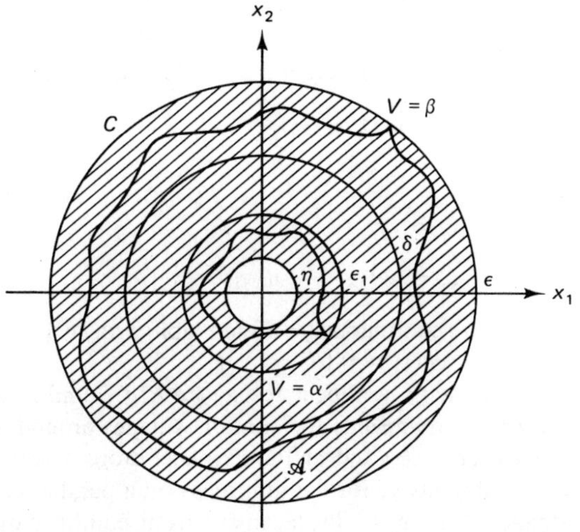

Figure 6.3 Illustration for the proof of stability theorems (1) and (2).

upper bound $-\gamma$, $\gamma > 0$ on the annular region $\mathscr{A}$ which is given by $\eta \leq \|\mathbf{x}\| \leq \epsilon$. It follows from $V(t) = V_0 + \int_0^t \dot{V}(\sigma)\, d\sigma$ that if $\mathbf{x}(t)$ remains inside $\mathscr{A}$, then $V(t) \leq V_0 - \gamma t$. But in such a case $V(t)$ would have to become negative, which contradicts the p.d. hypothesis. Therefore, $x(t)$ eventually must enter the disk of radius η, after which $V(x) < \alpha$ and $\|\mathbf{x}(t)\| < \epsilon_1$.

Example 6.4

The nonlinear system is given by a linear "marginally" stable portion plus a nonlinear power series, such that

$$\begin{aligned} \dot{x}_1 &= x_2 - x_1 f(x_1, x_2), \\ \dot{x}_2 &= -x_1 - x_2 f(x_1, x_2), \end{aligned} \tag{6.10}$$

where $f(\mathbf{x})$ is a convergent power series with $f(\mathbf{0}) = 0$. The linear portion, which is the same as the perturbation model at $\mathbf{0}$, has eigenvalues $\lambda_1, \lambda_2 = \pm j1$. Hence no conclusion can be made about stability of the $\mathbf{0}$ from Liapunov's first method as studied in Chapter 3.

As the first step of Liapunov's direct method, a simple quadratic form is chosen with

$$V(\mathbf{x}) = \tfrac{1}{2}(x_1^2 + x_2^2).$$

Then

$$\dot{V} = \left[\frac{\partial V}{\partial \mathbf{x}}\right]^{\mathrm{T}} \cdot \dot{\mathbf{x}} = [x_1 x_2]\begin{bmatrix} x_2 - x_1 f(\mathbf{x}) \\ -x - x_2 f(\mathbf{x}) \end{bmatrix}$$

or

$$\dot{V} = -(x_1^2 + x_2^2) f(\mathbf{x}).$$

Therefore, if

1. $f(\mathbf{x}) \geq 0$, arbitrarily close to $\mathbf{0}$, then the $\mathbf{0}$ is *stable*.
2. $f(\mathbf{x}) > 0$, arbitrarily close to $\mathbf{0}$, then the $\mathbf{0}$ is *asymptotically stable*.
3. $f(\mathbf{x}) < 0$, arbitrarily close to $\mathbf{0}$, then the $\mathbf{0}$ is *unstable*.

Recall that Example 4.9 is a special case of this example with $f(\mathbf{x})$ a quadratic function.

Example 6.5

Consider a special case of (5.25) that is different than (6.1) with

$$\ddot{x} + g(\dot{x}) + f(x) = 0,$$

or in canonical phase-variable form,

$$\begin{aligned} \dot{x}_1 &= x_2, \\ \dot{x}_2 &= -f(x_1) - g(x_2). \end{aligned} \tag{6.11}$$

Here it is assumed that $f(\cdot)$, $g(\cdot)$ are continuously differentiable with $f(0) = g(0) = 0$, and that $x_1 f(x_1) < 0$ for $x_1 \neq 0$ and $x_2 g(x_2) > 0$ for $x_2 \neq 0$.

For a passive *RLC* circuit with nonlinear resistance $R(\dot{q})$, nonlinear capacitance $C(q)$, and unit inductance, (6.11) represents the sum of voltages around a single loop. $x_1 = q$ charge, $x_2 = \dot{q} = i$ current; $f(\cdot)$ and $g(\cdot)$ are voltage drops across the capacitor and resistor, respectively. Analogously, for the unit mass on a parallel combination of spring and damper (dashpot), x_1 is mass displacement from equilibrium, $f(x_1)$ is the restoring force of the spring, and $g(x_2)$ is the damping force of the dashpot.

For the nonlinear, $f(\cdot)$ and $g(\cdot)$, characteristics that are assumed above, $\mathbf{x}_e = \mathbf{0}$ is the only equilibrium state (singular point in the phase plane). The energy stored in the capacitor, or the spring, is given by $V_c = \int_0^{x_1} f(\sigma)\, d\sigma > 0$. On the other hand, resistance or dashpot leads to energy dissipation at a rate $-g(x_2)x_2 = \dot{V}_r$ which is computed along a solution trajectory of (6.11). Compare this to (6.6) for system (6.2). As a result, complete stability is expected. Indeed, Liapunov's direct method verifies this with

$$V = \int_0^{x_1} f(\sigma)\, d\sigma + \frac{x_2^2}{2}$$

positive definite, and

$$\dot{V} = -x_2 g(x_2)$$

negative definite. At first glance $\dot{V}$ is only n.s.d., but with $\dot{V} = 0$ for $x_2 = 0$, and $\dot{x}_2 = 0$ for $x_2 = 0$ if $x_1 \neq 0$, trajectories must cross the x_1 axis except at $\mathbf{0}$. Hence the $\mathbf{0}$ is asymptotically stable. If it is assumed that $\int_0^{x_1} f(\sigma)\, d\sigma \to \infty$ as $|x_1| \to \infty$, then $V(\mathbf{x}) \to \infty$ as $\|\mathbf{x}\| \to \infty$, and the $\mathbf{0}$ is globally asymptotically stable. Since it is a unique equilibrium state, the system is said to be completely stable. Note that the condition $\int_0^{x_1} f(\sigma)\, d\sigma \to \infty$ as $|x_1| \to \infty$ merely guarantees that $f(x_1)$ does not converge to zero too rapidly as $|x_1| \to \infty$; for example, $1/x_1$ would be acceptable.

6.1.4 Linear Systems

The linear autonomous system is considered first with

$$\frac{d\mathbf{x}}{dt} = \mathbf{A}\mathbf{x}, \tag{6.12}$$

where $\mathbf{A}$ is an $n \times n$ constant matrix. Of course, it is well known that the equilibrium state of $\mathbf{0}$ for (6.12) is asymptotically stable if and only if the eigenvalues of $\mathbf{A}$ all have negative-real parts. The $\mathbf{0}$ is stable if the eigenvalues of $\mathbf{A}$ have nonpositive-real parts and every eigenvalue of $\mathbf{A}$ with zero-real part is nonrepeated.

It is of interest, however, to investigate the stability of (6.12) by the direct method. This result will be used in Vol. 2 as well.

For a Liapunov function of (6.12), choose the simple quadratic form given by (6.8), or

$$V(\mathbf{x}) = \mathbf{x}^T\mathbf{Q}\mathbf{x}, \tag{6.8}$$

where again, $\mathbf{Q}$ is a real symmetric matrix. Then

$$\dot{V}(\mathbf{x}) = \dot{\mathbf{x}}^T\mathbf{Q}\mathbf{x} + \mathbf{x}^T\mathbf{Q}\dot{\mathbf{x}}$$

or

$$\dot{V}(\mathbf{x}) = -\mathbf{x}^T\mathbf{P}\mathbf{x}, \tag{6.13}$$

where

$$-\mathbf{P} = \mathbf{A}^T\mathbf{Q} + \mathbf{Q}\mathbf{A}. \tag{6.14}$$

Equation (6.14) commonly is called Liapunov's equation, and the roles of $\mathbf{P}$ and $\mathbf{Q}$ are commonly exchanged.

If a matrix pair ($\mathbf{P}$, $\mathbf{Q}$) can be found which satisfies (6.14) such that both $\mathbf{P}$ and $\mathbf{Q}$ are positive-definite, then both $-\dot{V}$ and V are positive-definite functions. Therefore, the $\mathbf{0}$ of (6.12) is a globally asymptotically stable equilibrium point. On the other hand, if $\mathbf{P}$ is positive-definite and $\mathbf{Q}$ has at least one negative eigenvalue, $-\dot{V}$ is positive-definite and V assumes negative values arbitrarily close to the origin, which results in an unstable origin. Instability is seen directly from (6.14).

In the analysis of linear system stability by (6.14), it is desired to select a positive-definite $\mathbf{P}$ (e.g., $\mathbf{I}$ as a first try) so that $\mathbf{Q}$ is positive-definite for a given system matrix $\mathbf{A}$. Then the $\mathbf{0}$ is globally asymptotically stable. However, the success of this procedure hinges on our ability to find a unique $\mathbf{Q}$ solution to (6.14). Indeed, it can be shown [C5] that

1. Equation (6.14) has a unique solution for $\mathbf{Q}$ corresponding to every $n \times n$ real symmetric $\mathbf{P}$ if and only if $\lambda_i + \lambda_j^* \neq 0$, for all i, j where λ_i, $i = 1, \ldots, n$, are the eigenvalues of $\mathbf{A}$ (not necessarily distinct), and * denotes a complex conjugate. Also, the condition is satisfied for all stable $\mathbf{A}$ (i.e., $\mathbf{A}$ with all eigenvalues that have negative real parts).
2. If $\mathbf{A}$ is stable and $\mathbf{P}$ is positive-definite, $\mathbf{Q}$ also is p.d.

In summary, Liapunov [L7] showed that the $\mathbf{0}$ for (6.12) is globally asymptotically stable if and only if for any symmetric positive-definite matrix $\mathbf{Q}$ satisfies the $n(n + 1)/2$ scalar equations for (6.14). If the $\mathbf{0}$ of (6.12) is stable, there exists a unique Liapunov function that satisfies (6.8) and (6.14), where $\mathbf{P}$ is p.d.

Time-variant positive-definite scalar functions, $W(\mathbf{x}, t)$, were defined in Section 6.1.2. Such functions may be crucial to ascertain stability of time-variant systems. Even if a time-invariant Liapunov function $V(\mathbf{x})$ is chosen, in general its time derivative along a solution path will be a function of time $\dot{V}(\mathbf{x}, t)$. But in some cases time-variant $V(\mathbf{x}, t)$ may be required or even handier.

For time-variant systems, recall the definition of uniform asymptotic stability given in Section 6.3. Then it can be shown that $\mathbf{x}_e = \mathbf{0}$ for the linear time-variant system LTVS is stable if an appropriate time-variant Liapunov function is generated. Again, consider the unforced LTVS with

$$\frac{d\mathbf{x}}{dt} = \mathbf{A}(t)\mathbf{x}, \tag{6.15}$$

where $\mathbf{A}(t)$ is an appropriate nonsingular continuous matrix function. Then the zero equilibrium state is *uniformly asymptotically stable* if there exists a Liapunov function

$$V(\mathbf{x}, t) = \mathbf{x}^{\mathrm{T}}\mathbf{Q}(t)\mathbf{x}, \tag{6.16}$$

where $\mathbf{Q}(t)$ is defined by

$$\mathbf{Q}(t) = \int_t^{\infty} \mathbf{\Phi}^{\mathrm{T}}(\sigma, t)\mathbf{P}(\sigma)\mathbf{\Phi}(\sigma, t)\, d\sigma$$

which is a bounded function of t.

$\mathbf{\Phi}(\sigma, t)$ is the state-transition matrix of (6.15) that was studied in Chapter 4. $\mathbf{A}(t)$ is appropriately bounded in t. $\mathbf{P}(t)$ is a bounded, symmetric, p.d. matrix function of t such that there exists a finite constant γ with $\gamma\mathbf{x}^{\mathrm{T}}\mathbf{x} \le \mathbf{x}^{\mathrm{T}}\mathbf{P}(t)\mathbf{x}$ for all $\mathbf{x}$ and for $t \ge 0$.

From (6.16) and (6.15) the time-variant version of (6.14) becomes

$$\dot{\mathbf{Q}}(t) = -\mathbf{A}^{\mathrm{T}}(t)\mathbf{Q}(t) - \mathbf{Q}(t)\mathbf{A}(t) - \mathbf{P}(t), \tag{6.17}$$

with

$$\dot{V}(\mathbf{x}, t) = -\mathbf{x}^{\mathrm{T}}\mathbf{P}(t)\mathbf{x}.$$

Example 6.6

Consider the LTIS

$$\dot{\mathbf{x}} = \mathbf{A}\mathbf{x},$$

where

$$\mathbf{A} = \begin{bmatrix} -2 & -3 \\ 1 & 2 \end{bmatrix}.$$

Select $\mathbf{P} = \mathbf{I}$; then (6.14) yields

$$-4q_{11} + 2q_{12} = -1,$$
$$-3q_{11} + q_{22} = -0,$$
$$-6q_{12} + 4q_{22} = -1,$$

which do not yield a unique $\mathbf{Q}$, and in fact the eigenvalues of $\mathbf{A}$ are $\lambda_1 = -1$, $\lambda_2 = +1$, and the $\mathbf{0}$ is not stable.

If

$$\mathbf{A} = \begin{bmatrix} -2 & -3 \\ 1 & -2 \end{bmatrix}$$

and $\mathbf{P} = \mathbf{I}$, then

$$-4q_{11} + 2q_{12} = -1,$$
$$6q_{12} + 4q_{22} = 1,$$
$$3q_{11} - q_{22} + 4q_{12} = 0,$$

or $q_{11} = \frac{3}{14}$, $q_{22} = \frac{5}{14}$, $q_{12} = -\frac{1}{14}$. Therefore, the $\mathbf{0}$ is globally asymptotically stable. If one checks, it is seen that $\lambda_1, \lambda_2 = -2 \pm j\sqrt{3}$, which corresponds to a stable focus. In general, with the ith–jth element of $\mathbf{A}$ denoted by a_{ij}, where $i, j = 1, 2$, (6.14) reduces to

$$\begin{bmatrix} 2a_{11} & 2a_{21} & 0 \\ a_{12} & a_{11} + a_{22} & a_{21} \\ 0 & 2a_{12} & a_{22} \end{bmatrix} \begin{bmatrix} q_{11} \\ q_{12} \\ q_{22} \end{bmatrix} = \begin{bmatrix} -1 \\ 0 \\ -1 \end{bmatrix},$$

where it is assumed that $\mathbf{P} = \mathbf{I}$. Then it is seen that $\mathbf{Q}$ is positive-definite if

$$a_{11}a_{22} - a_{12}a_{21} > 0$$

and

$$a_{11} + a_{22} < 0.$$

These are the stability conditions for $\mathbf{A}$, which also result from its characteristic equation, $\det[\mathbf{I}\lambda - \mathbf{A}] = 0$.

6.1.5 Liapunov Function Generation

For linear systems it was seen in the preceding section that if the equilibrium state is asymptotically stable, a quadratic form which is a Liapunov function satisfies Liapunov's equation (6.14) or (6.17). In fact, the conditions are necessary and sufficient. Unfortunately, for nonlinear systems, no such methodology is available. First it can be difficult to find a satisfactory Liapunov function, and second it may yield a conservative estimate of stability since the conditions generally are sufficient but not necessary. Already, we have seen energy functions as well as more general quadratic forms utilized as Liapunov functions. Several other methods are available for their generation and are examined here.

First, consider *Aizerman's method*, which proceeds as follows to analyze stability of $\mathbf{0}$ for (6.9):

1. Linearize (6.9) at $\mathbf{0}$ to obtain

$$\dot{\delta\mathbf{x}} = \mathbf{A}(\mathbf{0})\,\delta\mathbf{x}, \qquad (6.18)$$

where

$$\mathbf{A}(\mathbf{0}) = \frac{\partial \mathbf{f}}{\partial \mathbf{x}}(\mathbf{0}).$$

2. Select a quadratic form for (6.18) which is p.d. so that

$$V = \mathbf{x}^{\mathrm{T}}\mathbf{Q}\mathbf{x} \qquad (6.8)$$

has unspecified $q_{ij} = q_{ji}$.

3. Select a n.d. $\dot{V}$ according to (6.14) and (6.18), which in turn specifies q_{ij} and therefore $\mathbf{Q}$ if (6.18) is asymptotically stable.
4. Now solve $\dot{V}$ along $\mathbf{x}(t)$ for (6.9) the nonlinear system, and recompute V from (6.8) and step 3.
5. Find the range of permissible parameter values for asymptotic stability.

Example 6.7

Consider the nonlinear feedback regulator shown by Fig. 6.4, or

$$\ddot{x} + 2\dot{x} + u = 0, \tag{6.19}$$

where $u = f(x)$. With $x_1 = x$, $x_2 = \dot{x}_1$,

$$\dot{x}_1 = x_2,$$
$$\dot{x}_2 = -2x_2 - f(x_1).$$

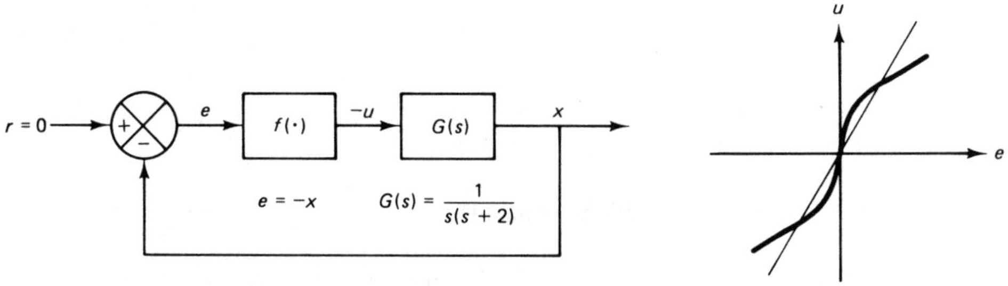

Figure 6.4 Nonlinear regulator for Example 6.7.

Then (6.18) provides

$$\delta\dot{x}_1 = \delta x_2,$$
$$\delta\dot{x}_2 = -2\delta x_2 - \frac{\partial f}{\partial f_1}(0)\,\delta x_1 .$$

Then select

$$V = q_{11}x_1^2 + q_{22}x_2^2 + q_{12}x_1x_2,$$

with

$$\dot{V} = 2q_{11}x_1\dot{x}_1 + 2q_{22}x_2\dot{x}_2 + q'_{12}(x_1\dot{x}_2 + \dot{x}_1x_2) = -\mathbf{x}^{\mathrm{T}}\mathbf{P}\mathbf{x}.$$

Or, with $\partial f/\partial x_1(0) = 2$,

$$\dot{V} = -2q_{12}x_1^2 + (2q_{11} - 4q_{22} - 2q'_{12})x_1x_2(q'_{12} - 4q_{22})x_2^2$$

and $\mathbf{P} = \mathbf{I}$. Then let $2q'_{12} = 1$ or $q'_{12} = 1/2$, $2q_{11} - 2q'_{12} - 4q_{22} = 0$, and $q'_{12} - 4q_{22} = -1$ or $q_{11} = \frac{5}{4}$, $q_{22} = \frac{3}{8}$, $q'_{12} = \frac{1}{2}$. Then computation of $\dot{V}$ for (6.19) yields

$$\dot{V} = \tfrac{5}{4}\,(2x_1\dot{x}_1) + \tfrac{1}{2}\,x_1\dot{x}_2 + \tfrac{1}{2}\,\dot{x}_1x_2 + \tfrac{3}{4}\,\dot{x}_2x_2$$

or

$$-\dot{V} = \left[\frac{1}{2}\frac{f(x_1)}{x_1}\right]x_1^2 + \left[\frac{3}{4}\frac{f(x_1)}{x_1} - \frac{3}{2}\right]x_1x_2 + x_2^2.$$

Assume that $f(x_1)/x_1 > 0$, so that $\dot{V} < 0$ if $f(x_1)/x_1 > 0$, and

$$\det[\mathbf{P}] = \det\begin{bmatrix} \dfrac{f(x_1)}{2x_1} & \dfrac{1}{2}\left[\dfrac{3}{4}\dfrac{f(x_1)}{x_1} - \dfrac{3}{2}\right] \\ \dfrac{1}{2}\left[\dfrac{3}{4}\dfrac{f(x_1)}{x_1} - \dfrac{3}{2}\right] & 1 \end{bmatrix} > 0.$$

Then the $\mathbf{0}$ for (6.19) is asymptotically stable if

$$0.573 < \frac{f(x_1)}{x_1} < 6.98.$$

Next consider the *variable-gradient method*, which again works from $\dot{V}$ back to V [S4]. Again, the autonomous nonlinear system is given by (6.9), for which the time rate of change of V is given by

$$\frac{dV}{dt} = \frac{\partial V}{\partial \mathbf{x}} \cdot \mathbf{f}(\mathbf{x}) = \boldsymbol{\nabla}^{\mathrm{T}} V\, \mathbf{f}(\mathbf{x}), \tag{6.20}$$

where $V(\mathbf{x})$ can be obtained from

$$V = \int_0^{\mathbf{x}} \boldsymbol{\nabla}^{\mathrm{T}} V(\boldsymbol{\sigma})\, d\boldsymbol{\sigma} \tag{6.21}$$

Then the following steps are used to generate $V(\mathbf{x})$:

1. Assume $\boldsymbol{\nabla} V$ to be an arbitrary column vector whose coefficients are functions of the state so that

$$\boldsymbol{\nabla} V = \mathbf{D}(\mathbf{x})\mathbf{x}, \tag{6.22}$$

where the ith–jth element of $\mathbf{D}(\cdot)$ is given by

$$\mathrm{d}_{\mathrm{ij}}(\mathbf{x}) = d_{ijc} + d_{ijv}(\mathbf{x});$$

d_{ijc} is a constant, and $i, j = 1, \ldots, n$. Since any constant $V(\mathbf{x})$ represents a closed surface, $\mathbf{D}(\cdot)$ is chosen independent of x_n by setting d_{nn} constant.

2. $V(\mathbf{x})$ from (6.21) is chosen independent of the path of integration so that

$$V(\mathbf{x}) = \int_0^{x_1} \nabla_1 V\, d\sigma_1 + \int_0^{x_2} \nabla_2 V\, d\sigma_2 + \cdots + \int_0^{x_n} \nabla_n V\, d\sigma_n,$$

where it is crucial that

$$\nabla_i V = \frac{\partial V(\mathbf{x})}{\partial x_i}, \qquad \frac{\partial \nabla_i V}{\partial x_j} = \frac{\partial \nabla_j V}{\partial x_i}, \qquad i \neq j,$$

$i, j = 1, \ldots, n.$

3. Check that $\dot{V}(\mathbf{x})$ is constrained to be at least semidefinite with opposite definiteness to $V(\mathbf{x})$.

Example 6.8

Consider

$$\dot{x}_1 = ax_2,$$
$$\dot{x}_2 = -bx_1 + \mu(c - x_1^r)^p x_2^q, \tag{6.24}$$

where r is a positive even integer, p and q positive odd integers, and a, b, c, μ positive constants. Recall that for Van der Pol's equation (5.26), $a = b = c = p = q = 1$ and $r = 2$. Then, from (6.23),

$$\frac{\partial f_1}{\partial x_1} = 0, \qquad \frac{\partial f_2}{\partial x_2} = \mu q(c - x_1^r)^p x_2^{q-1}.$$

Consequently, it is seen that $\partial f_1/\partial x_1 + \partial f_2/\partial x_2$ does not change sign in the strip $-(c)^{1/r} < x_1 < (c)^{1/r}$, and according to Bendixson's criterion (5.6), no limit cycle can exist in this strip.

Now let

$$\dot{V}(\mathbf{x}) = \left(\frac{\partial f_1}{\partial x_1} + \frac{\partial f_2}{\partial x_2}\right) P(\mathbf{x}),$$

where $P(\mathbf{x})$ is appropriately selected with sign identical to that of $\dot{V}(\cdot)$ since the first portion does not change sign in the strip given above, which contains the equilibrium point at the origin. If $P(\mathbf{x}) = x_2^2$,

$$\dot{V}(\mathbf{x}) = \mu q(c - x_1^r)^p x_2^{q+1},$$

which is p.s.d. for $-c^* < x_1 < c^*$. Then, from (6.20) and (6.23),

$$\dot{V}(\mathbf{x}) = \nabla_1 V(ax_2) + \nabla_2 V[-bx_1 + \mu(c - x_1^r)^p x_2^q]$$

or

$$\nabla_2 V = qx_2,$$

and

$$\nabla_1 V(ax_2) - \nabla_2 V(bx_1) = 0,$$

so that, as required,

$$\frac{\partial \nabla_1 V}{\partial x_2} = \frac{\partial \nabla_2 V}{\partial x_1}.$$

Finally,

$$V(\mathbf{x}) = \int_0^{x_1} \nabla_1 V\, d\sigma_1 + \int_0^{x_2} \nabla_2 V\, d\sigma_2$$
$$= \frac{q}{2}\left(\frac{bx_1^2}{a} + x_2^2\right),$$

which is p.d.; and $\dot{V}(\mathbf{x})$ is p.s.d. for $-c^{1/r} < x_1 < c^{1/r}$. Then the origin is unstable according to the instability theorem (3) of Section 6.1.3.

A general stability theorem due to Krasovskii [K3] and later simplified by Kalman and Bertram [K4] can be useful to generate a Liapunov function for the nonlinear autonomous system (6.9) with an equilibrium state at $\mathbf{0}$. All that is required here is to show that the matrix

$$\mathbf{F}(\mathbf{x}) = \frac{\partial \mathbf{f}}{\partial \mathbf{x}} + \left(\frac{\partial \mathbf{f}}{\partial \mathbf{x}}\right)^{\mathrm{T}} \tag{6.25}$$

is negative-definite for all $\mathbf{x}$. Then the origin is asymptotically stable. As a consequence,

$$V(\mathbf{x}) = \mathbf{f}^{\mathrm{T}}(\mathbf{x})\mathbf{f}(\mathbf{x}) \tag{6.26}$$

is a Liapunov function.

The proof of this result depends on asymptotic stability theorem (2) in Section 6.1.3. That is, substitution of (6.26) into (6.20) yields

$$\frac{dV}{dt} = \left(\frac{\partial V}{\partial \mathbf{x}}\right)^{\mathrm{T}} \mathbf{f}(\mathbf{x}) = 2\left[\left(\frac{\partial \mathbf{f}}{\partial \mathbf{x}}\right)^{\mathrm{T}} \mathbf{f}(\mathbf{x})\right]^{\mathrm{T}} \mathbf{f}(\mathbf{x})$$

$$= \left[\mathbf{f}^{\mathrm{T}}(\mathbf{x})\left(\frac{\partial \mathbf{f}}{\partial \mathbf{x}}\right)^{\mathrm{T}} \mathbf{f}(\mathbf{x})\right]^{\mathrm{T}} + \mathbf{f}^{\mathrm{T}}(\mathbf{x})\frac{\partial \mathbf{f}}{\partial \mathbf{x}}\mathbf{f}(\mathbf{x})$$

or, since this is a scalar,

$$\frac{dV}{dt} = \mathbf{f}^{\mathrm{T}}(\mathbf{x})\mathbf{F}(\mathbf{x})\mathbf{f}(\mathbf{x}),$$

which is assumed to be negative. Obviously, since $\mathbf{f}(\mathbf{0}) = \mathbf{0}$, from (6.26), $V(\mathbf{0}) = 0$, and $V(\mathbf{x}) > 0$ for $\mathbf{x} \neq \mathbf{0}$ if $\mathbf{0}$ is a unique equilibrium state. However, it can be shown that if (6.26) is negative-definite as assumed, then $\mathbf{0}$ is the only equilibrium state [K4]. The inherent problem with this method arises from the inability of satisfying $\mathbf{F}(\mathbf{x})$ negative-definite for all $\mathbf{x}$. However, it can be useful when combined with other methods.

Example 6.9

Consider an exothermic chemical reactor with perfect mixing, irreversible decomposition, and with equal volumetric feed and product rates both equal to constant F. Material and energy balances yield

$$\dot{y}_1 = 1 - y_1 - g(y_1, y_2),$$

$$\dot{y}_2 = \alpha - y_2 + g(y_1, y_2) - h(y_2),$$

where $y_1 = x_R/x_F$ is a dimensionless concentration ratio (i.e., mole fraction of material in reactor divided by mole fraction of material in the feed stream); $y_2 = c_pT/H_c$ is a dimensionless energy ratio (i.e., heat required to raise one mole of the reactor material temperature, T, divided by H_c, the heat generated in the chemical decomposition); and time is normalized by multiplication of t by the ratio of F to reactor volume v. Here c_p is again the reactor material specific heat. Also, it is determined [B5] that

$$g(y_1, y_2) = \frac{kvy_1}{F}\exp\left(\frac{-Ec_p}{Rx_fH_cy_2}\right),$$

$$h(y_2) = \frac{Q(T)}{F\rho x_F H_c},$$

$$\alpha = \frac{c_pT_F}{x_FH_c},$$

where k is a reaction rate constant, E the activation energy per mole, R the perfect gas constant, T the absolute reactor temperature, T_F the absolute feed system temperature, v the reactor volume, $Q(t)$ the heat-removal rate from the reactor, and ρ the reactor material density. Then a translation of variables locates the equilibrium state at the origin. Therefore, it follows that

$$\dot{x}_1 = -x_1 - G(x_1, x_2),$$

$$\dot{x}_2 = -x_2 + G(x_1, x_2) - H(x_2),$$

where

$$x_1 = y_1 - y_{1e},$$

$$x_2 = y_2 - y_{2e},$$

$$G(x_1, x_2) = g(x_1 + y_{1e}, x_2 + y_{2e}) - g(y_{1e}, y_{2e}),$$

$$H(x_2) = h(x_2 + y_{2e}) - h(y_{2e}),$$

$$1 - y_{1e} - g(y_{1e}, y_{2e}) = 0,$$

and

$$\alpha - y_{2e} + g(y_{1e}, y_{2e}) - h(y_{2e}) = 0.$$

Now examine the stability of the $\mathbf{0}$ from the linear approximation with

$$\mathbf{A} = \frac{\partial \mathbf{f}}{\partial \mathbf{x}}(\mathbf{0}) = \begin{bmatrix} -1 - \dfrac{\partial G}{\partial x_1}(0,0) & -\dfrac{\partial G}{\partial x_2}(0,0) \\ \dfrac{\partial G}{\partial x_1}(0,0) & -1 + \dfrac{\partial G}{\partial x_2}(0,0) - \dfrac{\partial H}{\partial x_2}(0) \end{bmatrix}.$$

Then from the conditions given by Example 6.6,

$$1 + \frac{\partial G}{\partial x_1}(0,0) - \frac{\partial G}{\partial x_2}(0,0) + \frac{\partial H}{\partial x_2}(0)\left[1 + \frac{\partial G}{\partial x_1}(0,0)\right] > 0$$

and

$$1 + \frac{\partial G}{\partial x_1}(0,0) - \frac{\partial G}{\partial x_2}(0,0) + \frac{\partial H}{\partial x_2}(0) > -1.$$

Next apply Krasovskii's result with (6.25) given by

$$\mathbf{F} = -\begin{bmatrix} 2\left(1 + \dfrac{\partial G}{\partial x_1}\right) & \dfrac{\partial G}{\partial x_2} - \dfrac{\partial G}{\partial x_1} \\ \dfrac{\partial G}{\partial x_2} - \dfrac{\partial G}{\partial x_1} & 2\left(1 - \dfrac{\partial G}{\partial x_2} + \dfrac{\partial H}{\partial x_2}\right) \end{bmatrix}.$$

Here, from (6.26),

$$V(\mathbf{x}) = [x_1 + G(x_1, x_2)]^2 + [x_2 - G(x_1, x_2) + H(x_2)]^2$$

will be p.d. if

$$1 + \frac{\partial G}{\partial x_1} > 0$$

and

$$4\left(1 + \frac{\partial G}{\partial x_1}\right)\left(1 - \frac{\partial G}{\partial x_2} + \frac{\partial H}{\partial x_2}\right) - \left(\frac{\partial G}{\partial x_2} - \frac{\partial G}{\partial x_1}\right)^2 > 0.$$

The more general result of Krasovskii [K3] states that the $\mathbf{0}$ for (6.9) is globally asymptotically stable if there exist constant symmetric positive-definite matrices $\mathbf{P}$ and $\mathbf{Q}$ such that

$$\mathbf{F}(\mathbf{x}) = \left(\frac{\partial \mathbf{f}}{\partial \mathbf{x}}\right)^{\mathrm{T}}\mathbf{Q} + \mathbf{Q}\frac{\partial \mathbf{f}}{\partial \mathbf{x}} + \mathbf{P}$$

is negative-definite for all $\mathbf{x}$. Here the Liapunov function is given by

$$V(\mathbf{x}) = \mathbf{f}^{\mathrm{T}}\mathbf{Q}\mathbf{f}.$$

The proof is similar to the previous case and will not be repeated here.

Zubov's method also selects a negative-definite $\dot{V}(\mathbf{x})$ to obtain a positive-definite $V(\mathbf{x})$ as a solution to (6.20) with $V(\mathbf{0}) = 0$. Let $P(\mathbf{x}) = \nabla^{\mathrm{T}}V(\mathbf{x})\mathbf{f}(\mathbf{x})$ in (6.20). If $\mathbf{0}$ is asymptotically stable, $V(\mathbf{x})$ will be p.d. in some neighborhood of $\mathbf{0}$. Then

$$V(\mathbf{x}(t; \mathbf{x}_0, t_0)) - V(\mathbf{x}_0) = \int_{t_0}^{t} P(\mathbf{x}(\sigma; \mathbf{x}_0, t_0))\, d\sigma$$

yields

$$V(\mathbf{x}_0) = -\int_{t_0}^{\infty} P(\mathbf{x}(t; \mathbf{x}_0, t_0))\, dt$$

within D, the domain of attraction of $\mathbf{0}$, for initial $\mathbf{x}_0$ that is positive unless $\mathbf{x}_0 = \mathbf{0}$. As $\mathbf{x}_0$ approaches the boundary of D, $V(\mathbf{x}_0) \to \infty$. To avoid computation of points of infinity, a new function $Z(\mathbf{x})$ is defined by

$$Z(\mathbf{x}) = 1 - \exp[-V(\mathbf{x})].$$

Now the partial differential equation for $Z(\mathbf{x})$ in terms of $P(\mathbf{x})$ becomes

$$\mathbf{f}^{\mathrm{T}}(\mathbf{x})\nabla Z(\mathbf{x}) = P(\mathbf{x})[1 - Z(\mathbf{x})],$$

where $Z(\mathbf{0}) = 0$ and the boundary of D is found from

$$Z(\mathbf{x}) = 1.$$

Example 6.10 [W2]

Consider

$$\dot{x} = x^3 - x.$$

With $P(\mathbf{x}) = -2x^2$, the above yields

$$\frac{dV}{dx} = -\frac{2x}{x^2 - 1}$$

and

$$\frac{1}{1 - Z}\frac{dZ}{dx} = -\frac{2x}{x^2 - 1}.$$

Consequently,

$$V(\mathbf{x}) = -\ln(1 - x^2)$$

and

$$Z(\mathbf{x}) = x^2,$$

with the boundary of D, the domain of attraction to 0, given by $|x| = 1$. Note that this result agrees with Theorem (5) in Section 6.1.3.

Example 6.11

A synchronous electric generator is connected to an infinite bus through a double transmission line with negligible resistance. Suppose that one of the transmission lines develops a ground fault and is isolated from the system in a time T by the circuit breakers at the two ends. The problem is to find a stable region of operation for the generator.

From Newton's second law, ϕ, the angle between the machine voltage and the infinite bus voltage is given by

$$M\ddot{\phi} + D(\phi)\dot{\phi} = P_i - P_o, \tag{6.27}$$

where P_i and P_o are the input and output power, M the inertial constant, and $D(\cdot)$ a damping term. Recall that (6.27) represents one part of the power system given in Example 3.4. Here, however, the power terms $P_i = \omega_0 T_m$ and $P_o = \omega_0 T_e$, and a damping term is incorporated. To simplify the problem, it is assumed that the machine voltages, and therefore the field-flux linkages, are constant. Furthermore, it is assumed that

$$P_o = P_{o1} \sin \phi(t) - P_{o2} \sin 2\phi(t),$$

where constants P_{o1}, P_{o2} are proportional to the product of machine voltage and the infinite bus voltage, and the square of the infinite bus voltage, respectively. Note that this is a generalization of the swing equation (5.23).

Now assume that P_i and D are nonnegative constants. Then in normalized canonical phase-variable form, (6.27) becomes

$$\dot{x}_1 = x_2,$$
$$\dot{x}_2 = -Dx_2 + f(x_1),$$

where

$$f(x_1) = P_i - P_{o1} \sin x_1 + P_{o2} \sin 2x_1.$$

At equilibrium, $x_2 = 0$ and $f(x_1) = 0$. For typical parameters

$$\mathbf{x}_{1e}^{\mathrm{T}} = [0.298 \quad 0]$$

and

$$\mathbf{x}_{2e}^{\mathrm{T}} = [2.53 \quad 0],$$

which are repeated periodically on the x_1 axis (every 2π, as in Example 5.5). For $D = 0$, $\mathbf{x}_{1e}$ is a center point and $\mathbf{x}_{2e}$ is a saddle point. For $D > 0$, $\mathbf{x}_{1e}$ is a stable focus and $\mathbf{x}_{2e}$ is a saddle point. By translation of x_1 so that the origin becomes an equilibrium corresponding to $x_{11e} = 0.298$,

$$\begin{aligned} \dot{x}_1 &= x_2, \\ \dot{x}_2 &= -0.279x_2 + f(x_1), \end{aligned} \tag{6.28}$$

where $f(x_1) = 0.3 - \sin(x_1 + 0.298) + 0.138 \sin(2x_1 + 0.596)$ and x_1, x_2 are now with respect to $x_{11e} = 0$. Then the energy type of Liapunov function is given by (6.3), or

$$V(x_1, x_2) = \frac{x_2^2}{2} + \int_0^{x_1} [-f(\sigma)]\, d\sigma,$$

or

$$V(x_1, x_2) = \frac{x_2^2}{2} - 0.3x_1 - \cos(x_1 + 0.298) + 0.069 \cos(2x_1 + 0.596).$$

Hence $V(\cdot)$ is p.d. in a region about the origin. Also,

$$\dot{V}(\mathbf{x}) = -0.279x_2^2$$

is n.s.d. for all $\mathbf{x}$. Since $V(\cdot)$ is a Liapunov function, the $\mathbf{0}$ is stable.

For the linearized model about the $\mathbf{0}$, it is found that

$$\mathbf{A} = \begin{bmatrix} 0 & 1 \\ -0.727 & -0.279 \end{bmatrix}.$$

Then following Aizerman's method,

$$V(\mathbf{x}) = q_{11}x_1^2 + 2q_{12}x_1x_2 + q_{22}x_2^2,$$

$$\dot{V} = -1.45q_{12}x_1^2 + (2q_{11} - 0.558q_{12} - 1.45q_{22})x_1x_2 + (2q_{12} - 0.558q_{22})x_2^2,$$

$q_{11} = 0.68$, $q_{12} - 0.69$, $q_{22} = 0.67$. Then

$$V(\mathbf{x}) = 0.68x_1^2 + 1.38x_1x_2 + 0.67x_2^2$$

and

$$\dot{V}(\mathbf{x}) = 1.38k(x_1)x_1^2 - g(x_1)x_1x_2 - x_2^2,$$

where $k(x_1) = f(x_1)/x_1$ and $g(x_1) = -1.35k(x_1) - 0.98$. $V(\cdot)$ is n.d. if the

$$\det \begin{bmatrix} -1.38k(x_1) & 0.5g(x_1) \\ 0.5g(x_1) & 1.0 \end{bmatrix} > 0$$

or

$$-4.37 < k(x_1) < -0.12. \tag{6.29}$$

Figure 6.5 shows the acceptable values with $k_2 = 4.37$, $k_1 = 0.12$.

The energy Liapunov function is a maximum at the saddle point (2.53, 0) as studied previously (see Example 5.5), and as can be verified here. If $D = 0$, the

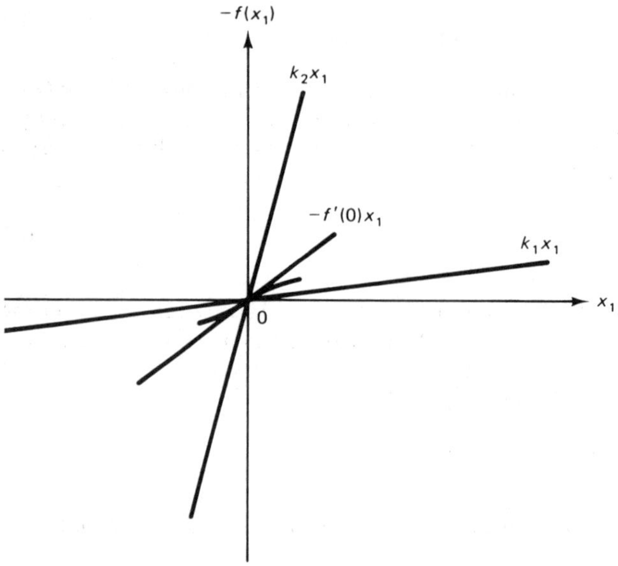

Figure 6.5 Stability region for Example 6.11.

trajectories about **0** are closed, with each corresponding to a constant energy $V(\mathbf{x})$. At some critical energy, however, the trajectory approaches the saddle point, and at higher energies the motion is unstable. As shown in Fig. 6.6, there exists some critical state $\mathbf{x}_c$ and time t_c such that the indicated transmission line, a ground-fault trajectory crosses the separatrix (trajectory to the saddle point) at $\mathbf{x} = \mathbf{x}_c$, which corresponds to a critical "swing" angle ϕ_c and time t_c. Then if the fault occurs at some $t = t_F$ and the required time to isolate the line is T, it is most important that

$$T < t_c - t_f. \tag{6.30}$$

Otherwise, the entire power system will become unstable.

If the input power is controlled by means of governor action (as normally done),

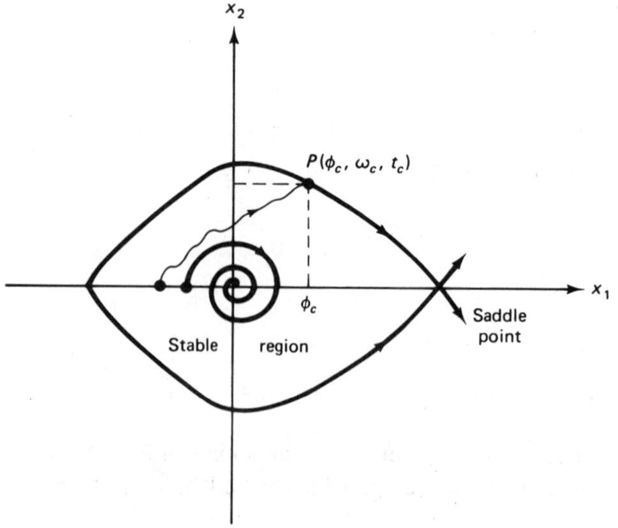

Figure 6.6 Critical angle ϕ_c and critical time t_c for Example 6.11.

the saddle point equilibrium occurs at a larger value and the stability region is enlarged. For example, consider the following system model [M13]:

$$\frac{d^3\phi}{dt^3} + a\frac{d^2\phi}{dt^2} + (b + \cos\phi)\frac{d\phi}{dt} + c\sin\phi - d = 0, \tag{6.31}$$

where a, b, c, and d are nonnegative constants. Then, with $a = 0.943$, $b = 1.197$, $c = 0.667$, and $d = 0.196$, (6.31) is at equilibrium with $\dot{\phi} = \ddot{\phi} = 0$ and $\phi = 0.298$, 2.843, . . ., which are again repeated periodically along the ϕ axis.

Again, a translation of coordinates is made to examine the stability of the first equilibrium state (now at the origin). Then, in phase variable form (6.31) becomes

$$\dot{x}_1 = x_2,$$

$$\dot{x}_2 = x_3, \tag{6.32}$$

$$\dot{x}_3 = -0.943x_3 - [1.197 + \cos(x_1 + 0.298)]x_2 - 0.667\sin(x_2 + 0.298) + 0.196.$$

While the Liapunov stability analysis can be made, the algebra becomes more burdensome. Instead, a computer simulation is made on an IBM-PC-AT with a Runge–Kutta fourth-order integration scheme. The system is stable for $x_1(0) \leq 2.5$. Figures 6.7 to 6.9 show typical state versus time and phase-plane (x_1 versus x_2) trajectories. Figure 6.7 shows the stable transient response with $x_1(0) = -2$, $x_2(0) = 0$, $x_3(0) = 0$. Figure 6.8 shows the lack of stability of $\mathbf{x} = \mathbf{0}$ with $x_1(0) = 2.7$, $x_2(0) = 0$, $x_3(0) = 0$. Finally, Fig. 6.9 shows the peculiar phase-plane projected trajectories for this third-order system. Figure 6.10 shows the typical decay of the energy Liapunov function for this example.

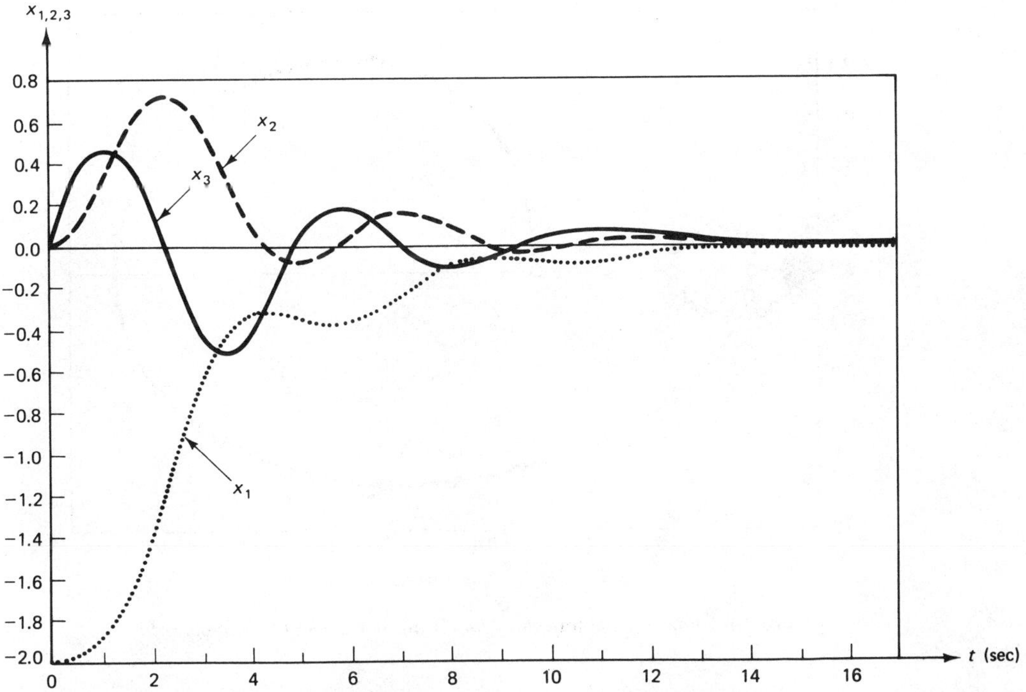

Figure 6.7 Stable transients for Example 6.11.

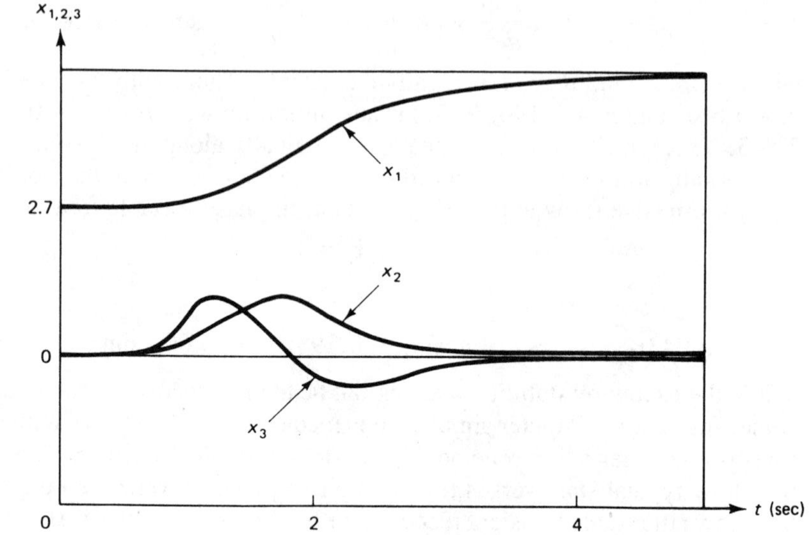

Figure 6.8 Lack of x_1 convergence to 0 for Example 6.11.

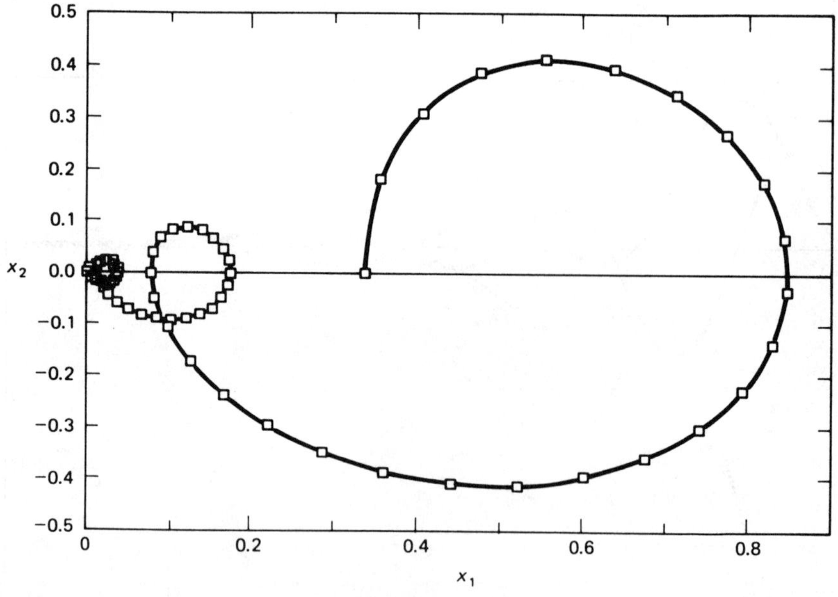

Figure 6.9 Phase-plane trajectory for third-order model of Example 6.11.

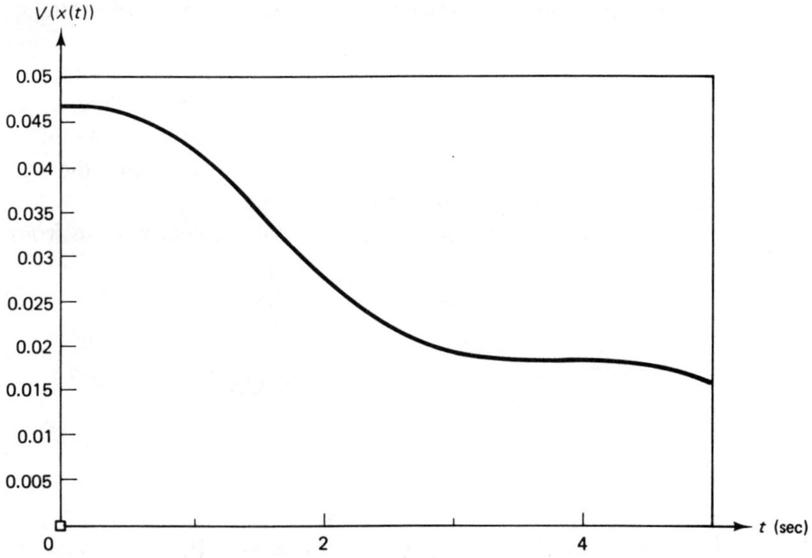

Figure 6.10 Liapunov function for Example 6.11.

6.2 LIAPUNOV TRANSIENT ANALYSIS AND DESIGN

Sometimes, Liapunov functions are useful to obtain bounds on system response time without solving the system differential equations. Then the method may be carried one step further to design an adaptive control system which adjusts control parameters according to the response time desired.

6.2.1 Transient Estimation

Transient response estimation by means of Liapunov functions proceeds from bounds on $V(x(t))$ itself. Since we can write

$$\dot{V}(x) = \frac{\dot{V}(x)}{V(x)} V(x),$$

then

$$\dot{V} \leq -\rho V \tag{6.33}$$

and

$$V(x(t)) \leq V(x_0)e^{-\rho(t-t_0)} \tag{6.34}$$

if the maximum value of $\dot{V}/V$ is never larger than $-\rho$ in a domain about the origin. Here $\rho \geq 0$. As a consequence of these relationships, the procedure may be as follows:

1. Determine a Liapunov function $V(\mathbf{x})$.
2. Maximize $-\dot{V}/V$ to obtain ρ.

3. For any motion from $\mathbf{x}(t_0)$ inside the region of asymptotic stability about the origin, compute $V(\mathbf{x}_0)$.

This provides an upper bound on $V(\mathbf{x})$ for all $t \geq t_0$. That is, for $t \geq t_0$, the state of the system will remain inside the contour $V(\mathbf{x}) = V(\mathbf{x}_0)\exp[-\rho(t - t_0)]$. Therefore, a large ρ is indicative of a fast transient response toward the origin. $1/\rho$ has the characteristic of a response time or time constant.

Consider the globally asymptotically stable linear system from Section 6.1.4:

$$\frac{d\mathbf{x}}{dt} = \mathbf{A}\mathbf{x}, \tag{6.12}$$

with

$$V(\mathbf{x}) = \mathbf{x}^{\mathrm{T}}\mathbf{Q}\mathbf{x}, \tag{6.8}$$

such that

$$\dot{V}(\mathbf{x}) = -\mathbf{x}^{\mathrm{T}}\mathbf{P}\mathbf{x}, \tag{6.13}$$

$$\mathbf{A}^{\mathrm{T}}\mathbf{Q} + \mathbf{Q}^{\mathrm{T}}\mathbf{A} = -\mathbf{P}. \tag{6.14}$$

Then

$$\rho = \min_{\mathbf{x}} \frac{\mathbf{x}^{\mathrm{T}}\mathbf{P}\mathbf{x}}{\mathbf{x}^{\mathrm{T}}\mathbf{Q}\mathbf{x}},$$

or the stationary point (necessary for a minimum) is obtained from the vector equation

$$\frac{\partial \rho}{\partial \mathbf{x}} = \frac{(2\mathbf{P}\mathbf{x})(\mathbf{x}^{\mathrm{T}}\mathbf{Q}\mathbf{x}) - 2(\mathbf{x}^{\mathrm{T}}\mathbf{P}\mathbf{x})(\mathbf{Q}\mathbf{x})}{(\mathbf{x}^{\mathrm{T}}\mathbf{Q}\mathbf{x})^2} = \mathbf{0}. \tag{6.35}$$

Equation (6.36) may result in numerous $\mathbf{x}$. For all such $\mathbf{x}$ that satisfy (6.36), let

$$\gamma = \frac{\mathbf{x}^{\mathrm{T}}\mathbf{P}\mathbf{x}}{\mathbf{x}^{\mathrm{T}}\mathbf{Q}\mathbf{x}}, \tag{6.36}$$

so that (6.36) can be written as

$$[\mathbf{P} - \gamma\mathbf{Q}]\mathbf{x} = \mathbf{0}$$

or

$$[\mathbf{Q}^{-1}\mathbf{P} - \gamma\mathbf{I}]\mathbf{x} = \mathbf{0}. \tag{6.37}$$

Therefore, γ is an eigenvalue of the matrix $\mathbf{Q}^{-1}\mathbf{P}$, and ρ, from (6.35), is the minimum eigenvalue, that is,

$$\rho = \min_i \lambda_i(\mathbf{Q}^{-1}\mathbf{P}), \tag{6.38}$$

where $i = 1, \ldots, n$.

Again, of course, $V(\mathbf{x})$ is not uniquely defined, and the estimate of the upper bound on response time may be quite conservative. To better ascertain this limit, different $\mathbf{P}$ may be selected, which result in different $\mathbf{Q}$. But in any case, the analysis would most likely be backed by computer simulation.

Example 6.12 [H1]

The equations of motion of a tumbling satellite with the tumbling rate much greater than the orbital rate is given by

$$
\begin{aligned}
I_x\dot{\omega}_x - (I_y - I_z)\omega_y\omega_z &= u_1, \\
I_y\dot{\omega}_y - (I_z - I_x)\omega_z\omega_x &= u_2, \\
I_z\dot{\omega}_z - (I_x - I_y)\omega_x\omega_y &= u_3,
\end{aligned}
\tag{6.39}
$$

where ω_x, ω_y, ω_z are angular rates about the principal body X, Y, Z axes. Also, the moments of inertia I_x, I_y, I_z and the thrust torques u_1, u_2, u_3 are taken about the same body axes.

Select the positive-definite square of the magnitude of the statellite angular momentum as a candidate for a Liapunov function. That is, let

$$V(\mathbf{x}) = M^2(\mathbf{x}) = (I_x x_1)^2 + (I_y x_2)^2 + (I_z x_3)^2, \tag{6.40}$$

where $x_1 = \omega_x$, $x_2 = \omega_y$, $x_3 = \omega_z$. Then

$$\dot{V}(\mathbf{x}) = 2(I_x x_1 u_1 + I_y x_2 u_2 + I_z x_3 u_3). \tag{6.41}$$

It is seen that (6.41) is negative-definite for all $\mathbf{x}$ if $u_i = -f_i(x_i)$ is a finite feedback control such that

$$
\begin{aligned}
& x_i f_i(x_i) > 0, \qquad \text{for } x_i \neq 0 \\
\text{and} \\
& \lim_{x_i \to \infty} \int_0^{x_i} f_i(\sigma)\, d\sigma = \infty,
\end{aligned}
\tag{6.42}
$$

$i = 1, 2, 3$. Since, from (6.40), $\lim V(\mathbf{x}) \to \infty$ as $\|\mathbf{x}\| \to \infty$, these conditions are sufficient for the $\mathbf{0}$ to be globally asymptotically stable according to Theorem (4) in Section 6.1.3. Here $\dot{V}(\mathbf{x}, \mathbf{u})$ is linear in $\mathbf{u}$ so that with $|u_i| \leq U_i$, $\dot{V}(\cdot)$ is minimized by the bang-bang thrust-torque control. Therefore,

$$u_i = -f_i(x_i) = -U_i \operatorname{sgn} x_i, \tag{6.43}$$

where

$$\operatorname{sgn} x_i = \begin{cases} 1, & x_i > 0, \\ -1, & x_i < 0, \end{cases}$$

$U_i > 0$, and $i = 1, 2, 3$. Then it is seen from these conditions that

$$|f_i(x_i)| \geq \frac{UI_i}{M(\mathbf{x})}|x_i|$$

for

$$|x_i| \leq \frac{M(\mathbf{x})}{I_i}$$

[always satisfied by (6.40)], where

$$U = \min_i U_i, \qquad i = 1, 2, 3$$

$$I_1 = I_x, \qquad I_2 = I_y, \qquad I_3 = I_z,$$

and $V^{1/2}(\mathbf{x})$ is the magnitude of the angular momentum of the satellite. Then

$$I_i x_i f_i(x_i) \geq \frac{UI_i^2}{M(\mathbf{x})} x_i^2 . \tag{6.44}$$

From (6.44) and (6.40),

$$\dot{V}(\mathbf{x}) = 2M(\mathbf{x})\dot{M}(\mathbf{x}) = 2\left[\sum_{i=1}^{3} I_i x_i f_i(x_i)\right] \geq 2U\left[\frac{V(\mathbf{x})}{M(\mathbf{x})} - M(\mathbf{x})\right].$$

Then

$$\dot{M}(\mathbf{x}) \leq -U$$

and

$$M \leq M(t_0) - U(t - t_0)$$

for

$$t_0 \leq t \leq t_0 + \frac{M(t_0)}{U}.$$

Consequently, the total angular momentum at t_0, $M(t_0)$, is transferred to the origin in finite time T such that

$$T \leq \frac{M(t_0)}{U}. \tag{6.45}$$

Since

$$M(t_0) = \left| \left\{ \sum_{i=1}^{3} [I_i^2 x_i(t_0)] \right\}^{1/2} \right|,$$

each $\mathbf{x}(t_0)$ corresponds to an $M(t_0)$ from which (6.45) provides an estimate of the upper bound on the time to reach equilibrium. If, for example, $x_1(t_0) = x_{10}$, $x_2(t_0) = x_3(t_0) = 0$, $M(t_0) = I_1 x_{10}$, and (6.45) yields

$$T \leq \frac{I_1 x_{10}}{U}.$$

This checks the physical reasoning that impulse, UT, equal to change in momentum, $I_1 x_{10}$, is required to bring the satellite to equilibrium. In this case (6.45) is a bit conservative in that the smallest $U_i = U$, $i = 1, 2, 3$, is used. In reality, small-duration impulsive corrections might be used so that the total impulse,

$$\sum_{i=1}^{3} U_i T_i = M(t_0),$$

in general, where T_i is the total duration of the pulses from the ith control.

6.2.2 Control Design

Here, consider the externally controlled, nonlinear time-invariant system

$$\frac{d\mathbf{x}}{dt} = \mathbf{f}(\mathbf{x}, \mathbf{u}), \tag{6.46}$$

where $\mathbf{u} \in \mathbb{R}^m$ is a control vector. As in the previous sections, assume that $V(\mathbf{x})$ is a Liapunov function for the autonomous nonlinear system (6.9) [i.e., (6.46) with $\mathbf{u} = \mathbf{0}$]. Furthermore, assume that the control objective is to drive $\mathbf{x}(t)$ for (6.46) to $\mathbf{0}$, and that the unforced system (6.9) is asymptotically stable at $\mathbf{0}$ with $\dot{V} < 0$.

It was shown in the preceding section that for a given $V(\mathbf{x})$, a maximum $-\dot{V}(\mathbf{x})$ for (6.9) would tend to minimize the system response to the origin. Therefore, it is assumed here that the objective of the control $\mathbf{u}(t)$ is to minimize $\dot{V}(\mathbf{x}, \mathbf{u})$ or to maximize $-\dot{V}(\mathbf{x}, \mathbf{u})$ for (6.46). Indeed, this was the policy that was used successfully in Example 6.12.

For linear systems the procedure is straightforward from the existence of a positive-definite quadratic form for a Liapunov function. For this purpose consider again the controlled linear system with

$$\frac{d\mathbf{x}}{dt} = \mathbf{Ax} + \mathbf{Bu}, \tag{6.47}$$

where $\mathbf{A}$, $\mathbf{B}$ are constant matrices of appropriate dimension, $n \times n$ and $n \times m$, respectively. Again, assume a Liapunov function

$$V(\mathbf{x}) = \mathbf{x}^T\mathbf{Qx}, \tag{6.8}$$

so that, from (6.47),

$$\begin{aligned} \dot{V}(\mathbf{x}, \mathbf{u}) &= 2\mathbf{x}^T\mathbf{Q}\dot{\mathbf{x}}, \\ &= 2\mathbf{x}^T\mathbf{QAx} + 2\mathbf{x}^T\mathbf{QBu}, \end{aligned}$$

or

$$\dot{V}(\mathbf{x}, \mathbf{u}) = -\mathbf{x}^T\mathbf{Px} + 2\mathbf{u}^T\mathbf{B}^T\mathbf{Qx}. \tag{6.48}$$

As before, the procedure is arbitrarily to select a symmetric positive-definite $\mathbf{Q}$, and then compute $\mathbf{P}$ according to (6.14).

Now, suppose that $\mathbf{u}$ is constrained so that

$$|u_k(t)| \leq U_k,$$

$k = 1, \ldots, m$. Then, to maximize $-\dot{V}(\mathbf{x}, \mathbf{u})$ [i.e., to minimize (6.48)], the control takes piecewise-constant, maximum allowable values as in Example 6.12. This so-called bang-bang control takes the following form:

$$u_k(t) = -U_k \operatorname{sgn}(\mathbf{B}^T\mathbf{Qx})_k, \tag{6.49}$$

where $(\cdot)_k$ denotes the kth component of the $(m \times 1)$ vector $\mathbf{B}^T\mathbf{Qx}$, and as before, sgn denotes the signum function.

Example 6.13

Consider the synchronous-machine Example 6.11 with control added to the assumed constant input power p_i. Then (6.28) becomes

$$\begin{aligned} \dot{x}_1 &= x_2, \\ \dot{x}_2 &= u - 0.279x_2 + f(x_1). \end{aligned}$$

With $V(\mathbf{x}) = 3.1x_1^2 + 1.67x_1x_2 + 4.78x_2^2$ for Aizerman's method, $\dot{V}(\mathbf{x}, \mathbf{u}) = \nabla^T(\mathbf{x})\mathbf{f}(\mathbf{x}, \mathbf{u})$ from (6.50) becomes

$$\dot{V}(\mathbf{x}, \mathbf{u}) = 1.67k(x_1)x_1^2 - g(x_1)x_1x_2 - x_2^2 + (1.67x_1 + 9.56x_2)u. \qquad (6.50)$$

Therefore, if negative p_i is admissible, by say mechanical, thermal, or electrical means, with symmetric constraints for convenience,

$$|u| \leq U.$$

Then (6.50) is minimized by

$$u = -U \text{ sgn } S(\mathbf{x}),$$

where $S(\mathbf{x}) = 1.67x_1 + 9.56x_2$.

If the control is constrained, say, to $\alpha \leq u \leq U$, (6.50) is minimized by

$$u = \begin{cases} \alpha & \text{for } S(\mathbf{x}) > 0, \\ U & \text{for } S(\mathbf{x}) < 0. \end{cases}$$

If negative p_i is not admissible, $\alpha = -0.3$ would be the minimum allowable $u(t)$ according to (6.28). It should be noted, however, that similar power dissipation by a so-called braking resistor has been utilized successfully by the Bonneville Power Administration for fault "correction" [M14].

The foregoing scheme of control generation to drive the state to the origin rapidly by maximizing $-\dot{V}(\mathbf{x}, u)$ for a given $V(\mathbf{x})$ can be useful for adaptive control as shown in Fig. 6.11. Here it is desired that the plant (controlled system) state $\mathbf{x}(t)$ behaves according to some desired system state $\mathbf{z}(t)$ relative to a reference input $\mathbf{r}(t)$. An error is generated and driven to zero by the method described above. Such controllers are especially useful for systems with some uncertainty due to changing plant parameters or unknown disturbances. While the process shows promise for nonlinear systems, its success has been demonstrated primarily for linear systems.

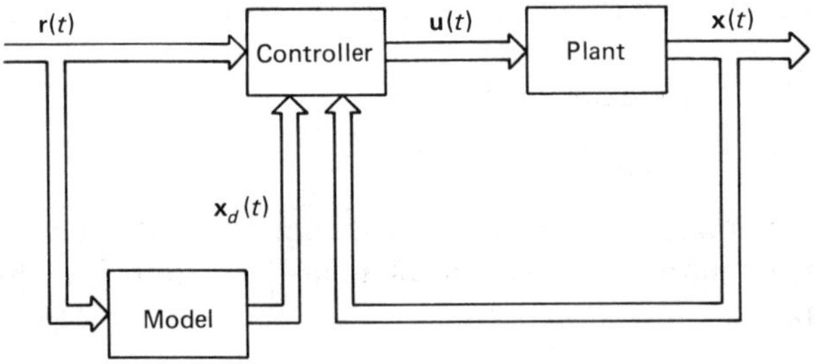

Figure 6.11 Model-reference adaptive controller.

Example 6.14

Consider the following time-variant linear plant from [G6]:

$$\dot{\mathbf{x}} = \mathbf{A}(t)\mathbf{x} + \mathbf{B}(t)\mathbf{u},$$

with reference model

$$\dot{\mathbf{z}} = \mathbf{A}_d\mathbf{z} + \mathbf{B}_d\mathbf{r},$$

where $\mathbf{A}_d$, $\mathbf{B}_d$ are appropriate constant reference matrices. The two systems are of the same order n. The reference model is asymptotically stable with $\mathbf{r} = \mathbf{0}$. Then there exists a p.d. Liapunov function

$$V_d(\mathbf{z}) = \mathbf{z}^T\mathbf{Q}\mathbf{z},$$

such that

$$\dot{V}_d(\mathbf{z}) = \mathbf{z}^T(\mathbf{A}_d^T\mathbf{Q} + \mathbf{Q}\mathbf{A}_d)\mathbf{z} = -\mathbf{z}^T\mathbf{P}\mathbf{z}$$

or

$$-\mathbf{P} = \mathbf{A}_d^T\mathbf{Q} + \mathbf{Q}\mathbf{A}_d,$$

where $\mathbf{P}$ is p.d. and $\dot{V}_d(\cdot)$ is n.d.

If $\mathbf{e} = \mathbf{z} - \mathbf{x}$, $\Delta\mathbf{A} = \mathbf{A}(t) - \mathbf{A}_d(t)$, and $\Delta\mathbf{B} = \mathbf{B}(t) - \mathbf{B}_d$,

$$\dot{\mathbf{e}} = \mathbf{A}_d\mathbf{e} - \Delta\mathbf{A}\mathbf{x} - \mathbf{B}(t)\mathbf{u} + \mathbf{B}_d\mathbf{r}.$$

Then let $V(\mathbf{e}) = \mathbf{e}^T\mathbf{Q}\mathbf{e}$, so that

$$\dot{V}(t;\, \mathbf{e},\, \boldsymbol{u}) = -\mathbf{e}^T\mathbf{P}\mathbf{e} - 2\mathbf{e}^T\mathbf{Q}[\Delta\mathbf{A}\mathbf{x} + \Delta\mathbf{B}\mathbf{r} + \mathbf{B}(\mathbf{u} - \mathbf{r})].$$

Now with $\mathbf{P}$ p.d., $\mathbf{u}(t)$ is selected so that

$$\mathbf{e}^T(t)\mathbf{Q}\{\Delta\mathbf{A}(t)\mathbf{x}(t) + \Delta\mathbf{B}(t)\mathbf{r}(t) + \mathbf{B}(t)[\mathbf{u}(t) - \mathbf{r}(t)]\}\begin{cases} = 0, & \Delta\mathbf{A} = \Delta\mathbf{B} = \mathbf{0}, \\ \geq 0, & \text{otherwise.}\end{cases}$$

6.3 SINGULAR PERTURBATION

The Van der Pol equation (5.26) is a good example of a system that changes its response drastically over the range of parameters μ. If $\mu \to 0$, the system resembles a linear system with a phase-plane portrait approaching that of a center point. Here there is a continuum of closed trajectories. On the other hand, for finite μ there exist unique limit cycles. If $\mu \to \infty$, the phase trajectories become discontinuous. Such discontinuities are typical of systems with an insufficient number of state variables. Indeed, as $\mu \to \infty$, (5.28) approaches a first-order system which naturally exhibits jump solutions in a phase plane. This is shown in some detail in [A3].

In general, consider the system of equations

$$\frac{d\mathbf{x}^1}{dt} = \mathbf{f}^1(\mathbf{x}^1, \mathbf{x}^2), \tag{6.51}$$

$$\epsilon\frac{d\mathbf{x}^2}{dt} = \mathbf{f}^2(\mathbf{x}^1, \mathbf{x}^2), \tag{6.52}$$

where $\mathbf{x}^1 \in \mathbb{R}^q$ and $\mathbf{x}^2 \in \mathbb{R}^{n-q}$. For any $\epsilon \neq 0$, (6.51)–(6.52) consists of n differential equations with system order of n. But if $\epsilon = 0$, the system order is only q. Sometimes a small ϵ can suggest a system order reduction of this type. Unfortunately, one must be careful, as the following analysis shows.

If it is assumed that $\epsilon = 0$, (6.52) reduces to

$$\mathbf{f}^2(\mathbf{x}^1, \mathbf{x}^2) = \mathbf{0}, \tag{6.53}$$

which may have a solution $\mathbf{x}^2(t) = \mathbf{x}^2(\mathbf{x}^1(t))$. Upon substitution into (6.51),

$$\frac{d\mathbf{x}^1}{dt} = \mathbf{f}^1(\mathbf{x}^1, \mathbf{x}^2(\mathbf{x}^1)), \tag{6.54}$$

which may drastically alter the character of the system. For this reason, $\epsilon = 0$ in (6.52) is referred to as a singular perturbation. Elements that may cause such structural changes in a physical system are said to be parasitic.

Example 6.15 [S5]

Consider the electrical circuit shown in Fig. 6.12(a) with the current–voltage characteristic for the diode shown in Fig. 6.13. Note the parasitic capacitor ϵ and its absence in the simplified circuit in Fig. 6.12(b). The general model is given by

$$\begin{aligned} \dot{x}_1 &= -x_1 + x_2 + E, \\ \epsilon\dot{x}_2 &= x_1 - g(x_2), \end{aligned} \tag{6.55}$$

where $x_1 = i$, the primary loop current, x_2 is the voltage across the parasitic capacitor, $g(x_2)$ represents the diode characteristic shown in Fig. 6.13, and E is an externally applied voltage. Here three possible operating equilibrium points are noted by a, b, c. Linearization about point b yields

$$\begin{aligned} \delta\dot{x}_1 &= -\delta x_1 - \delta x_2, \\ \epsilon\delta\dot{x}_2 &= \delta x_1 - g'(x_{2be})\,\delta x_2, \end{aligned} \tag{6.56}$$

where $\delta\mathbf{x} = \mathbf{x} - \mathbf{x}_{be}$, and

$$g'(x_{2e}) = \frac{dg}{dx_2}(x_{2be}).$$

Then the characteristic equation for the linearized system matrix is

$$\lambda^2 + \lambda\left[1 + \frac{g'(x_{2be})}{\epsilon}\right] + \frac{g'(x_{2be}) + 1}{\epsilon} = 0.$$

As shown in Fig. 6.13, $g'(x_{2be})$ is negative and, in fact,

$$g'(x_{2be}) < -1.$$

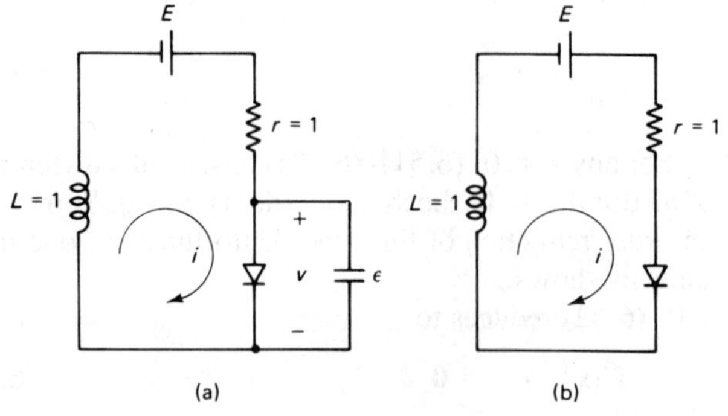

Figure 6.12 Parasitic electrical circuit for Example 6.15. (After Aggarwal [A3].)

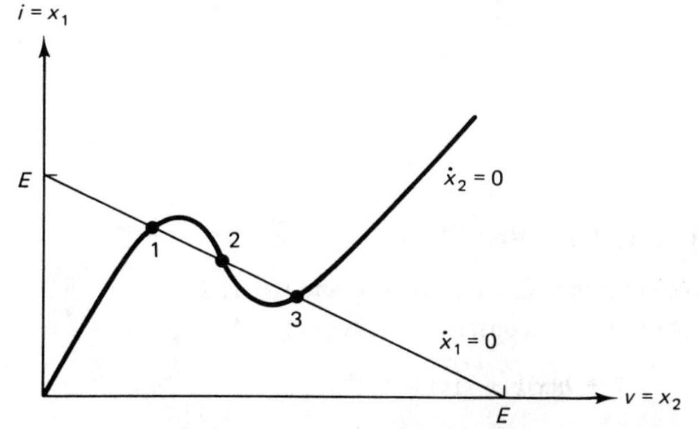

Figure 6.13 Diode characteristic for Example 6.15. (After Aggarwal [A3].)

Consequently, for sufficiently negative ϵ, $\mathbf{x}_{be}$ is unstable.

Without the parasitic capacitance

$$\dot{x}_1 = -x_1 - x_2 + E,$$

$$x_2 = g^{-1}(x_1).$$

On linearization at $\mathbf{x}_{be}$, $\delta x_1 = g'(x_{2be})\ \delta x_2$, and

$$\delta \dot{x}_1 = -\left[1 + \frac{1}{g'(x_{2be})}\right] \delta x_1,$$

which is stable if $g'(\cdot) < -1$, and of course, only of first order.

Consider the linear system

$$\begin{aligned} \frac{d\mathbf{x}^1}{dt} &= \mathbf{A}^1\mathbf{x}^1 + \mathbf{A}^{12}\mathbf{x}^2, \\ \epsilon \frac{d\mathbf{x}^2}{dt} &= \mathbf{A}^{21}\mathbf{x}^1 + \mathbf{A}^2\mathbf{x}^2, \end{aligned} \tag{6.57}$$

where again $\mathbf{x}^1$, $\mathbf{x}^2$ are q and $n - q$ vectors, respectively, and $\mathbf{A}^1$, $\mathbf{A}^2$, $\mathbf{A}^{12}$, $\mathbf{A}^{21}$ are appropriate constant matrices. With $\epsilon = 0$, $\mathbf{x}^2 = -(\mathbf{A}^2)^{-1}\mathbf{A}^{21}\mathbf{x}^1$, so that

$$\frac{d\mathbf{x}^1}{dt} = [\mathbf{A}^1 \quad -\mathbf{A}^{12}(\mathbf{A}^2)^{-1}\mathbf{A}^{21}]\mathbf{x}^1.$$

It is possible to deduce the stability of the zero equilibrium state for (6.57) by the following theorem [S5]: If the eigenvalues of $\mathbf{A}^2$ and of $[\mathbf{A}^1 - \mathbf{A}^{12}(\mathbf{A}^2)^{-1}\mathbf{A}^{21}]$ all have negative real parts, there exists an $\epsilon_0 > 0$ such that for every $\epsilon \in (0, \epsilon_0)$, the $\mathbf{0}$ of (6.57) is asymptotically stable. Although this result is not surprising, the proof does involve a little algebra, as given in [S5].

EXERCISES

6.1. Analyze the stability of the origin of the following nonlinear systems by Liapunov's second method. (See Exercise 3.2 for an analysis by the first method.) For cases of asymptotic stability, estimate the region of attraction.

(a) $\dot{x}_1 = -x_1^3 - 2x_2,$
$\dot{x}_2 = 2x_1 - 5x_2^3.$
(b) $\dot{x}_1 = x_1^2 + x_2,$
$\dot{x}_2 = x_1 + x_2^2.$
(c) $\dot{x}_1 = x_1^2 - x_2^2,$
$\dot{x}_2 = -x_1x_2.$
(d) $\dot{x}_1 = x_2,$
$\dot{x}_2 = a_1x_1 + a_2x_2 + (b_1x_1 + b_2x_2)^2x_2,$

where a_i are negative constants, b_i are constants, and $i = 1, 2$.

6.2. Consider Lienard's equation, the nonconservative system (5.24),

$$\ddot{x} + h(x)\dot{x} + f(x) = 0,$$

where $f(0) = 0$. Let $x_1 = x$, $x_2 = \dot{x} + \int_0^x f(\sigma)\, d\sigma$. Then estimate the domain of attraction for the zero solution and the possibility of limit cycles by solution of a Liapunov function.

6.3. Consider the following bilinear system with quadratic state feedback:

$$\dot{x}_1 = -x_1 + ux_1,$$

$$\dot{x}_2 = -x_2 + ux_2,$$

where $u = x_1^2 + x_2^2$.

(a) Compute the equilibrium states of the closed-loop system.
(b) Analyze the stability of the origin by means of a quadratic Liapunov function and thereby generate its region of attraction.
(c) Sketch the state-plane portrait.

6.4. Analyze the stability of the equilibrium state by Liapunov's direct method as dependent on $a_1(t)$ and $a_2(t)$ for

$$\dot{x}_1 = a_1(t)x_2 + a_2(t)x_1(x_1^2 + x_2^2),$$

$$\dot{x}_2 = -a_1(t)x_1 + a_2(t)x_2(x_1^2 + x_2^2).$$

6.5. For Exercise 3.4, compute a Liapunov function and show the region of attraction.

6.6. By the second method of Liapunov, analyze the stability of the origin for Exercise 6.4 with $a_1(t) = 1$ and $a_2(t) = a$, a constant. This system may be considered a harmonic oscillator with nonlinear damping.

6.7. Consider the simple first-order system $\dot{x} = x^2$ with $x(0) = 1$.
(a) Is the equilibrium stable?
(b) Does $x(t)$ exhibit a finite escape time?
(c) Compute $x(t)$.

6.8. Consider an R-L-C series network with

$$\dot{x}_1 = L^{-1}(t)x_2,$$

$$\dot{x}_2 = -C^{-1}(t)x_1 - R(t)L^{-1}(t)x_2,$$

where x_1 is the capacitor charge and $x_2 = L(t)i_2$, the inductor flux.
(a) Analyze stability of the network by means of an appropriate Liapunov function.
(b) Discuss constraints that are imposed on the network elements to guarantee stability.

6.9. Analyze stability of the equilibrium points of the nonlinear system with

$$\dot{x}_1 = x_2,$$

$$\dot{x}_2 = a_1 x_1 + a_2 x_2 - x_1^2,$$

where a_1, a_2 are nonpositive. Estimate the region of attraction of any asymptotically stable equilibria when $a_1 = -2$, $a_2 = -0.25$.

6.10. Consider the so-called gradient system

$$\dot{x}_i = f_i(\mathbf{x}) = -\frac{\partial \phi}{\partial x_i},$$

where $i = 1, \ldots, n$; $\mathbf{x} \in \mathbb{R}^n$, and $\partial f_i / \partial x_j = \partial f_j / \partial x_i$, for all $i, j = 1, \ldots, n$. Assume that $\phi(\mathbf{x})$ is a minimum at equilibria $\mathbf{x}_e$ such that the Hessian matrix of ϕ evaluated at $\mathbf{x}_e$ is nonsingular [L8].

(a) Show that $V(\mathbf{x}) - V(\mathbf{x}_e)$ is a Liapunov function, where $V(\mathbf{x}) = \phi(\mathbf{x})$.

(b) Note the special properties of these systems.

(c) Analyze the following system for global stability of $\mathbf{0}$:

$$\dot{x}_1 = -2x_1(1 + x_2^2),$$

$$\dot{x}_2 = -2x_2(1 + x_1^2).$$

6.11. Consider a representation of Lienard's equation (5.24) that is given by

$$\dot{x} = y - H(x),$$

$$\dot{y} = -f(x),$$

where $H(x) = \int_0^x h(\sigma)\, d\sigma$; $h(\cdot)$ is an even function, and $f(\cdot)$ is a monotone increasing function with $f(0) = 0$.

(a) Show that $V = y^2/2 + F(x)$, is an appropriate Liapunov function with $F(x) = \int_0^x f(\sigma)\, d\sigma$.

(b) Apply this result to the global stability analysis of $\mathbf{0}$ for the system

$$\ddot{x} + \dot{x} + x^3 = 0.$$

6.12. Show that condition (c) for p.d. functions (Section 6.1.2) is equivalent to $V(\mathbf{x}) \geq \omega(r)$ for all $\mathbf{x} \in D$, where $r = \|\mathbf{x}\|$, $\omega(r)$ is continuous with $\omega(0) = 0$, and $\omega(r)$ is a strictly increasing function.

6.13 Determine a region of attraction to the origin for the following system:

$$\dot{x}_1 = -4x_1 + x_2 + 4x_1^2,$$

$$\dot{x}_2 = x_1 - 4x_2 - x_1^2.$$

Use Liapunov's direct method.

CHAPTER

7

NONLINEAR STABILITY: FREQUENCY-DOMAIN ANALYSIS

In this chapter, stability of a feedback regulator is studied in more detail. First, the concept of absolute stability for a range of nonlinear feedback gains is introduced for the Luré regulator problem. Then a frequency-domain circle criterion is presented which resembles the classical Nyquist stability criterion. Whereas this method generally provides a conservative estimate of stability, the describing function method, which is presented last, can be a bit optimistic in its stability forecasts. On the other hand, describing functions are intuitively satisfying and easy to apply.

7.1 LURÉ REGULATOR PROBLEM

The basic problem, which is considered here, is the stability of the simple, scalar, feedback regulator given by Fig. 7.1. $G(s)$ is the transfer function for the linear part of the system. $\phi(\cdot)$ represents a nonlinear gain term. That is, $\phi(e)$ is a nonlinear operation on its input; it is not time variant and is nondynamical in nature. Recall that Example 6.11 [equation (6.29)] resulted in a system characterization of this kind. The problem is to find the bounds, that is, constant gains (slopes) k_1 and k_2 such that the system is stable for $0 \leq k_1 \leq \phi(e)/e \leq k_2$ (i.e., $\phi \in [k_1, k_2]$).

In the time domain, the system in Fig. 7.1 is described by

$$\frac{d\mathbf{x}}{dt} = \mathbf{A}\mathbf{x} + \mathbf{b}\phi(e), \tag{7.1}$$

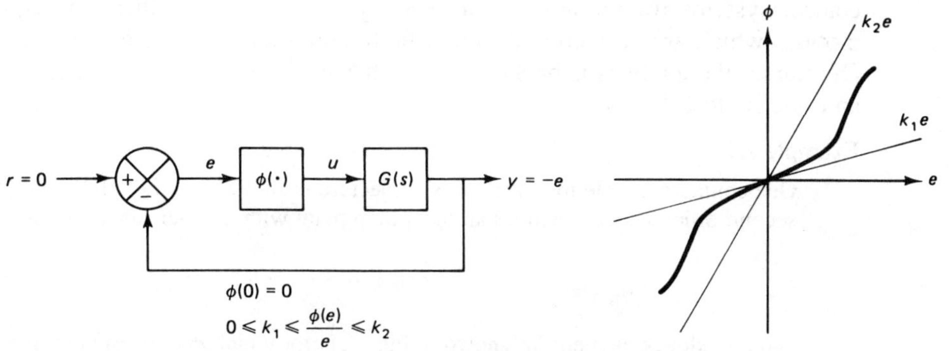

Figure 7.1 Feedback regulator with nonlinear gain characteristic.

where $\mathbf{A}$ is a stable matrix and

$$-e = y = \mathbf{c}^{\mathrm{T}}\mathbf{x} + du.$$

Here $\phi(0) = 0$ and $e\phi(e) \geq 0$. $\mathbf{b}$, $\mathbf{c}$, and d are appropriate constant vectors or scalars. Then the linear transfer function in Fig. 7.1 is

$$G(s) = \mathbf{c}^{\mathrm{T}}[\mathbf{I}s - \mathbf{A}]^{-1}\mathbf{b} + d. \tag{7.2}$$

As noted above, $G(s)$ has no poles in the open right-half plane.

This system does have physical significance. In many cases, the nonlinear characteristic may be separated from the system dynamics in this way; and more general multiple loop indirect control systems can be put into the form of Fig. 7.1 for stability analysis. Motivated by such applications, systems of this type have been studied since as early as the 1940s when *Aizerman* made his *conjecture*. That is, if the origin for system (7.1) is stable for all linear system approximations with corresponding constant gains k such that $k_1 \leq k \leq k_2$, the nonlinear system origin would be stable. Obviously, this would be a convenient result, in that linear system theory would be useful to ascertain the nonlinear system stability in the large. Unfortunately, the conjecture was shown to be invalid.

Then, Kalman conjectured in 1957 that if the system is stable for all linear approximations $d\phi(e)/de$ such that

$$k_0 \leq \frac{d\phi(e)}{de} \leq k_3,$$

the nonlinear system (7.1) origin would be stable. It is readily seen that for this entire class of nonlinearities $k_0 \leq k_1 \leq k_2 \leq k_3$, so that the Kalman conjecture is stronger than the Aizerman conjecture. Again, however, the conjecture was proven false.

Despite their failures, such conjectures may be useful when used with caution along with other methodologies and computer simulations. Indeed, similar conjectures have been used since the 1940s by control system designers in conjunction with nonlinear plants. Although involving a more complex system, stability of aircraft control systems traditionally is checked by a study of the linearized equations of motion, which are derived at different flight conditions over some desired envelope. Of course, the analysis is backed up by other methodologies, computer simulations, and finally, test flights.

Example 7.1

One counterexample to Aizerman's conjecture is studied in [H9], [F3], and [W3]. This second-order system includes an open-loop plant with transfer function

$$G(s) = \frac{s+1}{s^2}$$

and a feedback element (as shown in Fig. 7.1) for which $\phi(e = -y)$ becomes

$$\phi(-y) = \frac{-y}{[\exp(1) + 1]\exp(1)} \qquad \text{for } y \le 1$$

and

$$\phi(-y) = \frac{-\exp(-y)}{1 + \exp(y)} \qquad \text{for } y \ge 1.$$

Then the closed-loop system is described by $\ddot{y} = u + \dot{u}$ with $u = \phi(-y)$, or after some algebra by

$$\ddot{y} - \phi(-\dot{y} - y) = 0.$$

It is readily seen that by appropriate selection of initial conditions, $y(t)$ wil be greater than -1 and as $t \to \infty$, $\lim y(t) = \infty$. For example, if the initial conditions are given by

$$y(0) = 1 \;\; -\exp(-1)$$

and

$$\dot{y}(0) = \exp(-1),$$

then

$$\dot{y}(t) = \exp[-y(t) - \dot{y}(t)].$$

Consequently, $y(t)$ increases for all t, is unbounded, and does not converge to zero. Obviously, the zero equilibrium state is not stable in the large. Since $\phi \in (0, k_2)$, where

$$k_2 = \frac{1}{[\exp(1) + 1]\exp(1)} < 1,$$

it is seen that Aizerman's conjecture (from Nyquist's stability criterion) would falsely indicate stability.

It should be noted here that most references on the Luré problem and the Popov method (developed in Section 7.2) assume that the nonlinear element $\phi'(y)$ operates directly on the output y prior to the sign change of the regulator. That is, $\phi'(\cdot)$ is in the feedback rather than the forward path. In most control theory and practice, however, such nonlinear elements appear in the forward path as shown by Figs. 7.1 and 7.2. It is seen, however, for the first- and third-quadrant class of nonlinearities which are

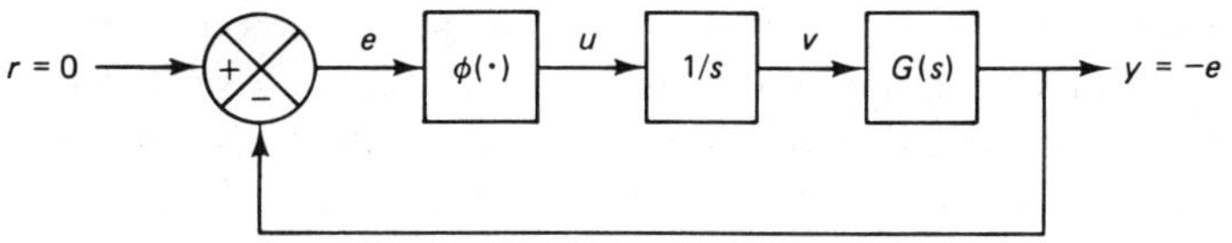

Figure 7.2 Feedback regulator structure for Popov criterion.

assumed here, $\phi(e)$ with $e = -y$ is identical, with the sign inversion on y, to $-\phi'(y)$, so that in both cases $\phi \in (k_1, k_2)$ would be the open sector of interest. In the example above, the equivalent closed-loop equation with $\phi'(y)$ in the feedback branch would become

$$\ddot{y} + \phi'(\dot{y} + y) = 0.$$

If $G(s) = N(s)/D(s)$ is a ratio of polynomials with the order of $D(\cdot)$ no less than the order of $N(s)$, then Fig. 7.1 can be represented by the equation

$$D(p)y(t) = N(p)[\phi(-y)],$$

where the operator $p \equiv d(\cdot)/dt$. Alternatively, with $\phi'(y)$, the closed-loop regulator model in this form becomes

$$D(p)y(t) = N(p)[-\phi'(y)].$$

Throughout this book the nonlinearity is presented as shown in Figs. 7.1 and 7.2, which is more consistent with its role in practice.

7.1.1 Absolute Stability

If the origin for the system (7.1) and Fig. 7.1 is globally asymptotically stable for all gains on the sector (k_1, k_2), the system is said to be *absolutely stable,* or stable on the sector (k_1, k_2). Although it is not always stated, the results presented here use the following assumptions:

1. $\phi(e)$ satisfies a global Lipschitz condition so that the solution to (7.1) is unique.
2. Either (a) $\mathbf{A}$ is Hurwitz [i.e., it has all eigenvalues with negative real parts as assumed with (7.1)], or (b) $\mathbf{A}$ has all eigenvalues with nonpositive real parts but with only a simple zero eigenvalue, as used in the next section.
3. $(\mathbf{A}, \mathbf{b})$ is controllable.
4. $(\mathbf{A}, \mathbf{c})$ is observable.
5. As stated above, $\phi(0) = 0$, $e\phi(e) \geq 0$.

7.2 POPOV METHOD

Numerous researchers, starting with Aizerman [A2] and Luré and Postnikov ([L9] and [L10]) in the USSR, followed by Popov [P3] in Romania and Lefschetz [L11], LaSalle [L12], and Kalman [K5] in the United States, pioneered the investigation of this regulator problem. Obviously, the results are numerous with various specific conditions and can be quite confusing to the reader. Consequently, only the most basic criterion is presented in Section 7.2.1. Also, this is the simplest to apply. A slightly more general circle criterion is presented in Section 7.2.2. Then several generalizations are studied in Section 7.2.4.

7.2.1 Basic Stability Criteria

To make the problem slightly more general than that associated with the autonomous nonlinear system (7.1), allow the addition of a free integrator [assumption 2(b)] so that $u = dv/dt$. Now (7.1) can be written as shown in Fig. 7.2 with $\phi(e) = \phi(-y)$, so that

$$\frac{d\mathbf{x}}{dt} = \mathbf{Ax} + \mathbf{b}\phi(e),$$
$$-e = y = \mathbf{c}^T\mathbf{x} + dv, \tag{7.3}$$
$$\frac{dv}{dt} = \phi(e).$$

Here

$$G(s) = \mathbf{c}^T[\mathbf{I}s - \mathbf{A}]^{-1}\mathbf{b} + \frac{d}{s}. \tag{7.4}$$

However, by simply defining an addended state variable, v, (7.4) is basically the same as (7.1) with

$$\begin{bmatrix} \dot{\mathbf{x}} \\ \dot{v} \end{bmatrix} = \begin{bmatrix} \mathbf{A} & \mathbf{0} \\ \mathbf{0} & 0 \end{bmatrix}\begin{bmatrix} \mathbf{x} \\ v \end{bmatrix} + \begin{bmatrix} \mathbf{b} \\ 1 \end{bmatrix}u, \tag{7.5}$$

$$y(t) = [\mathbf{c}^T \quad d]\begin{bmatrix} \mathbf{x} \\ \mathrm{V} \end{bmatrix}$$

and

$$u(t) = \phi(-y(t)).$$

Now (7.5) puts (7.4) into the structure of (7.1) with assumption 2; that is, a simple zero eigenvalue is admitted, with all the others in the open left-half plane. Then, since $\phi \in (0, \infty)$, it can readily be shown that it is necessary d be positive for (7.4) to be absolutely stable [A2].

The central *Popov theorem* for our interst here now may be stated as follows: System (7.3) is absolutely stable for all nonlinearities $\phi(e) \in (0, K)$ if there exists a real number q such that

$$\text{Re}\,[(1 + j\omega q)G(j\omega)] + \frac{1}{K} > 0 \tag{7.6}$$

for all real ω. The proof is somewhat lengthy and is given in a very readable form in [A2], [L10], [H1], and [V1]. Only a brief outline of the proof is given here, which follows [V1]. Popov used a Liapunov function of the following form to prove this crucial theorem:

$$V = \mathbf{x}^{\mathrm{T}}\mathbf{Q}\mathbf{x} + \alpha' v^2 + \beta \int_0^e \phi(-\sigma)\, d\sigma, \tag{7.7}$$

where $\mathbf{Q}$ is a p.d. matrix, $\alpha' \geq 0$, and $\beta \geq 0$. Here we can let $\alpha' v^2 = (\alpha/d^2)(y - \mathbf{c}^{\mathrm{T}}\mathbf{x})^2$. The proof can be shortened somewhat by a result of Lefschetz [L11]. Here we are given a stable matrix $\mathbf{A}$ and a vector $\mathbf{b}$ such that $(\mathbf{A}, \mathbf{b})$ is a controllable pair. Also, we are given a real vector $\mathbf{w}$, scalars $\gamma \geq 0$, $\epsilon > 0$, a p.d. matrix $\mathbf{P}$, and a vector $\mathbf{p}$ which satisfy

$$\mathbf{A}^{\mathrm{T}}\mathbf{Q} + \mathbf{Q}\mathbf{A} = -\mathbf{p}\mathbf{p}^{\mathrm{T}} - \epsilon\mathbf{P},$$

$$\mathbf{Q}\mathbf{b} - \mathbf{w} = \gamma^{1/2}\mathbf{p}$$

if and only if ϵ is small enough, and the scalar function $h(s) \equiv \gamma + 2\mathbf{w}^{\mathrm{T}}[\mathbf{I}s - \mathbf{A}]^{-1}\mathbf{b}$ satisfies

$$\mathrm{Re}\,[h(j\omega)] > 0 \tag{7.8}$$

for all real ω.

With regard to (7.4) and (7.7),

$$\begin{aligned} \mathbf{w} &\equiv \tfrac{1}{2}\beta\mathbf{A}^{\mathrm{T}}\mathbf{c} + \alpha\, d\mathbf{c}, \\ \gamma &\equiv \beta(\mathbf{c}^{\mathrm{T}}\mathbf{b} + d). \end{aligned} \tag{7.9}$$

This result is a convenient representation of the Kalman–Yakubovich lemma ([K5] and [V1]). With $q = \beta/(2\alpha d)$ and $K = \infty$, (7.6) is in a form amenable to this lemma. Thus

$$\beta d + \mathrm{Re}\,[(2\alpha d + j\omega\beta)\mathbf{c}^{\mathrm{T}}(j\omega\mathbf{I} - \mathbf{A})^{-1}\mathbf{b}] > 0$$

for all ω. The crucial relation that evolves is given by

$$\gamma + 2\,\mathrm{Re}\,[\mathbf{w}^{\mathrm{T}}[j\omega\mathbf{I} - \mathbf{A}]^{-1}\mathbf{b}] > 0 \tag{7.10}$$

for all real ω. Consequently, it follows that

$$\dot{V} = -\epsilon\mathbf{x}^{\mathrm{T}}\mathbf{P}\mathbf{x} - (\mathbf{x}^{\mathrm{T}}\mathbf{p} + \gamma^{1/2}\phi)^2 - 2\alpha\, d\phi\, e \tag{7.11}$$

which is n.d. in $\mathbf{x}$ and v since $\mathbf{P}$ is p.d. and $\alpha > 0$. Since $e\phi(e) \geq 0$ for all e, $\dot{V}$ is always nonpositive for $\phi \in (0, K)$. It can further be shown that $\dot{V} < 0$ if $\mathbf{x}$ or v are nonzero. Therefore, (7.10) is sufficient for absolute stability. Then substitution of (7.9) into (7.10) yields

$$\mathrm{Re}\,[(1 + j\omega q)G(j\omega)] > 0, \tag{7.12}$$

where $G(j\omega) = d/j\omega + \mathbf{c}^{\mathrm{T}}[j\omega\mathbf{I} + \mathbf{A}]^{-1}\mathbf{b}$ and $q = \beta/2\alpha d$, which completes the proof outline for $K = \infty$.

Now consider Fig. 7.3, which makes a convenient loop transformation. If $\phi \in (0, K)$, then (7.4) with $\phi \in (0, \infty)$ can be transformed to the system of Fig. 7.3, where the new forward transfer function becomes

$$G_1(s) = G(s) + \frac{1}{K}, \tag{7.13}$$

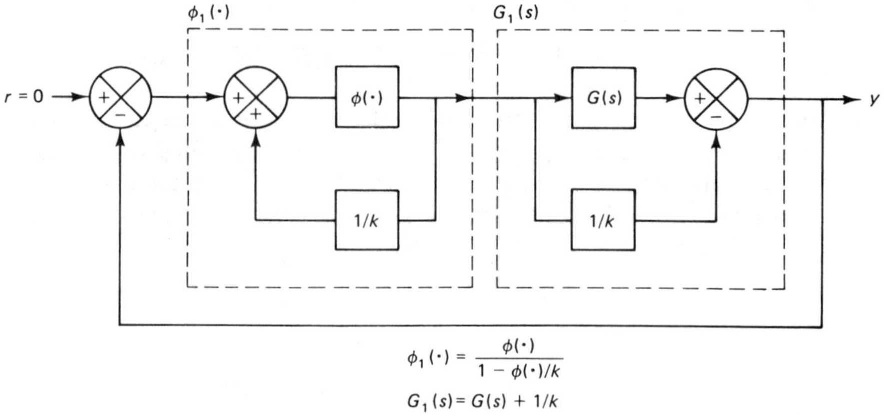

Figure 7.3 Loop transformation for a finite sector.

and the new feedback nonlinearity becomes

$$\phi_1 = \phi\left(1 - \frac{\phi}{K}\right)^{-1}, \tag{7.14}$$

where the inverse is well defined with $\phi \in (0, K)$. Now $\phi_1 \in (0, \infty)$. As a consequence, a sufficient condition for absolute stability of this sytem is that there exists a nonnegative constant q such that

$$\text{Re}\,[(1 + j\omega q)G_1(j\omega)] > 0, \tag{7.15}$$

for all real ω, or that

$$\text{Re}\,[(1 + j\omega q)G(j\omega)] + \frac{1}{K} > 0, \tag{7.6}$$

for all real ω.

While the aguments above only establish a sufficient condition for the existence of a Liapunov function, Popov shows that (7.6) is necessary for the existence of a quadratic Liapunov function (7.7) for system (7.4). However, the reader should be reminded that since the criterion is based on Liapunov's direct method, (7.6) is only a sufficient condition for absolute stability of (7.4). As such, it can be somewhat conservative. However, when combined with linear stability analysis and the conjectures above, the criterion can be quite useful.

Figure 7.4 provides a geometric interpretation of the Popov criterion (7.6). This is seen by rewriting (7.6) as

$$\text{Re}\,G(j\omega) > -\frac{1}{K} + q\omega\ \text{Im}\,G(j\omega), \tag{7.16}$$

where q is a nonnegative real constant. In other words, for each frequency ω, the Nyquist (polar-locus) plot of $G(j\omega)$ must lie to the right of a straight line that is given by

$$\text{Re}\,G(j\omega) = -\frac{1}{K} + q\omega\ \text{Im}\,G(j\omega). \tag{7.17}$$

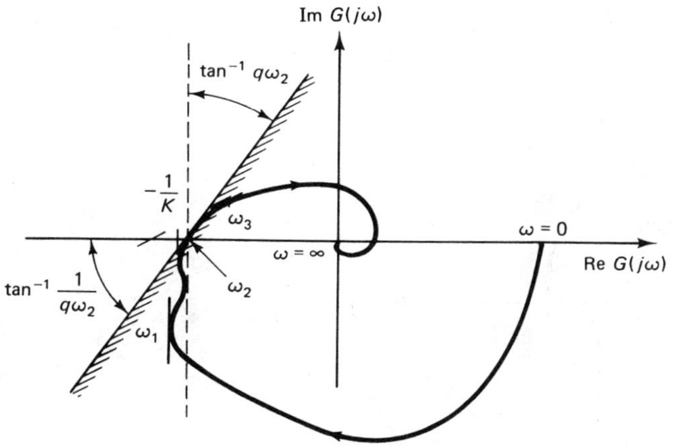

Figure 7.4 Illustration of Popov criterion for $G(j\omega)$.

Unfortunately, as shown in Fig. 7.4, the slope of this line is equal to $1/q\omega$. The line as shown for three frequencies ω_1, ω_2, ω_3 indicates that the system is absolutely stable for $\phi \in (0, K)$.

A simpler representation of Popov's criterion is derived by definition of $\mathbf{G}^*(j\omega)$ such that

$$\text{Re } G^*(j\omega) = \text{Re } G(j\omega)$$

and

$$\text{Im } G^*(j\omega) = \omega \text{ Im } G(j\omega).$$

That is,

$$G^*(j\omega) = \text{Re } G(j\omega) + j\omega \text{ Im } G(j\omega). \tag{7.18}$$

Now the frequency dependence of the line slope is removed, as shown in Fig. 7.5. Thus the *Popov criterion,* (7.6) or (7.16), becomes

$$\text{Re } G^*(j\omega) > -\frac{1}{K} + q \text{ Im } G^*(j\omega), \tag{7.19}$$

or *the polar-locus plot of* $G^*(j\omega)$ *must lie to the right of the straight line of nonnegative slope* $1/q$.

In Fig. 7.5 a case is shown that satisifes Aizerman's conjecture since the Popov line is tangent to the $G^*(j\omega)$ locus at the point where the locus intercepts the negative real axis, and the $G^*(j\omega)$ locus lies completely to the right of the Popov line. That is, K is the maximum admissible value for the linear gain k, such that $\phi(e) = u = ke$. Figure 7.6 shows an example where by the use of Nyquist's stability criterion, Aizerman's conjecture would suggest absolute stability for $k < k_a$. However, the more conservative (but sufficient) Popov criterion shows absolute stability for $\phi(e)/e < K$, where obviously $K < k_a$. Such a situation is seen to arise for

$$G(s) = \frac{s^2}{(s^2 + 1)(s^2 + 9) + \alpha(a_3 s^3 + a_2 s^2 + a_1 s + b)},$$

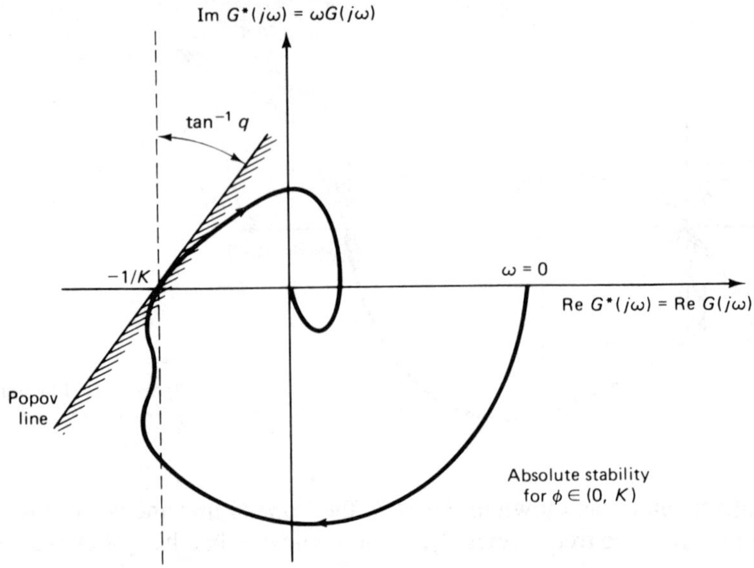

Figure 7.5 Popov criterion for $G^*(j\omega)$.

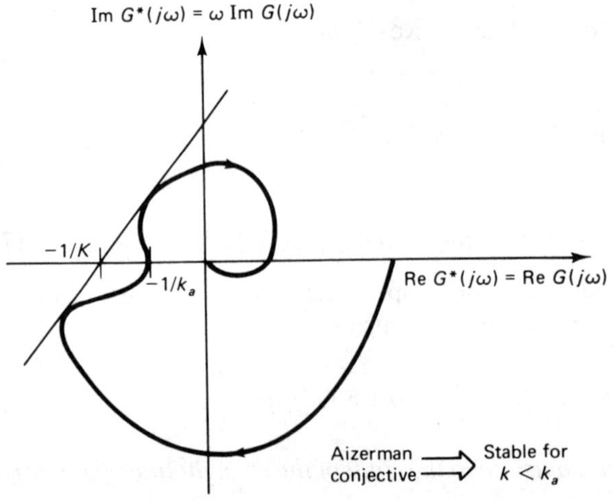

Figure 7.6 Illustration of difference between Popov criterion and Aizerman conjecture.

which is studied in some detail in [H9] and [F3]. It can be shown that if $a_3 = \frac{3}{16}$, $a_2 = \frac{29}{16}$, $a_1 = \frac{15}{16}$, and $b = \frac{117}{16}$, for example, with $\phi(e) = \alpha e^3$ for $|e| \leq 1.88$, Aizerman's conjecture is satisfied for sufficiently small α. However, Popov's criterion is not satisfied.

As shown in [A2], the system (7.1) is absolutely stable for $\phi \in [0, K]$ if the Popov criterion is satisfied for the principal case [i.e., (7.1) with **A** Hurwitz]. It is absolutely stable for $\phi \in [\epsilon, K]$ if the Popov criterion is satisfied for the particular case [i.e., (7.3), where a simple open-loop pole at the origin is allowed]. Here ϵ is an arbitrarily small positive constant. The proof above simplifies somewhat for the principal case with the latter two terms dropped in (7.7) and $1 + Kd > 0$ a necessary stability assumption. Furthermore, it can be shown that the Popov condition for the particular case (7.3) can be weakened as follows [A2]:

$$\text{Re}\,[(1 + j\omega q)G(j\omega)] + \frac{1}{K} \geq 0 \tag{7.20}$$

and

$$G(j\omega) \neq -\frac{1}{K}$$

for all $\omega \geq 0$. Without any loss in generality (7.6) can be written as

$$\text{Re}\,[(1 + j\omega q)G(j\omega)] + \frac{1}{K} \geq \delta > 0, \tag{7.21}$$

where δ is an arbitrarily small positive constant. The latter condition probably is the one most used in practice, where some margin of error is accounted for.

The Popov criterion can be generalized to include a linear plant with imaginary poles. Our discussion of this case follows [A2]. First, $G(s)$ is called a *strictly positive real function* of complex s if for real s, $G(s)$ is real and if in the closed right-half plane ($\text{Re}\ s \geq 0$)

$$\text{Re}\ G(s) > 0. \tag{7.22}$$

It can be shown that $G(s)$ is strictly positive if and only if

1. For real values of s, $G(s)$ is real.
2. $G(s)$ has no poles in the open right-half plane.
3. Any poles on the imaginary axis are simple with positive residues.
4. $\text{Re}\ G(j\omega) > 0$ for all ω except at poles on the imaginary axis.

The index μ of $G(s)$, a rational function of s, is defined as the total number of its poles that are on the imaginary axis. Then it is shown in [A2] that $G(j\omega)$ with index μ satisfies the Popov criterion (7.6) if $W(s) = (1 + qs)G(s)$ is strictly positive (i.e., positive real with positive residues) as follows. Here, in terms of residues,

$$G(s) = \frac{c_0}{s} + \sum_{i=1}^{\mu} \frac{d_i + je_i}{s - j\beta_i} + G_0(s),$$

where all the poles of $G_0(s)$ lie in the open left-half plane. Then the residue of $W(s)$ at the pole $s = 0$ is c_0 and at $s = j\beta_i$, the residues are given by

$$(1 + j\beta_i q)(d_i + e_i) = d_i - qe_i\beta_i + j(e_i + qd_i\beta_i)$$

or

$$c_0 > 0, \qquad e_i + qd_i\beta_i = 0, \qquad d_i - qe_i\beta_i > 0, \qquad i = 1, \ldots, \mu.$$

If q is finite, d_i, e_i cannot simultaneously be zero; then $d_i \neq 0$ and

$$q = \frac{-e_i}{d_i \omega_i}.$$

Consequently, from the last inequality,

$$d_i + \frac{e_i^2}{d_i} > 0, \tag{7.23}$$

hence $d_i > 0$. Then since d_i and c_0 are both positive residues, the requirement is satisfied to ensure stability in the limit.

Example 7.2 [W3]

Given

$$G(s) = \frac{1}{s(s^2 + as + b)},$$

where, a, b are positive constants, Aizerman's conjecture would predict that the nonlinear Luré regulator should be stable if $\phi(0) = 0$ and $0 < \phi(e)/e < ab$ for all $e \neq 0$, where $e = -y$. That is, the linear systems for gains on the range $(0, ab)$ would be stable. For the modified transfer function

$$G^*(j\omega) = \frac{-a\omega + j(\omega^2 - b)}{a^2\omega^2 + (\omega^2 - b)^2},$$

it is seen from Fig. 7.7 that Aizerman's conjecture is validated with $K = 1/ab$ and $q = a/b$ for the Popov criterion.

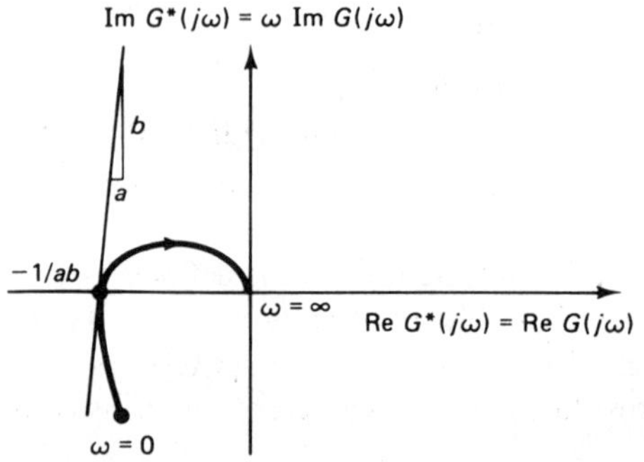

Figure 7.7 Popov stability criterion for Example 7.2.

7.2.2 Circle Criterion

It is desired to examine stability on a sector $\phi \in (k_1, k_2)$ as shown in Fig. 7.8. Here the transformed nonlinearity is

$$\phi_T(e) = \phi(e) - k_1 e$$

or

$$\phi_T(-y) = \phi(-y) + k_1 y.$$

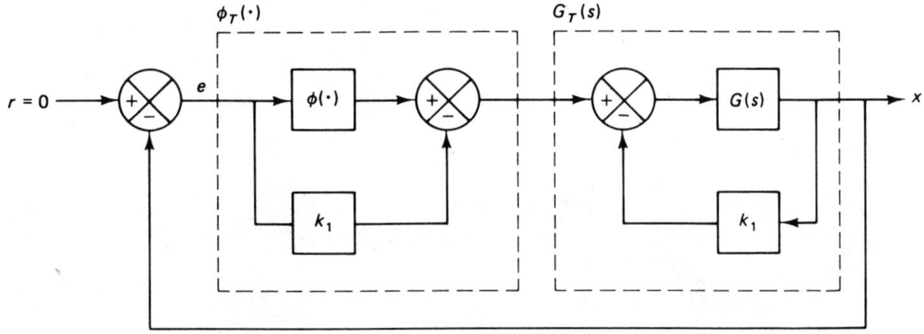

Figure 7.8 Sector transformation from $[k_1, k_2]$ to $[0, k_2 - k_1]$.

Also,

$$\phi_T(y) = \phi(y) - k_1 y$$

if $\phi(-y) = -\phi(y)$ and $\phi_T(-y) = -\phi_T(y)$. Hence, if $\phi \in (k_1, k_2)$, then $\phi_T \in (0, k_2 - k_1)$. As for K above, k_2 is assumed positive for convenience. Before the criterion of Popov above can be applied to the new sector, however, it is necessary to show that the transformed transfer function has the structure

$$G_T(s) = \mathbf{c}_T^T[\mathbf{I}s - \mathbf{A}_T]^{-1}\mathbf{b}_T + d_T, \tag{7.24}$$

where $\mathbf{A}_T$ is an asymptotically-stable matrix [i.e., $(\mathbf{A}_T, \mathbf{b}_T)$ is a controllable pair, and the pair $(\mathbf{A}_T, \mathbf{c}_T)$ is observable]. Then the eigenvalues of $\mathbf{A}_T$ are identical to the poles of $\mathbf{G}_T(s)$. With $k_1 \neq 0$, the Nyquist stability criterion can be used to check that $1 + k_1 G(s)$ has no zeros in the right-half plane.

From the Popov criterion given by (7.6), the criterion for the transformed system (7.23)–(7.24) or Fig. 7.8 becomes

$$\text{Re}\,[(1 + j\omega q)G_T(j\omega)] + \frac{1}{k_2 - k_1} > 0. \tag{7.25}$$

The sector transformation (7.23) results in pole shifting with

$$G_T(s) = \frac{G(s)}{1 + k_1 G(s)}, \tag{7.26}$$

or where, of course, for (7.25) $G_T(j\omega)$ is used. Such a transformation maps the Popov line of (7.20) or (7.21) into a circle for (7.25) as shown in Fig. 7.9. If $k_2 > k_1$ as usually assumed, the region to the right of the Popov line maps into the region that is exterior to the critial circle in the $G(j\omega)$ plane. Then from (7.25) and (7.26), the critical circle determines stability. Consequently, (7.1) is absolutely stable for $\phi \in (k_1, k_2)$ with $\mathbf{A}$ Hurwitz if there exists a real constant q such that for all real nonnegative ω, the following condition is satisfied:

$$\left| G(j\omega) + \frac{(k_2 + k_1) - j\omega q(k_2 - k_1)}{2k_1k_2} \right|^2 - \left(\frac{k_2 - k_1}{2k_1k_2}\right)^2 (1 + q^2\omega^2) > 0, \tag{7.27}$$

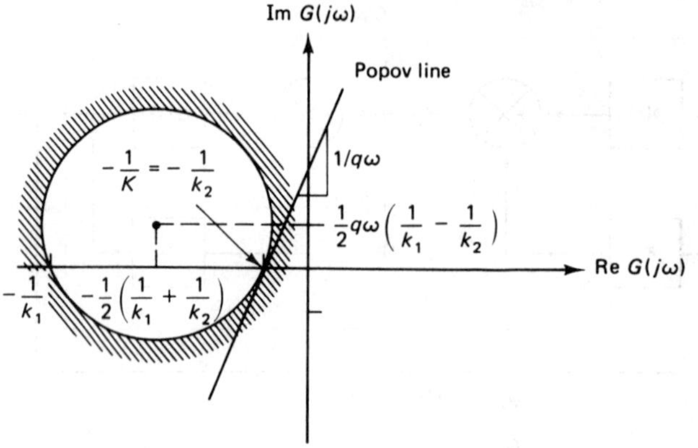

Figure 7.9 Critical circle illustration with $(1/k_1) > (1/k_2)$.

if $1/k_1 > 1/k_2$, or

$$\left| G(j\omega) + \frac{(k_2 + k_1) - j\omega q(k_2 - k_1)}{2k_1k_2} \right|^2 - \left(\frac{k_2 - k_1}{2k_1k_2}\right)^2 (1 + q^2\omega^2) < 0, \tag{7.28}$$

if $1/k_1 < 1/k_2$. In both cases it is assumed that $k_1 < k_2$. For example, if $k_2 = 2$ and $k_1 = 1$ or $k_2 = \frac{1}{2}$ and $k_1 = \frac{1}{4}$, then (7.27) should be satisfied. If $k_1 < 0 < k_2$, however, (7.28) should be satisfied. Here (7.27) and (7.28) are synonymous with equal and opposite signs of k_1 and k_2 respectively.

Equation (7.27) may be interpreted geometrically for absolute stability of (7.1) with $\phi \in (k_1, k_2)$. If $1/k_1 > 1/k_2$, for each $\omega \geq 0$, the polar-locus plot $G(j\omega)$ must lie outside the critical circle whose center is located at

$$X + jY = -\frac{1}{2}\left(\frac{1}{k_1} + \frac{1}{k_2}\right) + \frac{1}{2}j\omega q\left(\frac{1}{k_1} - \frac{1}{k_2}\right), \tag{7.29}$$

where X, Y represent real and imaginary components. Again, this is a sufficient condition only and corresponds to the case with k_1, k_2 of identical signs. If $1/k_1 < 1/k_2$, the corresponding locus for $G(j\omega)$ must remain inside this critical circle to satisfy the Popov criterion, and corresponds to k_1, k_2 of opposite signs.

Suppose that **A** has P eigenvalues in the open right-half plane; that is, $G(s)$ has P poles with positive real parts. Then, from (7.26), the application of Nyquist's stability criterion shows that $1 + k_1G(s)$ has no right-half plane zeros [i.e., no right-half-plane poles of $G_T(s)$] if and only if the Nyquist plot $G(j\omega)$ encircles the $-1/k_1$ point P net times. Here positive is taken in the counterclockwise direction with increasing frequency, such that $G(j\omega)$ does not travel through the $-1/k_1$ point. The circle criterion above, (7.27) or (7.28), with the point $-(1/k_1)$ encircled, implies that one of the following strips is encircled in the same manner: $(-(1/k_1), (1/k_2))$ if $1/k_1 > 1/k_2$, or $(-\infty, -(1/k_2))$ and $(-(1/k_1), \infty)$ if $1/k_1 < 1/k_2$.

If we limit the stability region somewhat by letting $q = 0$, (7.25) reduces to

$$\text{Re}\left[1 + (k_2 - k_1)\frac{G(j\omega)}{1 + k_1G(j\omega)}\right] > 0 \tag{7.30}$$

or

$$\mathrm{Re}\left[\frac{1 + k_2 G(j\omega)}{1 + k_1 G(j\omega)}\right] > 0 \tag{7.31}$$

for all real ω. Then the critical circle C with center given by (7.29), with $q = 0$, has radius $(k_2 - k_1)/2k_1k_2$ if $k_1 > 0$, and is centered at the point $-(k_1 + k_2)/2k_1k_2$. Figure 7.10 shows this condition. Equation (7.30) indicates that the plot $G(j\omega)$ never crosses the critical circle C. Then if $1/k_1 > 1/k_2$, $G(j\omega)$ remains exterior to C and encircles the point $-1/k_1$ P times, it must encircle C P times and vice versa for complete stability if the open-loop linear system $G(s)$ has P right-half-plane poles.

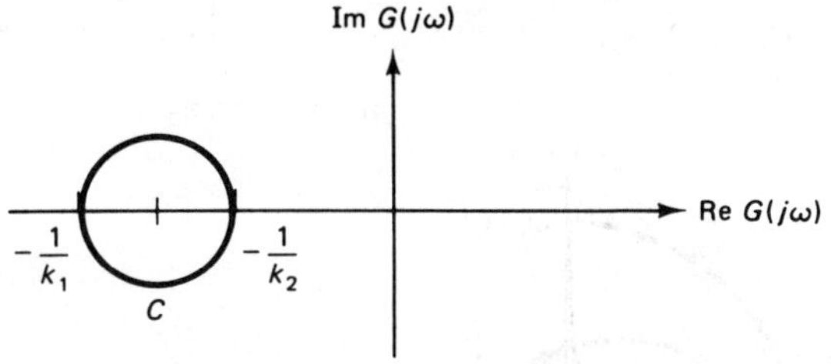

Figure 7.10 Critical circle with $q = 0$ and $(1/k_1) > (1/k_2)$.

If $k_1 < 0 < k_2$, it is necessary for absolute stability that $\mathbf{A}$ is Hurwitz. [Check this by setting $\phi(\cdot) \equiv 0$.] In this case, the criterion remains as before; that is, $G(j\omega)$ should not encircle the $-1/k_1$ point or pass through it. Indeed, the $G(j\omega)$ locus should remain interior to the critical circle C. If $k_1 = 0$, the criterion shows that $G(j\omega)$ must remain to the right of $X = -1/k_2$. Again, X designates the real axis of the complex plane. Figure 7.11 demonstrates each of the the three cases. Note that if k_1 and k_2 are both negative, their roles are reversed and the criterion is applied to $-G(j\omega)$.

7.2.3 Stability Region

It was shown in Section 6.1.3 by Theorem (5) that a region of asymptotic stability including $\mathbf{x}_e = \mathbf{0}$ can be generated from a Liapunov function, $V(\mathbf{x})$, which is defined over that region. Hence the Liapunov function (7.7) which was used in the Popov criterion above can be utilized in those cases that are asymptotically stable but are not globally so. Suppose, for example, that $G(s)$ satisfies the Popov criterion with $\phi(\cdot) \in (0, K)$ only for some finite range of y that corresponds to a finite range of u in Figs. 7.1 and 7.2 which is given by $-\alpha < u < \beta$, where α and β are positive constants. Then assume that the Liapunov function (7.7) with addended variables is given by a quadratic in $\mathbf{x}$, v plus a quadratically bounded term so that

$$V(\mathbf{x}, v) = q_a(\mathbf{x}, v) = q_a(\mathbf{x}_a),$$

where $\mathbf{x}_a^T = [\mathbf{x} \quad v]$. Then examination of the inner product of the gradient of V, $\nabla V(\mathbf{x}_a) = \partial V(\mathbf{x}_a)/\partial \mathbf{x}_a$, with the state velocity shows that

$$\dot{\mathbf{x}}_a \cdot \frac{\partial V}{\partial \mathbf{x}_a} = [\mathbf{A}\mathbf{x}_a + \mathbf{b}u(\mathbf{x}_a)]^T \frac{\partial V}{\partial \mathbf{x}_a}.$$

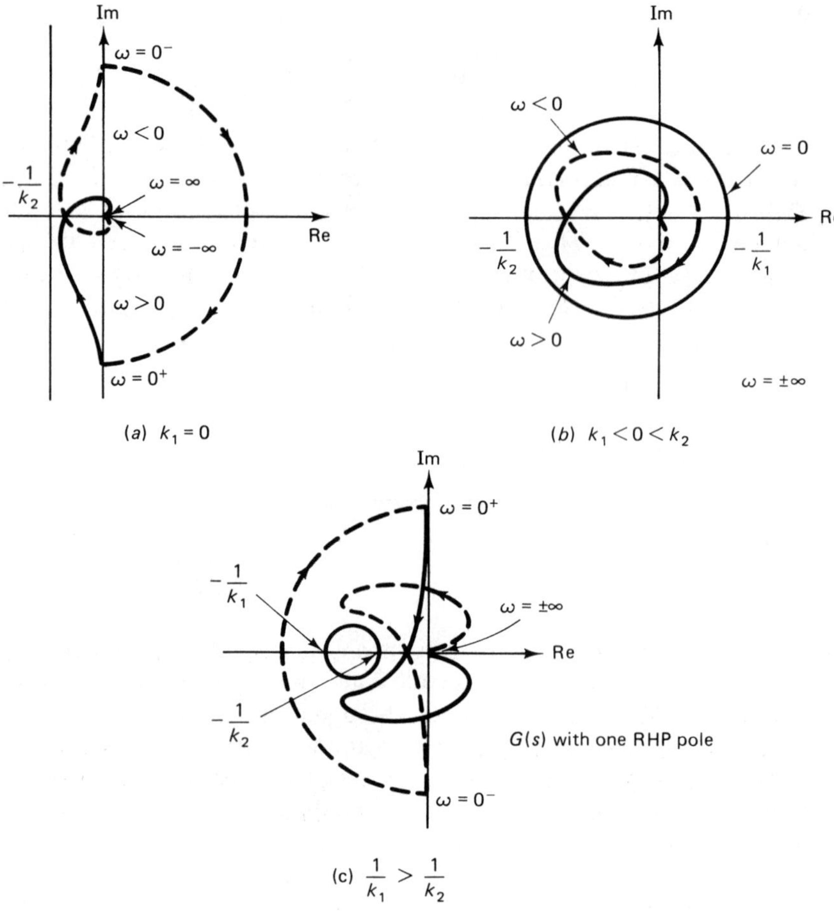

Figure 7.11 Demonstration of the circle criterion for three cases.

Therefore, the first term on the right-hand side is a quadratic function in $\mathbf{x}_a$, which is designated $p_a(\mathbf{x}_a)$, and the latter term with u $\in (-\alpha, \beta)$ is bounded by a linear function of $\mathbf{x}_a$, labeled $l_a(\mathbf{x}_a)$. The preceding equation thus becomes

$$\dot{\mathbf{x}}_a \cdot \frac{\partial V}{\partial \mathbf{x}_a} \leq p_a(\mathbf{x}_a) + l_a(\mathbf{x}_a).$$

Now it is seen that since $p_a(\cdot)$ is negative definite, for some $V(\mathbf{x}_a) = V_m$,

$$\dot{\mathbf{x}}_a \cdot \nabla V_m(\mathbf{x}_a) < 0.$$

Consequently, $\dot{\mathbf{x}}_a$ will always point into the region D of asymptotic stability, as shown in Fig. 7.12. Here a simple second-order case is illustrated with $|\alpha| = |\beta|$ such that with $u = -\alpha$ and $u = \beta$ the hypersurfaces H_1 and H_2, respectively, are generated. $V(\mathbf{x}) = V_m$ at the points of tangency of H_1, H_2 with S, the closed set of states such that $V(\mathbf{x}) = V$. For states on H_1, H_2, $V(\mathbf{x}) \geq V_m$. Obviously, all states $\mathbf{x}(t)$ with $\mathbf{x}(t_0) \in D$ will not leave D and are asymptotically stable relative to the origin since for all $\mathbf{x} \in D$:

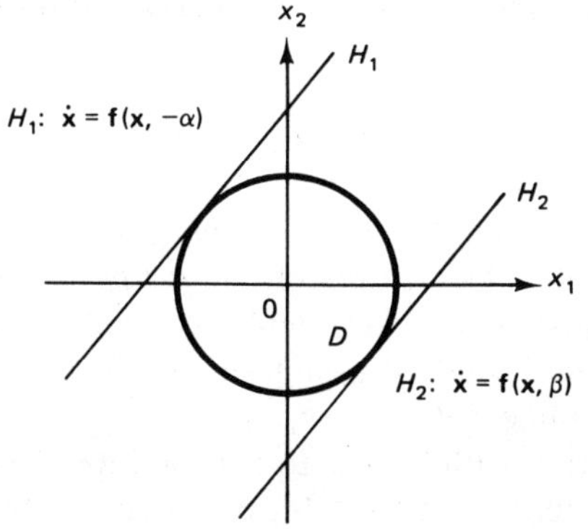

Figure 7.12 Asymptotic stability region for the origin.

1. $V(\mathbf{x})$ is positive definite.
2. $V(\mathbf{x}) < V_m$.
3. $\dot{V}(\mathbf{x})$ is negative definite.

7.2.4 Generalized Criteria and Comments

Throughout the discussion above it was assumed that the nonlinear nondynamical element $\phi(\cdot)$ was time invariant. Now let $\phi(e, t) \in (k_1, k_2)$. Then it can be shown from a simple quadratic Liapunov function $V(\mathbf{x}) = \mathbf{x}^T\mathbf{Q}\mathbf{x}$ that $\dot{V}(\mathbf{x}, t) \leq 0$ for system (7.1) with **A** Hurwitz [B6]. After some algebra it is seen that (7.31) results. Consequently, system (7.1) with the required conditions (1) to (5) satisfied, including 2(a), is *absolutely stable* for $\phi(e, t) \in (k_1, k_2)$ if

$$\operatorname{Re}\left[\frac{1 + k_2G(j\omega)}{1 + k_1G(j\omega)}\right] > 0. \tag{7.31}$$

Consequently, the interpretations given for the circle criterion such as with respect to Fig. 7.11 are valid for a *time-variant nonlinear element* as well as a constant one. Again, the presence of open-loop poles in the right-half plane leads to the requirement of encirclements of the critical circle as shown by Fig. 7.11. Of course, for $k_1 = 0$ and $k_2 = K$, (7.31) becomes

$$\operatorname{Re}\,[1 + KG(j\omega)] > 0, \tag{7.32}$$

which was an early basic result for the Luré regulator problem. Note that (7.32) is identical to (7.6) with $q = 0$ and $K = \infty$. Obviously, (7.32) provides a more conservative sufficient condition for absolute stability than does (7.6).

Sometimes it is convenient in practice to define a weaker concept of stability than that of the state as formulated by Liapunov and studied in the preceding four chapters and used with the Popov method. For example, the output may be the primary consideration. With this in mind, system (7.1) (Fig. 7.1) is said to be *output asymptotic of degree* α if for every set of initial conditions, $\mathbf{x}(t_0)$, there exists a real α such that

$$\int_0^\infty [\exp(\alpha t)e(t)]^2 dt < \infty. \tag{7.33}$$

Similarly, this regulator is *control asymptotic of degree* α if for every $\mathbf{x}(t_0)$ there exists a real α such that

$$\int_0^\infty [\exp(\alpha t)u(t)dt]^2\, dt < \infty. \tag{7.34}$$

Then the concept of absolute stability may be relaxed somewhat by the following definitions: Regulator system (7.1) is *absolutely output asymptotic of degree* α for $\phi(e, t) \in (k_1, k_2)$ if (7.33) is satisfied for every nonlinear element $\phi(\cdot) \in (k_1, k_2)$. Similarly, system (7.1) is *absolutely control asymptotic of degree* α for $\phi(e, t) \in (k_1, k_2)$ if (7.34) holds for every nonlinear $\phi(\cdot) \in (k_1, k_2)$.

Another useful definition with respect to the stability of $y(t)$ in Fig. 7.1, system (7.1), is given next. The system is output stable of degree α if there exists a real α such that for every $\mathbf{x}(t_0)$, the unit-impulse response $g(t)$ of the linear system, $G(s)$, and the initial condition response $y_0(t)$ satisfy the following for $t \in [0, \infty) : |e^{\alpha t}y(t)| < \infty$, and

$$\int_0^\infty [e^{\alpha t}g(t)]^2 dt, \qquad \int_0^\infty e^{\alpha t}|\,g(t)\,|\,dt,$$

$$\int_0^\infty [e^{\alpha t}y(t)]^2\,dt, \qquad \int_0^\infty [e^{\alpha t}\dot{y}(t)]^2\,dt,$$

are all finite integrals, that is, that $e^{\alpha t}g(t)$, $e^{\alpha t}y(t)$, $e^{\alpha t}\dot{y}(t)$ are square-integrable functions. If $\alpha = 0$, the linear system $G(s)$ in Fig. 7.1 is said to be *output stable*. Thus the linear system is output stable of degree α if all its poles of $G(s)$ are in the zone of $\text{Re}[s] < -\alpha$, a degree of damping.

It can be shown [H1] that if $G(s)$ in Fig. 7.1 is output asymptotic of degree α and the system is control asymptotic of degree α, then

$$\lim_{t\to\infty} \exp(\alpha t)e(t) = 0 \tag{7.35}$$

if for every $t \geq t_0$ and every real positive e_m there exists a finite positive real u_m such that

$$|u(t)| \leq u_m \leq \infty, \tag{7.36}$$

where $|e(t)| \leq e_m$. Equation (7.36) holds for almost all practical nonlinear elements $\phi(e, t)$, including relays, hysteresis, polynomials, and so on.

Figure 7.13 shows examples of both passive and active hysteresis that do satisfy this condition. Passive hysteresis can be transformed into active hysteresis by the change of variables $\phi_a(e) = ke - \phi(e)$ as shown in Fig. 7.13. Then it can be shown that the system in Fig. 7.1 is both *absolutely output and control asymptotic with active or passive hysteresis* $\phi(e) \in (0, K)$ if the Popov criterion (7.6) is satisfied [H1] with q nonpositive for the passive case and nonnegative for the active case ([Y1] and [H1]). In terms explained in Fig. 7.13, a nonlinearity is said to be passive hysteresis in general if

$$-\left(\int_{e_1}^{e_2}\gamma\right)ude \geq -\left(\int_{e_1}^{e_2}\gamma_{12}\right)ude = \left(\int_0^{e_2}\gamma_1\right)ude - \left(\gamma_1\int_0^{e_2} + \gamma_{12}\right)ude$$

is valid for every pair (e_1, e_2), where γ is the path of integration over $\phi(e(t))$. γ_1, γ_{12}

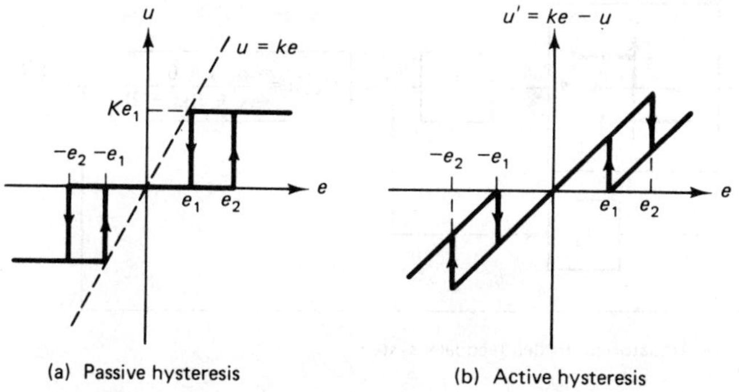

Figure 7.13 Hysteresis characteristic examples.

are single paths (not closed) on the characteristic $\phi(\cdot)$ between 0 and e, and e_1 and e_2, respectively. Similarly, a generalized active hysteresis exists if

$$-\left(\int_{e_1}^{e_2}\gamma\right)ude \le -\left(\int_{e_1}^{e_2}\gamma_{12}\right)ude = \left(\int_0^{e_2}\gamma_1\right)ude - \left(\gamma_1\int_0^{e_2} + \gamma_{12}\right)ude$$

holds for every part (e_1, e_2). This extends the class of nonlinearities which are covered by the foregoing hysteresis case of the Popov criterion ([Y1] and [H1]).

Relay nonlinear characteristics with and without sliding motion have been incorporated into the Popov criterion with $\phi \in (0, \infty)$ and $q > 0$ [Y1]. Consider the following examples.

Example 7.3 [H1].

Consider Fig. 7.14, the closed-loop control system with a relay, around which an internal actuator closed loop is shown such that it is desired to determine the allowable actuator loop gain k_a for absolute stability. The original indirect-control system of Fig. 7.14(a) is transformed into Fig. 7.14(b), the direct form with

$$G(s) = \frac{1}{s}[G_1(s) + k_a],$$

where

$$G_1(s) = \frac{s + 6}{(s + 2)(s + 3)}.$$

The initial condition response of the error is given by

$$e_0(t) = e_{10} + e_{20}e^{-2t} + e_{30}e^{-3t} + r,$$

where $e_0(0) = e_{10}$, and the input r is a constant that can be combined with e_{10}. Then

$$G^*(j\omega) = H^*(j\omega) - jk_a,$$

where

$$H^*(j\omega) \equiv -\frac{24 + \omega^2}{36 + 13\omega^2 + \omega^4} - j\frac{36 - \omega^2}{36 + 13\omega^2 + \omega^4}.$$

With $K = \infty$, the Popov criterion (7.6) becomes

$$\text{Re } G^*(j\omega) - q \text{ Im } G^*(j\omega) = \text{Re } H^*(j\omega) - q[\text{Im } H^*(j\omega) - k_a] > 0,$$

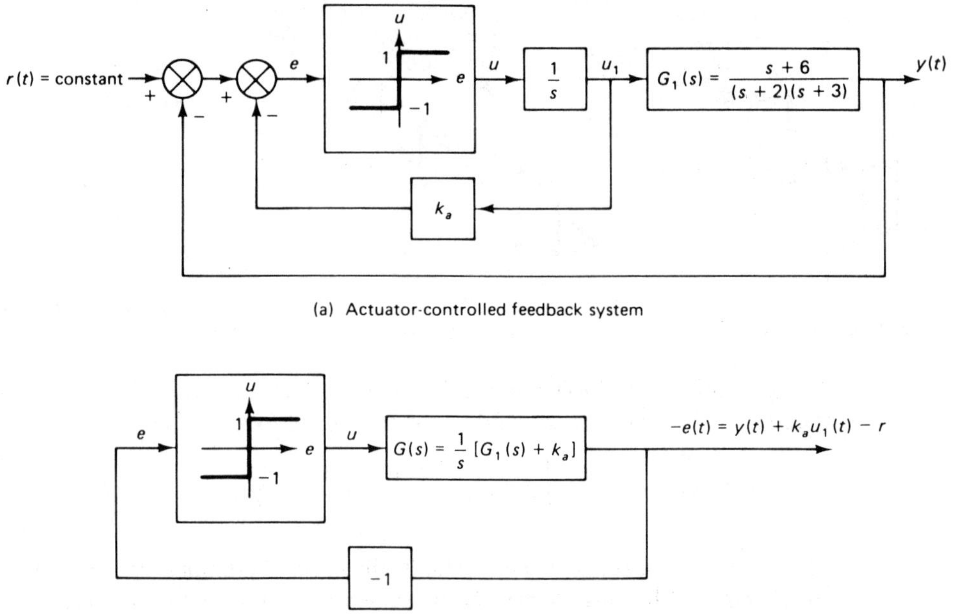

(a) Actuator-controlled feedback system

(b) Modification of (a)

Figure 7.14 Block diagram for Example 7.3.

where q is a positive constant for the relay problem ([Y1] and [H1]). The polar-locus plot shown in Fig. 7.15 has Im $H^*(J\omega)$ with an approximate maximum value of 0.059. Consequently, for

$$k_a > 0.059,$$

it is absolutely control and output asymptotic with $\lim_{t\to\infty} e(t) = 0$ and $\lim_{t\to\infty} y(t) = r/(k_a + 1)$. Further, if there is no pole–zero cancellation so that the system is controllable and observable, it is absolutely stable.

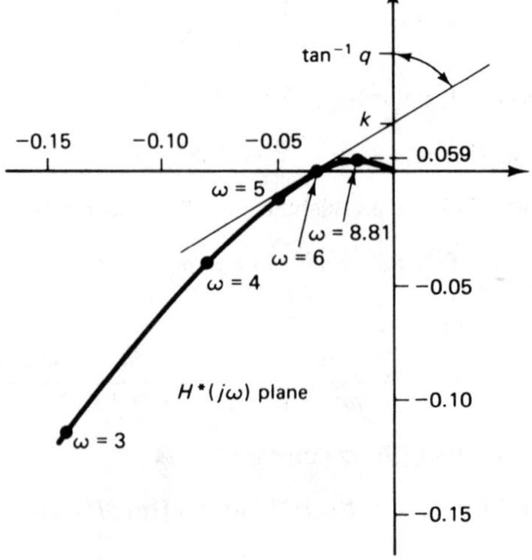

Figure 7.15 Popov condition for Example 7.3. (After Hsu and Meyer [H1].

Example 7.4 [P4]

Consider the following third-order nonlinear regulator with the nonlinear element, $\phi(\cdot)$, in the feedback path:

$$\dot{x}_1 = -cx_1 + x_2 - \phi(x_1),$$
$$\dot{x}_2 = -x_1 + x_3,$$
$$\dot{x}_3 = -cx_1 + b\phi(x_1),$$

where $b > 0$, $c > 0$, $b < c^2$, and $\phi(x)$ is a single-valued piecewise-continuous nonlinear function with $\phi(0) = 0$ and $0 < \phi(x) < Kx$, for $x \neq 0$. Then it can be shown from linear stability analysis that the equivalent linear system would be stable for $0 < K < c/b$. With $y = x_1$, and $u = -\phi(y)$ as shown in Fig. 7.16,

$$G(s) = \frac{s^2 - b}{s^3 + cs^2 + s + c} = \frac{s^2 - b}{(s^2 + 1)(s + c)}, \tag{7.37}$$

which has poles at $\lambda_{1,2} = \pm j$, $\lambda_3 = -c$. As shown above, (7.22) must be satisfied (i.e., the residues at the imaginary poles must have positive real parts). Here

$$d + je = \lim_{s \to j} [(s - j)G(s)] = \frac{1}{2}\frac{b + 1}{c^2 + 1}(1 + jc).$$

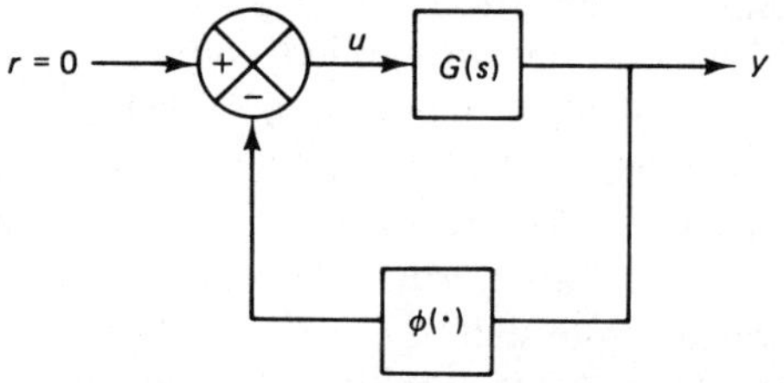

Figure 7.16 Conventional Luré regulator for Example 7.4.

So $d > 0$ satisfies the stability in the limit condition.

Now let $w = \omega^2$, so that

$$X = \text{Re } G(j\omega) = \text{Re } G^*(j\omega) = \frac{c(w + b)}{(w - 1)(w + c^2)}$$

and

$$Y = \omega \text{ Im } G(j\omega) = \text{Im } G^*(j\omega) = \frac{-w(w + b)}{(w - 1)(w + c^2)}.$$

Therefore,

$$X - KY = -c + \frac{c(c^2 - b)}{w + c^2}.$$

and unless $K = -c$, $X - KY$ has a simple pole near $w = 1$ and the Popov criterion (7.20) cannot be satisfied for finite K. Then check $K = -c$, or

$$X - KY = -c + \frac{c(c^2 - b)}{w + c^2}.$$

Case 1: $b < c^2$. Here

$$\frac{c(c^2 - b)}{w + c^2} > 0$$

and

$$X - KY + \frac{1}{K} = 0 \qquad \text{for } \frac{1}{K} = c.$$

Thus $\phi(x_1) \in (0, 1/c)$ ensures absolute stability.

Case 2: $b > c^2$. In this case the Popov sector is $\phi(x_1) \in (0, c/b)$.

Case 3: $b = c^2$. Here the Popov criterion is satisfied on the sector $(0, 1/c)$.

Consequently, it is seen that Aizerman's conjecture violates the bounds of the Popov criterion. In [P4] the existence of oscillatory solutions is derived to disavow Aizerman's conjecture for this example.

7.2.5 Summary of Results

The *Popov criterion* provides sufficient conditions for the Luré regulator to be absolutely stable with a large class of nonlinearities. The general criterion can be stated in terms of a *Popov line* for nonlinear elements $\phi(\cdot) \in (0, K)$, and in terms of a *critical circle* for $\phi(\cdot) \in (k_1, k_2)$. Both criteria are applied conveniently in the frequency domain to the polar-locus plot of the open-loop linear portion $G(j\omega)$. The line criterion in its simplest form (7.19), however, uses the modified polar-locus plot of $G^*(j\omega)$ from (7.18). The circle criterion, (7.27)–(7.29), is more cumbersome to apply since each circle is dependent on ω. In the basic form, the constant parameter q may take on any real value to determine the absolute stability sector $(0, K)$ or (k_1, k_2), as shown by (7.19) or (7.27)–(7.29), respectively. If $q = 0$, the allowable sector generally is reduced, but the circle criterion is much simpler to apply. If the open-loop linear system has poles in the open right-half plane (not on the imaginary axis) and k_1, k_2 are of identical signs, absolute stability can conveniently be ascertained by counting net encirclements of the critical circle [(7.27)–(7.29)] with $q = 0$, similar to that of the -1 point for the Nyquist stability criterion. If k_1, k_2 are of opposite signs, $\mathbf{A}$ must be Hurwitz for absolute stability of the system.

Also, the criterion can be extended to relay and hysteresis (finite memory) types of nonlinear elements. Weaker definitions of output stability, output asymptoticity, and control asymptoticity were introduced for cases where it may not be required that all the states are globally asymptotically stable to the origin. In particular, hidden modes [pole–zero cancellation in $G(s)$ due to a lack of controllability and/or observability] may be tolerated if the mode is not divergent or even if it is very slowly divergent in some cases. Popov's method, of course, is sufficient for the regulator to be absolutely output and absolutely control asymptotic on the prescribed sector. In some cases, the Liapunov function (7.7) may lead to stability for some region of state space. As in Chapter 5, it may be possible to determine such a region of stability by an analysis of the inner product of the state velocity vector and the spatial gradient of the Liapunov function, as shown by Section 7.2.3.

Advantages of the Popov method include:

1. Its ease of application to high-order systems by a frequency response that may be obtained for a system without knowledge of its state equations in some cases
2. Its validity for a very large class of memoryless nonlinear elements
3. Its sufficiency, which guarantees absolute stability

Disadvantages include the following:

1. Being sufficient only, it can lead to a conservative estimate of stability conditions.
2. It does not use any known structure of the nonlinear element to extend realistically the sector of absolute stability.

In the basic cases it was assumed that either $G(s)$ had no poles on the imaginary axis (**A** Hurwitz), the so-called "principal case," or that $G(s)$ has only a simple pole at the origin, the so-called "particular case." If $G(s)$ has simple poles on the imaginary axis, it is sufficient that the corresponding residues of $W(s) = (1 + qs)G(s)$ have positive real parts, where $G(s)$ is a rational function that has no poles in the open right-half plane. The Popov criterion is extended to time-variant nonlinearities $\phi(e, t) \in (0, \infty)$ with $q = 0$ as indicated for $K = \infty$. Again, this represents an apparent relaxation that reduces the admissible range of linear systems $G(s)$ which will lead to stability by (7.31). Note that only the principal case of $G(s)$ was considered here. However, the circle criterion as appropriate for (7.31) can be used for both the stable and unstable cases of the open-loop linear system $G(s)$.

Numerous attempts have been made to utilize the Popov criterion or to extend it in conjunction with other Liapunov functions so that the nonlinear structure is utilized through computation of the time rate of change of the Liapunov function. Examples of this are presented in [D2], [Y1], [B7], and [Z2]. Apparently, Aizerman's conjecture and Kalman's conjecture, in conjunction with the Popov method, can be quite useful. Since both conjectures satisfy classes of functions belonging to sectors $(0, K)$ or (k_1, k_2) as for the Popov criterion, they can be considered as necessary conditions. Generally, these conjectures will give a somewhat optimistic range of stability, whereas the Popov method may be pessimistic. References [B8] and [F4] study this problem in some detail for third- and fourth-order systems. Table 7.1 provides sufficient conditions for these conjectures to be fulfilled for several systems.

In the next section, another frequency-domain method of stability analysis, describing functions, is studied. The method is very useful for feedback regulator systems of the type presented above. Similar to Aizerman's conjecture, it generally provides a simplistic and optimistic estimate of the realm of stability, but in practice it has proven to be among the most successful methods.

Example 7.5

This example is intended to summarize the use of the Popov line criterion, the pole transformation, the circle criterion, and the use of constant-M circles with this methodology when using a personal computer.

TABLE 7.1 SUFFICIENT CONDITIONS FOR AIZERMAN AND KALMAN CONJECTURES TO BE TRUE

	Parameter Conditions		
$G(s)$	Aizerman Conjecture	Kalman Conjecture	Relations
$\dfrac{b_0 s^2 + b_1 s + b_2}{s(s^2 + a_1 s + a_2)}$	$b_0 = 0$ $b_0 > 0,\ b_2 \geq 0$ $b_0 > 0,\ b_2 < 0,\ A \geq 0$ $b_0 < 0,\ b_2 \geq 0,\ B \geq 0$ $b_0 < 0,\ \dfrac{b_2}{b_0} > a_2,\ A \geq 0$	$b_0 > 0,\ b_2 < 0$ $b_0 < 0,\ b_2 \geq 0$	$A = a_2^2 b_2(a_2 b_0 - b_2) + a_1^2 a_2 b_2^2$ $B = a_2(b_0 a_2 - b_2)^2 + a_1^2 a_2 b_0 b_2$
$\dfrac{b_0 s^2 + b_2}{s(s^3 + a_1 s^2 + a_2 s + a_3)}$	$b_0 = 0,\ b_2 > 0,\ A \geq 0$ $b_0 = 0,\ b_2 < 0,\ a_1^2 > 2a^2$ $B > 0$ $b_0 = 0,\ b_2 < 0,$ $4B > \dfrac{(a_1^2 - 2a_2)^2 a_3}{a_1}$ $a_3/a_1 > \dfrac{b_2}{b_0} > 0$		$A = a_1^2 + 2a_3/a_1 - 2a_2$ $B = \dfrac{a_3(a_2^2 - a_1 a_3)}{a_1}$
$\dfrac{b_0 s^2 + b_1 s + b_2}{(s^2 + \omega_0^2)(s + a_1)}$	$b_0 = b_1 = 0,\ b_2 < 0$ $b_0 = 0,\ b_1 > 0,\ a_1 b_1 > b_2 \geq 0$ $b_0 = 0,\ b_1 > 0,\ b_2 < 0,$ $b_1 \omega_0^2 < -a_1 b_2$ $b_0 = 0,\ b_1 < 0,\ b_2 < 0,$ $a_1 b_1 > b_2$	$b_0 < 0,\ b_1 = 0$ $0 > b_2 > \omega_0^2 b_0$ $b_0 > 0,\ b_2 < 0,\ b_0 > b_1,$ $a_1 b_1 > b_2,$ $b_1(a_1 b_0 + b_2 - a_1 b_1) > 0$ $b_0 > 0,\ b_1 > b_0,\ b_2 < 0$	

	$b_0 > 0,\ b_1 = 0,\ b_2 < 0$ $a_1^2 b_0 + b_2 < 0$	$b_0 = b_1 > 0,\ b_2 < 0$	
	$b_0 < 0,\ b_1 = 0,\ b_2 < 0,$ $b_0 \omega_0^2 > b^2$	$b_0 > 0,\ b_0 > b_1,$ $a_1 b_1 \geq b_2 > 0,$ $b_2 < (b_0 - b_1)\omega_0^2$	
	$b_0 > 0,\ a_1 > b_0,$ $0 < b_2 - a_1 b_1 < \omega_0^2 b_0$		
	$b_1 > b_0 > 0,\ a_1 b_1 \geq b_2 > 0$		
	$b_1 > b_0 > 0,\ b_2 < 0,$ $b_2 > (b_0 - b_1)\omega_0^2$		
	$b_0 < 0,\ b_2 < 0,\ b_1 > b_0$ $a_1 b_1 > b_2,$ $\omega_0^2 < \min\left(\dfrac{b_2}{b_0 - b_1}, \dfrac{b_2 - a_1 b_1}{b_0}\right)$		
	$b_0 = b_1 > 0,$ $\omega_0^2 b_0 > b_2 - a_1 b_1 > 0$		
	$b_0 = b_1 > 0,\ a_1 b_1 > b_2,$ $a_1^2 + \dfrac{(a_2 - a_1 b_1)}{b_0} < \omega_0^2$		
	$b_0 = b_1 < 0,\ a_1 b_1 > b_2$		
$\dfrac{b_0 s + b_1}{(s^2 + \omega_0^2)(s^2 + a_1 s + a_2)}$	$b_0 = 0,\ b_1 < 0,\ a_1^2 \geq 2a_2$	$b_0 = 0,\ b_1 < 0,$ $a_2 - \omega_0^2 \geq (a_1^2 - 2a_2)^2$	$A = \dfrac{a_2 b_1}{b_1 - a_1 b_0}$
	$b_1 > a_1 b_0 > 0,\ a_2 b_0 > a_1 b_1,$ $B > \omega_0^2 > A$	$b_1 > a_1 b_0 > 0,$ $a_2 b_0 > a_1 b_1,\ A > \omega_0^2$	$B = \dfrac{a_2 b_0 - a_1 b_1}{b_0}$
	$b_1 < 0,\ B > \omega_0^2$ $A(B - \omega_0^2)(a_1^2 - 2a_2) \geq a_2^2(\omega_0^2 - A)$		

Source: After Fuji and Shoji [F4].

(a) Consider the following third-order system as part of Fig. 7.1:

$$G(s) = \frac{1}{(s + 1)(s + 2)(s + 3)}. \tag{7.38}$$

The application of the Popov criterion (7.6) is illustrated by Fig. 7.17, which shows a personal-computer graph of $G(j\omega)$ relative to the Popov line which has a maximum sector gain of $K = 60.33$ with $q = 0.55$ and $\omega = 3.33$. Therefore, for the conventional time-invariant $\phi(\epsilon) \in (0,\ 60.33)$, the regulator is absolutely stable. For a time-variant non-linearity $\phi(e,\ t) \in (0,\ 27.96)$, it also is stable as shown with $q = 0$ and $K' = 27.96$.

This method has the difficulty that the line ωq Im $G(j\omega)$ must lie to the left of the $G(j\omega)$ locus for all ω. Hence given a q it must be tested for all ω. A simple solution is to

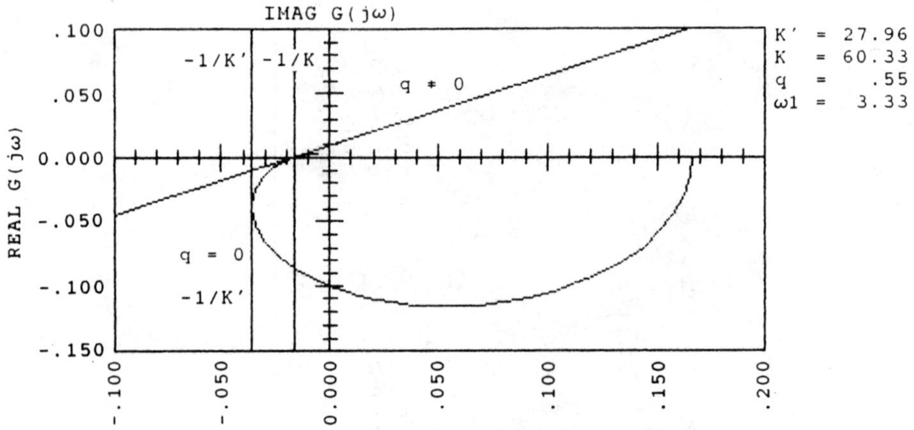

(a) Computer plot of Popov's line

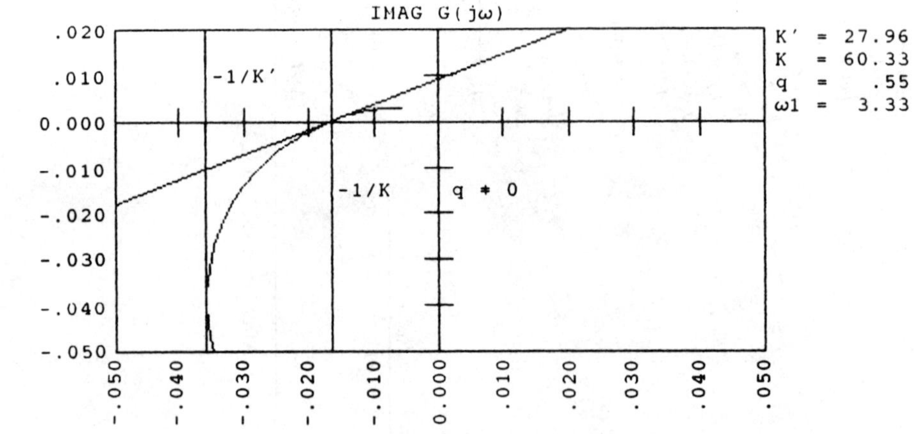

(b) Enlarged plot at point of contact

Figure 7.17 $G(j\omega)$ illustration of the Popov criterion, Example 7.5(a).

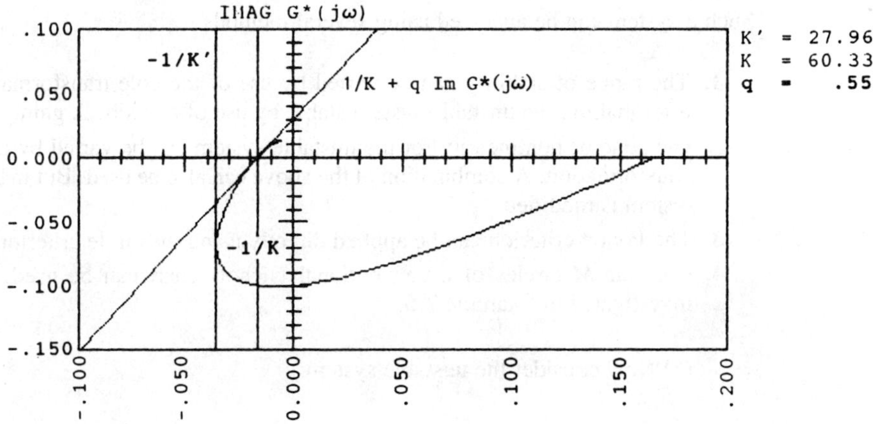

(a) Plot of $G^*(j\omega)$ and Popov line

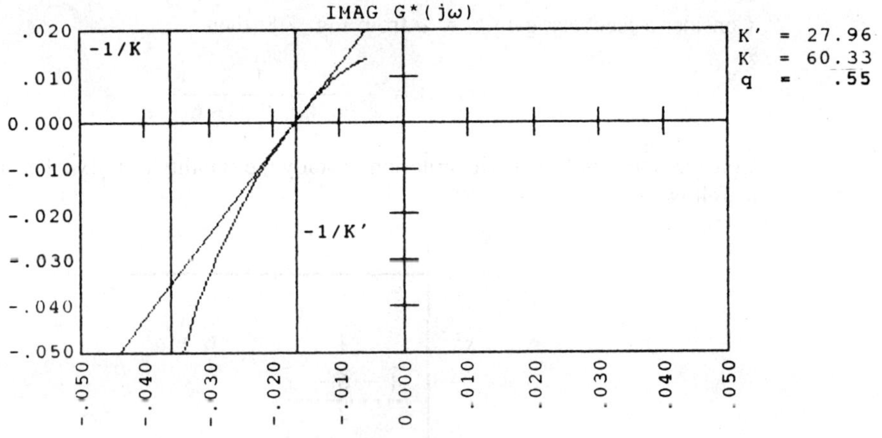

(b) Enlarged plot

$$G(s) = \frac{1}{(s+1)(s+2)(s+3)}$$

$$G^*(j\omega) = \text{Re}\, G(j\omega) + j\omega \,\text{Im}\, G(j\omega)$$

Figure 7.18 $G^*(j\omega)$ illustration of the Popov criterion, Example 7.5(a).

plot the locus $G^*(j\omega)$ defined by (7.18) so that the Popov theorem is modified to (7.19). A computer plot for $G^*(j\omega)$ is given by Fig. 7.18. Values of q and K are computed again, and as expected, they are the same as that obtained for the $G(j\omega)$ plot in Fig. 7.17.

We next study the application of the Popov criterion to an unstable system with the bounded nonlinearity $\phi(e)$ such that

$$\frac{u}{e} \in (a, b).$$

Such a system can be analyzed using several methods.

1. The range of stability can be varied by use of the pole transformation approach after making the unstable system stable by use of a feedback gain.
2. The zone of nonlinearity for the unstable system can be varied by using the zero transformation. A combination of the above can also be used. But in both cases the system is modified.
3. The Popov criterion can be applied directly using the circle criterion.
4. Constant-M circles of a conventional Nichols chart can be used. This will be investigated in Example 7.6.

(b) Now consider the unstable system:

$$G(s) = \frac{1}{(s-1)(s+2)(s+3)}, \tag{7.39}$$

$$= \frac{1}{s^3 + 4s^2 + s - 6}.$$

Consider a feedback gain $k_1 = a$ from Fig. 7.8; then

$$G_T(s) = \frac{1}{s^3 + 4s^2 + s - 6 + a}$$

Consider the Routh–Hurwitz criterion to study the stability of $G_T(s)$. The array becomes as follows:

s^3	1	1
s^2	4	$-6 + a$
s^1	$\dfrac{4 + 6 - a}{4}$	0
s^0	$-6 + a$	

For stability, $10 \geq a \geq 6$. So for the pole transformation we use $a = 6$, and the sector [6, 10] becomes [0, 4]. Then

$$G_T(s) = \frac{1}{s^3 + 4s^2 + s} = \frac{1}{s(s^2 + 4s + 1)}.$$

The pole-transformed plots $G(j\omega)$ are shown by Fig. 7.19 with the stability points and corresponding nonlinear sectors from the Popov criterion. Again, the Popov criteria (7.6) and (7.19) are used to show stability by Fig. 7.19 for $G(j\omega)$ and Fig. 7.20 for $G^*(j\omega)$. As shown, the compensated (pole-shifted) system is now stable for $\phi(e) \in (0, 4.0)$ with $q = 4.0$, and for $\phi(e, t) \in (0, 0.25)$ with $q = 0$. Next the circle criterion (7.27) is utilized by application to the $G(s)$ above with $k_1 = a = 6$ and $K_2 = 10$. Again it is seen that the system is absolutely stable for $\phi(e) \in (6, 10)$, as shown

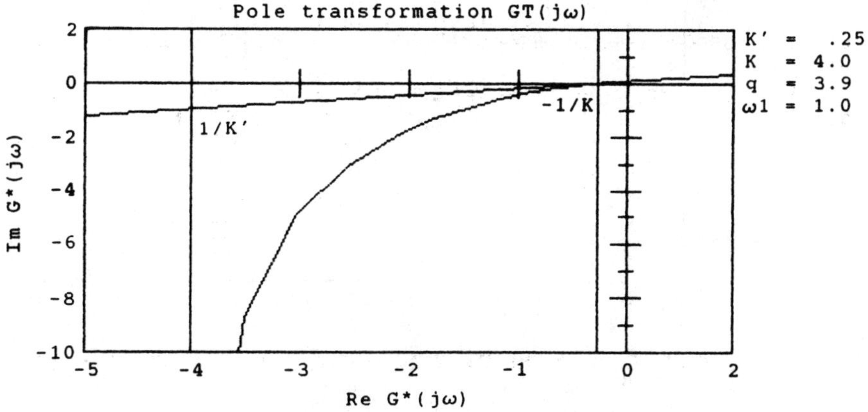

(a) Pole shifting for an unstable system

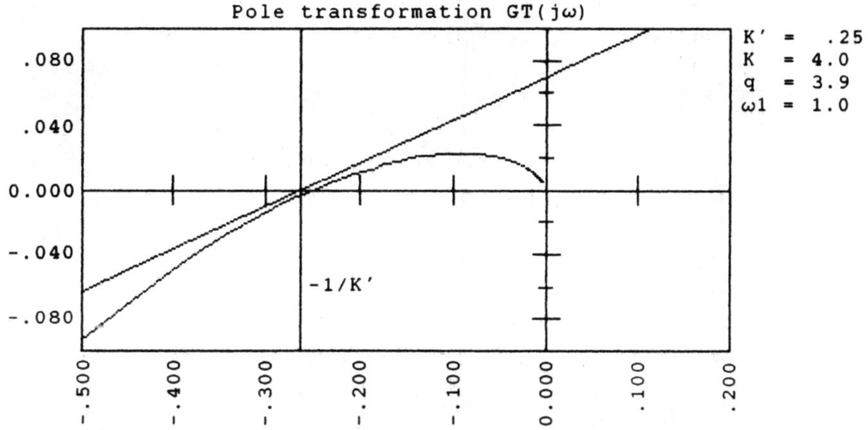

(b) Enlarged plot

$$G(s) = \frac{1}{s(s^2 + 4s + 1)}$$

Figure 7.19 $G_T(j\omega)$ illustration of the Popov criterion, Example 7.5(b).

on Fig. 7.21. Finally, the most convenient, but also the most conservative criterion is applied: that is, the circle criterion of (7.31) which results for (7.27)–(7.29) with $q = 0$. Recall that if $G(s)$ has P open right-half-plane poles, then $G(j\omega)$ must encircle, but not enter, the critical circle P net times in the counterclockwise direction with increasing frequency.

For $q = 0$ all of the critical circles, C, of the circle criterion will lie on the X axis [Re $G(j\omega)$]. Then, having plotted the $G(j\omega)$ locus, there are an infinite set of ranges that can be selected for the nonlinear element so that C is encircled $+1$ times since there exists an open-loop right-half-plane pole. For this example, (I), (II), (III), . . . are all possible Popov sectors centered on the real $a(j\omega)$ axis (Fig. 7.22).

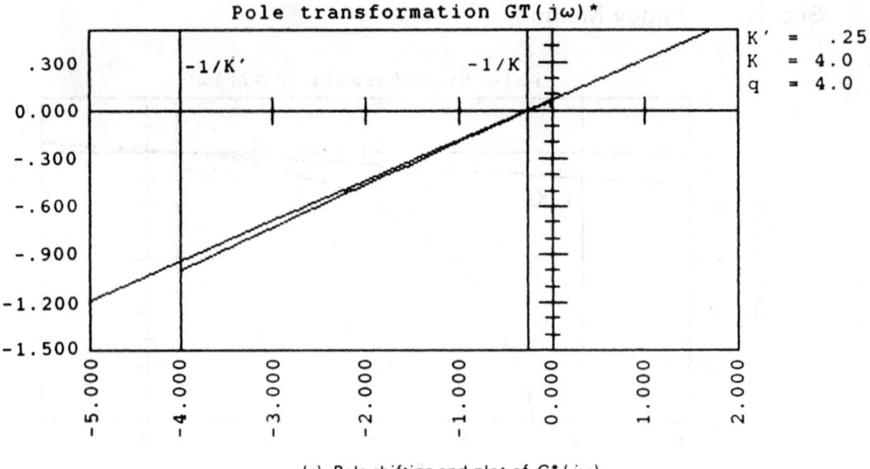

(a) Pole shifting and plot of $G_T^*(j\omega)$

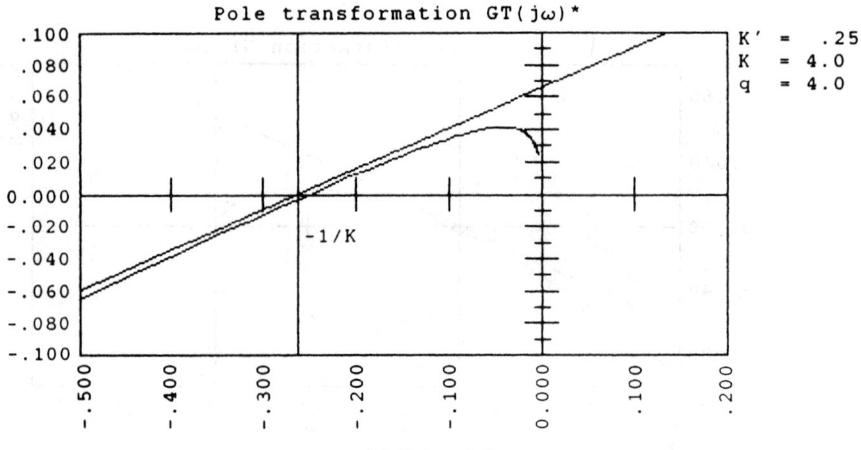

(b) Enlarged plot

Figure 7.20 $G_T^*(j\omega)$ illustration of the Popov criterion, Example 7.5(b).

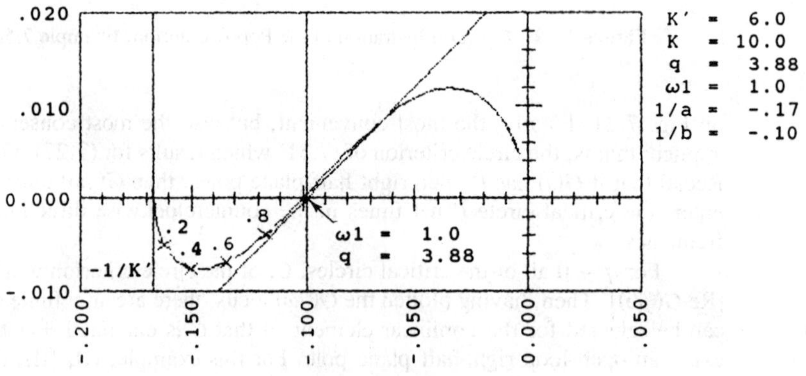

(a) Unstable system

Figure 7.21 Circle criterion for Example 7.5(b).

$$G(s) = \frac{1}{(s-1)(s+2)(s+3)}, \; a = 6, \; b = 10$$

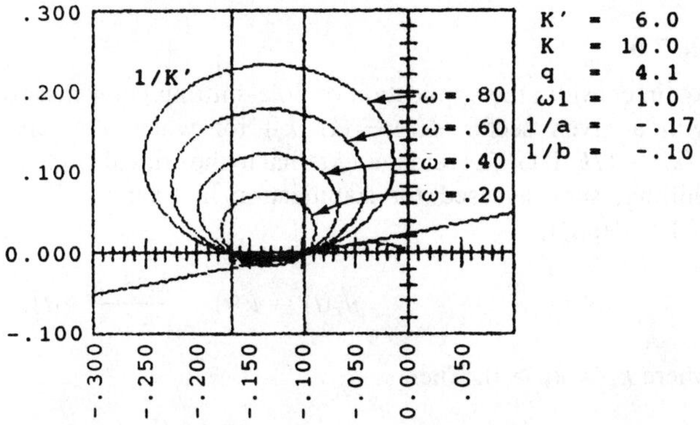

(b) Circle criterion applied

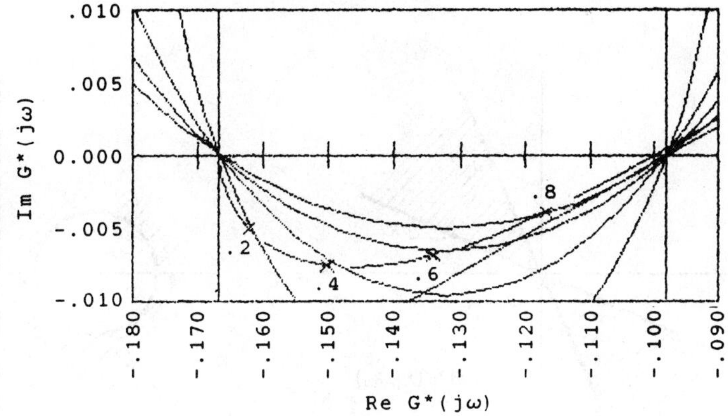

(c) Enlarged plot

Figure 7.21 Circle criterion for Example 7.5(b). (cont.)

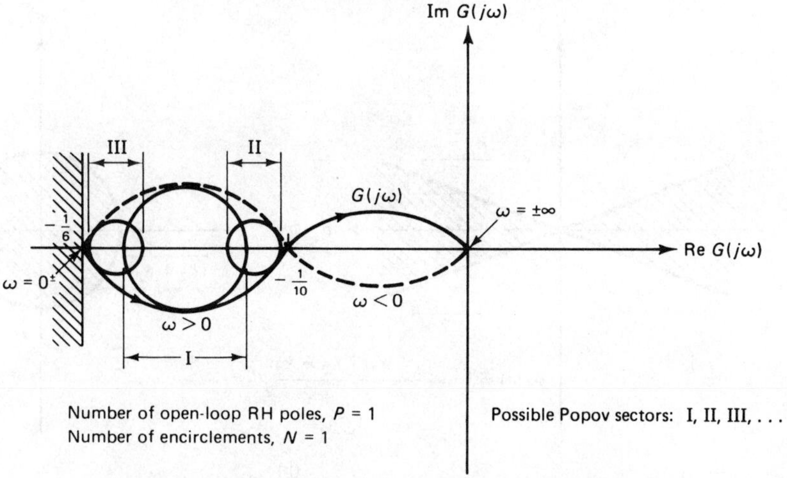

Figure 7.22 Circle criterion with $q = 0$ for Example 7.5.

Example 7.6

Again consider the application of pole-shifting compensation. In general, $G(s)$ is given with a given sector $\phi(e) \in (k_1, k_2)$ for which the circle criterion is satisfied for $1/k_1 > 1/k_2$ if $G(j\omega)$ remains external to the critical circle as shown in Fig. 7.23(a). Pole shifting, such as used for stabilization in Example 7.5(b), is used, such that with $u(t) = \phi(e(t))$,

$$u_T(t) = u(t) - \frac{k_1 + k_2}{2} e(t), \tag{7.40}$$

where $k_2 > k_1 > 0$. Then

$$G_T(s) = \frac{G(s)}{1 + [(k_1 + k_2)/2]G(s)}, \tag{7.41}$$

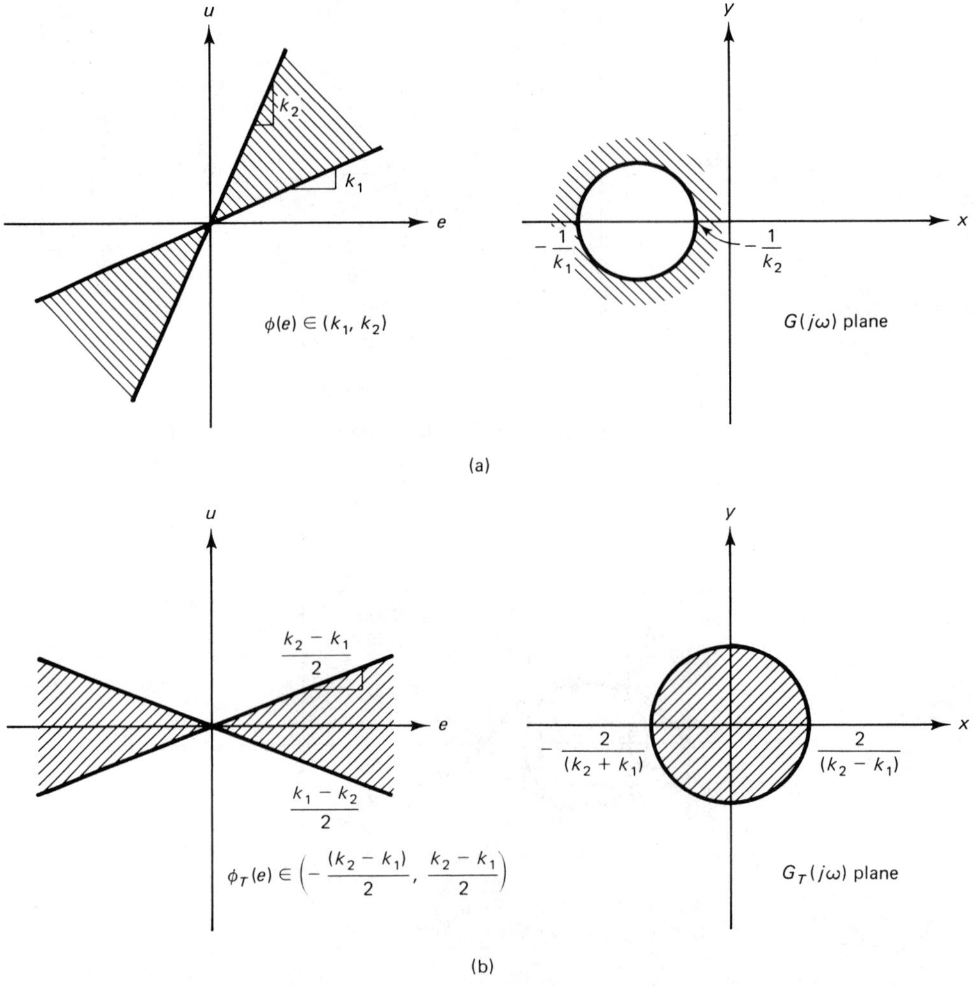

Figure 7.23 Pole-shifting effect on sector and stability criterion.

with

$$\phi_T(e) \in \left(-\frac{k_2 - k_1}{2}, \frac{k_2 - k_1}{2}\right)$$

such that the $G(j\omega)$ locus must remain internal to the critical circle with $1/k_1 > 1/k_2$. The corresponding sector and critical circle are shown on Fig. 7.23(b).

In order to find the critical circle, obtain the lines of constant $|G_T(j\omega)| = 2/(k_2 - k_1)$ such that

$$\left|\frac{G(j\omega)}{1 + [(k_1 + k_2)/2]G(j\omega)}\right| = \frac{2}{k_2 - k_1}$$

or

$$\left|\frac{[(k_1 + k_2)/2]G(j\omega)}{1 + [(k_1 + k_2)/2]G(j\omega)}\right| = \left|\frac{k_1 + k_2}{k_2 - k_1}\right|. \tag{7.42}$$

This represents the boundary of the admissible region for stability in the $[(k_1 + k_2)/2]G(j\omega)$ plane, which is obtained similar to that with constant-M circles in classical negative unity-feedback linear systems. In this manner a Nichols chart can readily be used to obtain the contours of constant $|(k_1 + k_2)/(k_2 - k_1)|$ in the $[(k_1 + k_2)/2]G(j\omega)$ plane. In order to satisfy the circle criterion the $[(k_1 + k_2)/2]G(j\omega)$ locus must remain outside this contour on the Nyquist (polar-locus) chart, or outside this contour in decibels on the Nichols chart.

Let

$$M = \left|\frac{k_1 + k_2}{k_2 - k_1}\right|,$$

so that with $G(j\omega) = X + jY$,

$$M = \frac{|X + jY|}{|1 + X + jY|} = \left|\frac{k_1 + k_2}{k_2 - k_1}\right|.$$

Then

$$\left(X + \frac{M^2}{M^2 - 1}\right)^2 + Y^2 = \frac{M^2}{(M^2 - 1)^2} \tag{7.43}$$

is the equation of a circle with radius

$$r = \left|\frac{M}{M^2 - 1}\right|$$

and center at

$$X = -\frac{M^2}{M^2 - 1}, \qquad Y = 0.$$

Consequently, circles of constant $M = |(k_1 + k_2)/(k_2 - k_1)|$ may be mapped on the $G(j\omega)$ plane. Then, the next step is to plot $[(k_1 + k_2)/2]\ G(j\omega)$. At the points of tangency of the $|(k_1 + k_2)/(k_2 - k_1)|$ and $[(k_1 + k_2)/2]\ G(j\omega)$ loci; the values of k_1 and k_2, corresponding to possible Popov sectors, can be found. That is, for a given $M = |(k_1 + k_2)/(k_2 - k_1)|$, a value of $(k_1 + k_2)/2$ can be found such that the $G(j\omega)$ locus is tangential to the constant-M locus. This is demonstrated below by means of a computer program. Given a $G(j\omega)$ plot and constant-M circles, $(k_1 + k_2)/2$ can be varied until the

$[(k_1 + k_2)/2]\,G(j\omega)$ locus is tangential to the given M circles. Thus $|(k_1 + k_2)/(k_2 - k_1)|$ and $|(k_1 + k_2)/2|$ can be solved simultaneously to yield a possible Popov sector $u/e \in (k_1, k_2)$.

Again consider Example 7.5(a) with

$$G(s) = \frac{1}{(s+1)(s+2)(s+3)}. \tag{7.38}$$

$G(j\omega)$ is graphed along with the standard Popov line in Fig. 7.17. Then Fig. 7.24 shows the M loci which are tangential to $[(k_1 + k_2)/2]G(j\omega)$ and correspond to various values of $(k_1 + k_2)/2$ given as follows:

M	1.2	1.6	2.2	3.0	4.0	5.2	6.8
$(k_1 + k_2)/2$	17	22	27	32	37	42	47

For $M = (k_1 + k_2)/(k_2 - k_1) = 4$,

$$k_1 + k_2 = 74,$$

$$k_2 - k_1 = \frac{k_1 + k_2}{4} = 18.5.$$

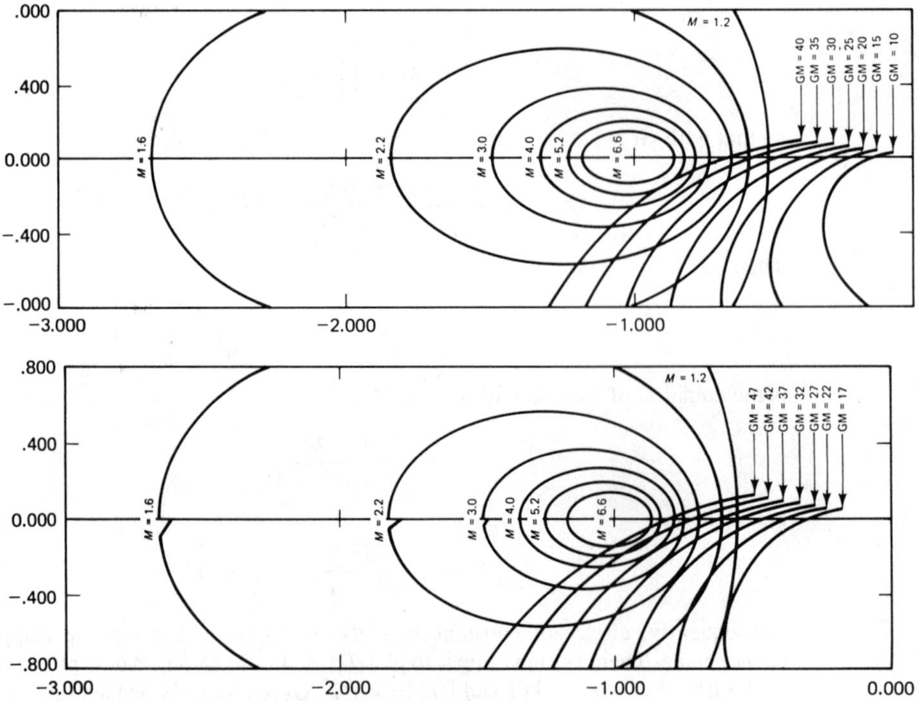

Gain margin, GM $\equiv \frac{k_1 + k_2}{2}$ changed to get the tangency condition for

$M = \frac{k_2 + k_1}{k_2 - k_1}$ and $\frac{k_1 + k_2}{2} G(j\omega)$

Figure 7.24 Constant-M circles for Example 7.5.

Therefore, $k_2 = 46.25$ and $k_1 = 27.75$. Other cases may be solved similarly. It is noted that the Nichols-chart method, although similar, is somewhat more cumbersome here.

7.3 DESCRIBING-FUNCTION METHOD

Nonlinear elements of the type studied in Section 7.2 are studied by means of a quasilinear approximation in this section. The method uses an optimum first-harmonic representation of the output of the nonlinear element for a sinusoidal input. Again, the method is found to be useful in conjunction with the Nyquist stability criterion and frequency-response characteristics.

7.3.1 Quasi-linear Optimal Approximation

Consider Fig. 7.1, where the output of the nonlinear element, for an input $e(t)$, is $u(t) = \phi(e(t))$. Now, compare this to the output $u^*(t)$ of a reference linear model which has an input $e^*(t)$, an impulse response $h^*(t)$, and an output $u^*(t)$. We consider $u^*(t)$ to be an approximation of $u(t)$ which is given by

$$u^*(t) = h * e^*(t) = \int_0^t h(t-\tau)e^*(\tau)\, d\tau. \tag{7.44}$$

In other words, compute $h * e^*(t)$ to minimize the mean-square error (m.s.e.)

$$\epsilon(h(t)) = \lim_{T\to\infty} \frac{1}{T}\int_0^t [\phi(e^*(t)) - h * e^*(t)]^2\, dt. \tag{7.45}$$

The optimal m.s.e. $h(t) \equiv h^*(t)$. Assume that

(i) $e(t)$ is a continuous bounded function, that is, $e(t) \subset C[0, \infty)$ and $|e(t)| \le b$, $|u(t)| \le m_0(b)$ such that m_0 is a finite constant that depends on every finite constant b.

(ii) $\int_0^\infty |h(t)|\, dt < \infty$, so the linear system is BIBO stable.

Since $h^*(t)$ is optimum

$$\epsilon(h) \ge \epsilon(h^*)$$

or

$$\epsilon(h) - \epsilon(h^*) \ge 0 \tag{7.46}$$

for all $h(\cdot)$. Then, from (7.45) and (7.46),

$$\begin{aligned}\Delta\epsilon(h) &= \epsilon(h) - \epsilon(h^*)\\ &= \lim_{T\to\infty} \frac{1}{T}\int_0^T \{[\phi(e^*(t)) - h * e^*(t)]^2 - [\phi(e^*(t)) - h^* * e^*(t)]^2\}\, dt.\end{aligned}$$

Then with $\Delta e^*(t) \equiv h * e^*(t) - h^* * e^*(t)$,

$$\Delta\epsilon(h) = \lim_{T\to\infty} \frac{1}{T} \int_0^T \{[\Delta e^*(t)]^2 - 2[\Delta e^*(t)][h * e^*(t) - \phi(e^*(t))]\}\, dt. \quad (7.47)$$

Therefore, $\Delta\epsilon(h) \geq 0$ if and only if the linear term in Δe^* is identically zero, or that

$$\lim_{T\to\infty} \frac{1}{T} \int_0^T \left\{\int_0^T [h(t-\sigma) - h^*(t-\sigma)]e^*(\sigma)\, d\sigma\right\}[h^* * e^*(t) - \phi(e^*(t))]\, dt = 0.$$

Then a change of variable $\tau = t - \sigma$, an interchange in the order of integration, and interchanging the limit and integration yields

$$\lim_{T\to\infty} \frac{1}{T} \int_\tau^T e^*(t-\tau)[h^* * e^*(t) - \phi(e^*(t))]\, dt = 0 \quad (7.48)$$

for all $\tau \geq 0$. Then in terms of the time-averaged cross-correlation functions $R_{e^*h^*}(\tau)$, $R_{e^*\phi}(\tau)$,

$$R_{e^*h^*}(\tau) = R_{e^*\phi}(\tau) \quad (7.49)$$

for all $\tau \geq 0$, where

$$R_{e^*h^*}(\tau) = \lim_{T\to\infty} \frac{1}{T} \int_\tau^T e^*(t-\tau)[h^* * e^*(t)]\, dt$$

and

$$R_{e^*\phi}(\tau) = \lim_{T\to\infty} \frac{1}{T} \int_\tau^T e^*(t-\tau)\phi(e^*(t))\, dt.$$

Here it should be noted that for a given reference $e^*(t)$, (7.48) does not necessarily yield a unique $h^*(t)$, and even if $h^*(t)$ is a unique solution, it may not be optimal for another class of $e^*(t)$. Thus the term "quasi-linearization" is natural.

If the input class of interest are *constants*,

$$e^*(t) \equiv e_0$$

for all $t \geq 0$. Then

$$R_{e^*\phi}(\tau) = e_0\phi(e_0)$$

and

$$R_{e^*h^*}(\tau) = e_0\phi(e_0) \quad (7.50)$$

for all $\tau \geq 0$. An obvious solution is the impulse response function

$$h^*(t) = \frac{\phi(e_0)}{e_0}\, \delta(t).$$

Therefore, for a constant input the "best" linear equivalence is a constant gain $k = u/e_0 = \phi(e_0)/e_0$.

Suppose that the class of inputs are sine waves such that

$$e^*(t) = A \sin \omega t, \tag{7.51}$$

where $A \geq 0$ is constant. If $u(t) = \phi(e^*(t))$, then $u(t)$ is periodic with a Fourier series

$$\phi(e^*(t)) = a_0 + \sum_{k=1}^{\infty} a_k \cos k\,\omega t + b_k \sin k\,\omega t. \tag{7.52}$$

If $\phi(\cdot)$ is odd, $\phi(-e) = -\phi(e)$, and a_o, a_k are zero, so that

$$u^*(t) = \phi(e^*(t)) = \sum_{k=1}^{\infty} b_k \sin k\omega t. \tag{7.53}$$

Then, from (7.48), it can be shown from the orthogonality of the sine and cosine bases that

$$R_{e^*\phi}(t) = \lim_{T\to\infty} \frac{1}{T} \int_{\tau}^{T} A \sin \omega(t - \tau) b_1 \sin \omega t \; dt$$

for all $\tau \geq 0$. Thus

$$h^* * e^*(t) = b_1 \sin \omega t,$$

or

$$h^*(t) = \frac{b_1}{A}\,\delta(t). \tag{7.54}$$

Now the quasi-optimal gain for sinusoidal input is given by

$$N(A, \omega) = \frac{b_1(A\omega)}{A},$$

where $\phi(\cdot)$ is odd, and the Fourier coefficient b_1 is generally a function of A and ω. Then the output of the nonlinear element is represented by

$$u^*(t) = \frac{b_1}{A} \sin \omega t. \tag{7.55}$$

This match of (7.55) to (7.53) is referred to as the principle of harmonic balance.

This representation, however, assumed an odd function with no bias. If it is assumed that any bias term may be so represented by a constant or subtracted out, then, in general, the quasi-optimal gain from (7.48) is derived to be

$$N(A, \omega) = \frac{b_1(A, \omega) + ja_1(A, \omega)}{A} \tag{7.56}$$

$$= \gamma(A, \omega) \exp\,[j\theta(A, \omega)], \tag{7.57}$$

where $\gamma(\cdot) = [a_1^2(\cdot) + b_1^2(\cdot)]^{1/2}/A$, $\theta(\cdot) = \tan^{-1}[a_1(\cdot)/b_1(\cdot)]$,

$$a_1(\omega) = \frac{\omega}{\pi} \int_0^{2\pi/\omega} u(t) \cos \omega t \; dt$$

and (7.58)

$$b_1(\omega) = \frac{\omega}{\pi} \int_0^{2\pi/\omega} u(t) \sin \omega t \, dt.$$

This particular representation (7.56) of the nonlinearity is called the *describing function*.

7.3.2 Foundations of Describing Functions

Equation (7.56) defines the describing function of the nonlinear time-invariant nonlinearity as the fundamental (first) component of the output Fourier series divided by the sinusoidal input reference. That is, the describing funtion is merely the fundamental Fourier-coefficient complex representation of the output divided by the amplitude of the assumed sinusoidal input. It is reasoned that any nonlinearity-generated higher-harmonic components are filtered out in traversing the feedback loop, which classically is composed of low-frequency band components. Again, refer to Fig. 7.1 for the classical regulator application. Also, it is assumed that the nonlinearity generates no subharmonics. Then it is reasonable to expect that for a sinusoidal reference signal with $e = A \sin \omega t$, the following approximation is valid:

$$u(t) \approx A\gamma(A, \omega) \sin[\omega t + \theta(A, \omega)], \tag{7.59}$$

where $\gamma(\cdot)$, $\theta(\cdot)$ are given by (7.57) and (7.58). Equation (7.59) is a generalization of (7.55) which assumed that the nonlinear element was an odd function so that $\phi(-e) = -\phi(e)$. Also, it can be shown from (7.58) and (7.56) that $N(\cdot)$ is nonnegative and independent of ω *if* $\phi(\cdot)$ is odd. Consequently, *for odd nonlinearity* characteristics the describing function is *real* such that

$$N(A, \omega) = \frac{b_1(A\omega)}{A} = \gamma(A) \tag{7.60}$$

is nonnegative and $\theta(\cdot) = 0$. Of course, in general the describing function is given by (7.57), which is repeated for convenience as

$$N(A, \omega) = \gamma(A, \omega) \exp[j\theta(A, \omega)], \tag{7.57}$$

where $\gamma(\cdot) = [a_1^2(\cdot) + b_1^2(\cdot)]^{1/2}/A$, and $\theta(\cdot) = \tan^{-1}[a_1(\cdot)/b_1(\cdot)]$. The describing function $N(A, \omega)$ for $\phi(e)$ is also called its *equivalent gain*.

In Fig. 7.1, if $u(t)$ is approximated by (7.59) for $e(t) = A \sin \omega t$, then $y(t)$ is approximated by

$$y(t) \approx A\gamma(A, \omega)|G(j\omega)| \sin[\omega t + \theta(A, \omega) + \theta_G(\omega)],$$

where $G(j\omega) = |G(j\omega)| \exp[j\theta_G(\omega)]$; or

$$\frac{y(j\omega)}{e(j\omega)} \approx N(A, \omega)G(j\omega). \tag{7.61}$$

Also, $e(j\omega) = -y(j\omega)$ for $r = 0$ in Fig. 7.1. Therefore, oscillations are sustained approximately by

$$1 + N(A, \omega)G(j\omega) = 0, \tag{7.62}$$

which effectively becomes an equivalent characteristic equation. Indeed, the approximation works real well if $|G(j\omega)|$ decreases rapidly as $\omega \to \infty$. A more rigorous justification of the describing function as a practical methodology of stability analysis is given in [B9]. In essence, (7.62) can be applied to the *Nyquist stability criterion* with $N(\cdot)$ replacing the -1 point. Of course, in general this point will not be on the real axis and may depend on ω as well as A. If the nonlinear characteristic is odd as assumed in (7.60), $\phi(\cdot)$ is a first- and third- quadrant function as assumed for the Luré problem and the Popov criterion in Section 7.2. In this case $N(\cdot)$ is real and nonnegative so that $-1/N(\cdot)$ is on the negative real axis.

Figure 7.25 shows the closed-loop regulator with the describing function $N(A, \omega)$ in place of $\phi(\cdot)$ given by Fig. 7.1. Then Fig. 7.26 shows a typical $G(j\omega)$ plot along with the graph for a general $-1/N(A, \omega)$. Note that the intersections between $G(j\omega)$ and $-1/N(A, \omega)$ represent the solutions to (7.62) and should thus correspond to limit-cycle solutions. Limit-cycle stability can be determined roughly by noting the stability of a perturbation from the limit-cycle solution. Take, for example, the limit-cycle solution (I) in Fig. 7.26. A positive perturbation $A_I + \delta A$ from (I) places the entire $G(j\omega)$ curve to the right of the point $-1/N(A_I + \delta A, \omega)$ and thus represents a stable configuration. Similarly, a negative perturbation places the $G(j\omega)$ curve to the left of the point $-1/N(A_I - \delta A, \omega)$ and thus reprsents an unstable configuration point or encircling the $(-1, j0)$ point in the Nyquist plot in linear system theory. Therefore, any perturbations from (I) converge to (I), so (I) is a stable limit cycle. In a similar manner, positive perturbations from (II) are unstable and negative perturbations are stable. Thus (II) represents an unstable limit-cycle solution.

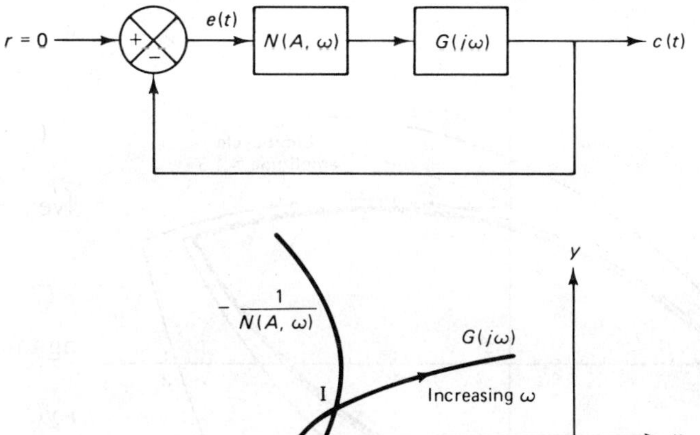

Figure 7.25 Closed-loop regulator with describing function.

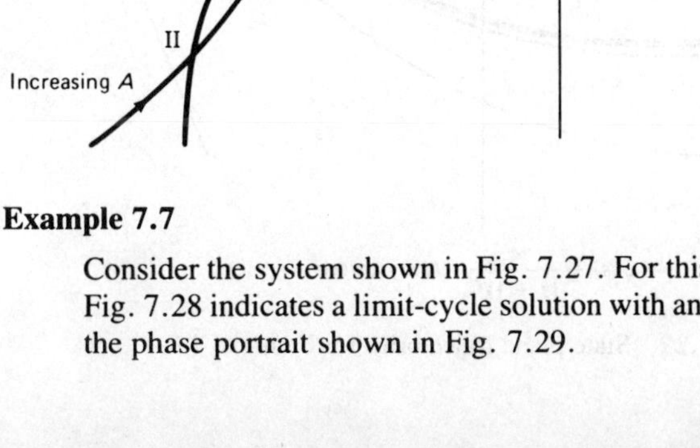

Figure 7.26 Nyquist criterion with a describing function.

Example 7.7

Consider the system shown in Fig. 7.27. For this system the gain-phase plot as shown in Fig. 7.28 indicates a limit-cycle solution with an amplitude of 1.24. This is confirmed by the phase portrait shown in Fig. 7.29.

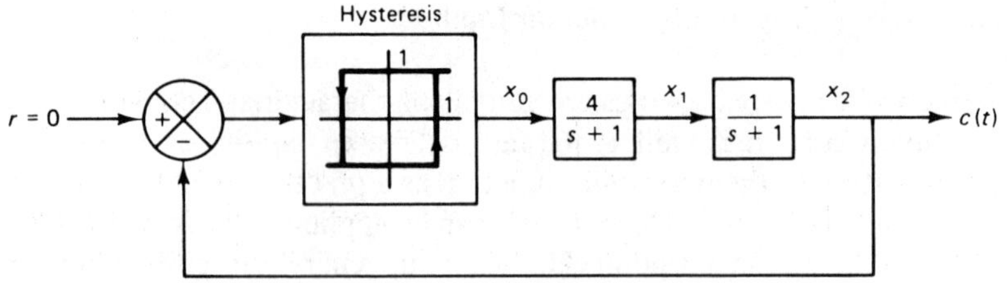

Figure 7.27 Closed-loop regulator for Example 7.7.

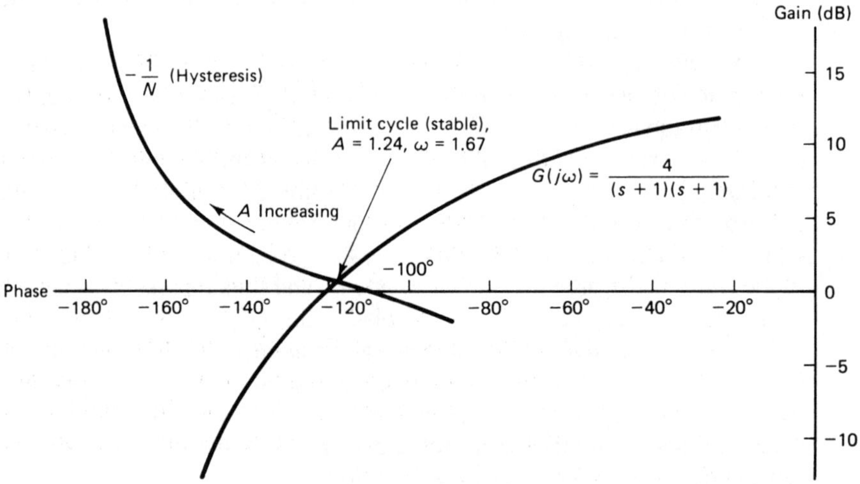

Figure 7.28 Gain-phase plot for Example 7.7.

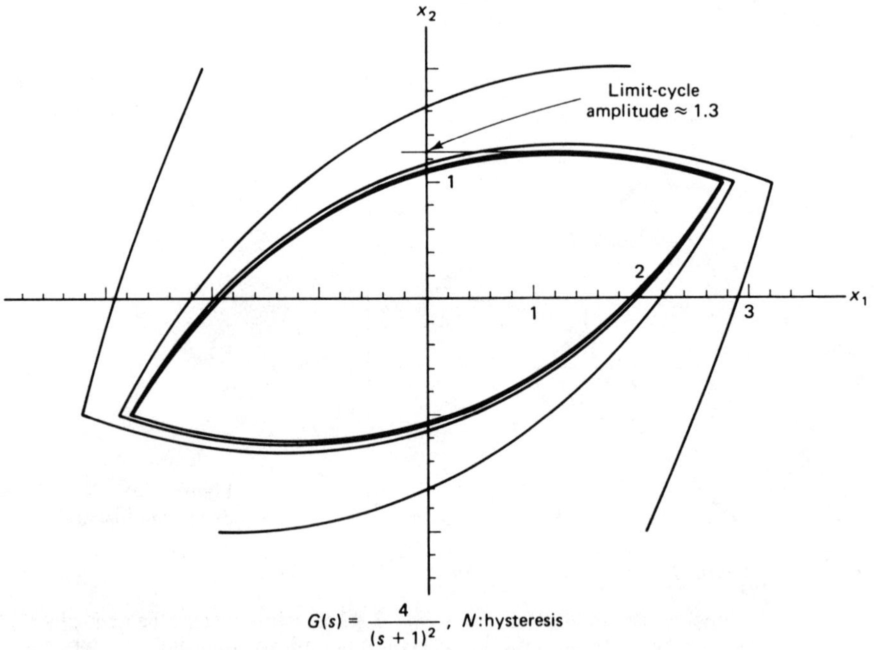

Figure 7.29 State plane trajectories for Example 7.7.

These solutions were obtained with a personal computer. The hysteresis describing function is computed as shown in the next example.

Example 7.8

Consider the *hysteresis* element (relay with hysteresis) as shown with input and output in Fig. 7.30. Assume that $u(0) = -M$. Then for $e(t) = A \sin \omega t$ for $\omega t \in [0, 2\pi]$,

$$u(t) = -M \qquad \text{for } -\pi + \sin^{-1}\frac{d}{A} < \omega t < \sin^{-1}\frac{d}{A}$$

and

$$u(t) = M \qquad \text{for } \sin^{-1}\frac{d}{A} < \omega t < \pi + \sin^{-1}\frac{d}{A}.$$

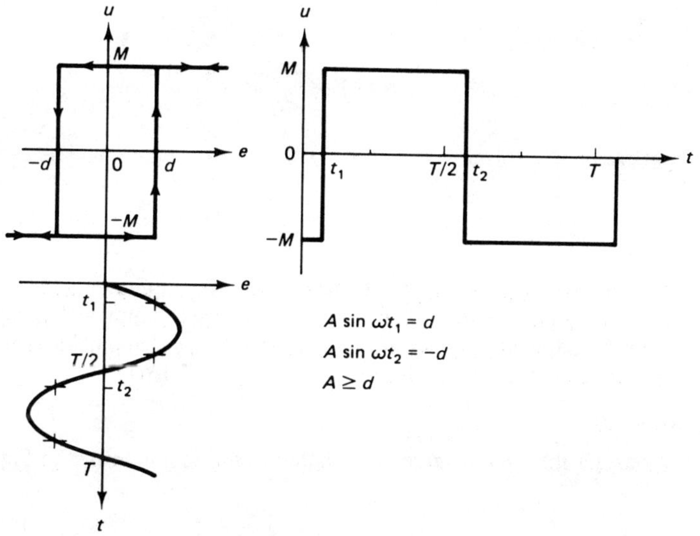

Figure 7.30 Input-output for a relay with hysteresis.

Then the Fourier series for $u(t)$ is given by

$$u(t) = a_0 + a_1 \cos \omega t + b_1 \sin \omega t + \cdots,$$

where $a_k = 0,\ k = 0, \ldots,\ b_{2l} = 0,\ l = 1, \ldots,$

$$a_1 = \frac{M}{\pi}\left[-\int_{-\pi + \sin^{-1}(d/A)}^{\sin^{-1}(d/A)} \cos \omega t \, d\omega t + \int_{\sin^{-1}(d/A)}^{\pi + \sin^{-1}(d/A)} \cos \omega t \, d\omega t\right]$$

or

$$a_1 = -\frac{4Md}{\pi A} \tag{7.63}$$

and

$$b_1 = \frac{M}{\pi}\left[-\int_{-\pi+\sin^{-1}(d/A)}^{\sin^{-1}(d/A)} \sin \omega t \, d\omega t + \int_{\sin^{-1}(d/A)}^{\pi+\sin^{-1}(d/A)} \sin \omega t \, d\omega t\right]$$

or

$$b_1 = -\frac{4M}{\pi}\left(1 - \frac{d^2}{A^2}\right)^{1/2}. \tag{7.64}$$

Here it is assumed that $A > d$.

Therefore, the describing function (7.56) or (7.57) is given by

$$N(A) = \frac{4M}{\pi A}\left[-\frac{d}{A} + j\left(1 - \frac{d^2}{A^2}\right)^{1/2}\right]$$

or

$$N(A) = \frac{4M}{\pi A} e^{j\theta(A)}, \tag{7.65}$$

where

$$\theta(A) = -\sin^{-1}\frac{d}{A}.$$

Notice that the phase shift $\theta(A)$ decreases with amplitude, which is typical of a nonlinear element with memory such as the hyseteresis exhibits. The sign of the phase shift, however, does not depend on whether the memory element is active or passive [recall active and passive hystresis (Section 7.2.4 and Fig. 7.13)].

Example 7.9

Consider the simple *saturation* characteristic that is shown in Fig. 7.31. Here

$$u = \phi(e) = \begin{cases} \frac{M}{d}e, & e \in (-d, d) \\ M, & d \leq e \\ -M, & -d \geq e. \end{cases} \tag{7.66}$$

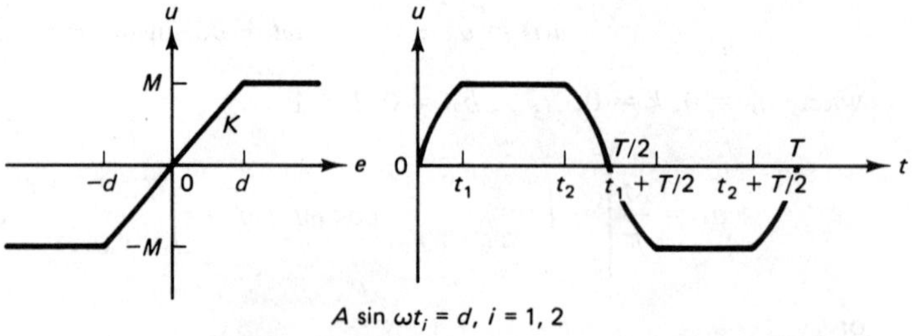

Figure 7.31 Saturation characteristic.

Apparently, on the linear range,

$$N(A) = \frac{M}{d}, \qquad e \in (-d, d). \tag{7.67}$$

Let $\theta_1 = \omega t_1$ and $\theta_2 = \omega t_2 = k\pi - \theta_1$ in Fig. 7.31 such that $\sin \theta_1 = d/A$. Then

$$u(t) = \begin{cases} \frac{M}{d} A \sin \omega t, & k\pi - \theta_1 < \omega t < k\pi + \theta_1 \\ M, & 2k\pi + \theta_1 \le \omega t \le (2k+1)\pi - \theta_1 \\ -M, & (2k-1)\pi + \theta_1 \le \omega t \le 2k\pi - \theta_1, \end{cases}$$

$k = 0, \pm 1, \pm 2, \cdots$.

The Fourier series of $u(t)$ with $\phi(\cdot)$ again odd becomes

$$u(t) = b_1 \sin \omega t + b_3 \sin 3\omega t + \cdots.$$

Then, performing the required integrations over one period and dividing by A for $d \le A$, it is seen that

$$N(A) = \frac{M}{\pi}\left[\left(\frac{2}{d}\theta_1 - \frac{1}{d}\sin 2\theta_1\right) + \frac{4}{A}\cos\theta_1\right], \tag{7.68}$$

where $\theta_1 = \sin^{-1}(d/A)$. In other words,

$$N(A) = K = \frac{M}{d}, \qquad \text{for all } A \le d$$

and

$$N(A) = \frac{2K}{\pi}\left[\sin^{-1}\left(\frac{d}{A}\right) + \frac{d}{A}\left(1 - \frac{d^2}{A^2}\right)^{1/2}\right] \qquad \text{for } A > d. \tag{7.69}$$

Obviously, with $\phi(\cdot)$ again odd, $N(A)$ is real, nonnegative, and independent of ω. If M is fixed in the limit as $d \to 0$, with L'Hospital's rule applied to (7.68), the *relay describing function* becomes

$$N(A) = \frac{4M}{\pi A}. \tag{7.70}$$

The describing function for the unit limiter [i.e., (7.69) with $M = d = 1$] is given by

$$N_u(A) = \begin{cases} 1, & \text{for } A \le 1, \\ \frac{2}{\pi}\left[\sin^{-1}\frac{1}{A} + \frac{1}{A}\left(1 - \frac{1}{A^2}\right)^{1/2}\right] & \text{for } A \ge 1. \end{cases} \tag{7.71}$$

Then it is seen that for (7.69),

$$N\left(\frac{A}{d}\right) = \begin{cases} K & \text{for } \frac{A}{d} \le 1, \\ KN_u\left(\frac{A}{d}\right) & \text{for } \frac{A}{d} \ge 1. \end{cases} \tag{7.72}$$

The unit-limiter describing function $N_u(A)$ is a basic function for the generation of other describing functions.

7.3.3 Properties of Describing Functions

The analysis in the preceding section shows that describing functions for odd characteristics are real, nonnegative, and independent of ω. Three such describing functions—saturation, relay, and a relay with hysteresis—were computed in Examples 7.8 and 7.9. Other typical characteristics and their describing functions are given in Table 7.2. Notice the basic role of $N_u(A)$, the unit limiter, in constructing many other describing functions.

If two characteristics appear in parallel with $u(t) = \phi_1(e) + \phi_2(e)$, it is readily seen from (7.56) that their describing functions, DF, are additive. That is,

$$N(A, \omega) = N_1(A, \omega) + N_2(A, \omega),$$

where $N_1(\cdot)$, $N_2(\cdot)$ are the DF for $\phi_1(\cdot)$ and $\phi_2(\cdot)$, respectively. For example, consider Fig. 7.32. A computation of $N(\cdot)$ directly (see Table 7.2) yields

$$N\left(\frac{A}{d}\right) = \begin{cases} K, & \dfrac{A}{d} \le 1, \\ K_1 + (K - K_1)N_u\left(\dfrac{A}{d}\right), & \dfrac{A}{d} \ge 1, \end{cases}$$

where $N_u(\cdot)$ is the DF of the unit limiter. Similarly, in Fig. 7.32,

$$N_1\left(\frac{A}{d}\right) = \begin{cases} 0, & \dfrac{A}{d} \le 1, \\ K_1\left[1 - N_u\left(\dfrac{A}{d}\right)\right], & \dfrac{A}{d} \ge 1 \end{cases}$$

and

$$N_2\left(\frac{A}{d}\right) = \begin{cases} K, & \dfrac{A}{d} \le 1, \\ KN_u\left(\dfrac{A}{d}\right), & \dfrac{A}{d} \ge 1. \end{cases}$$

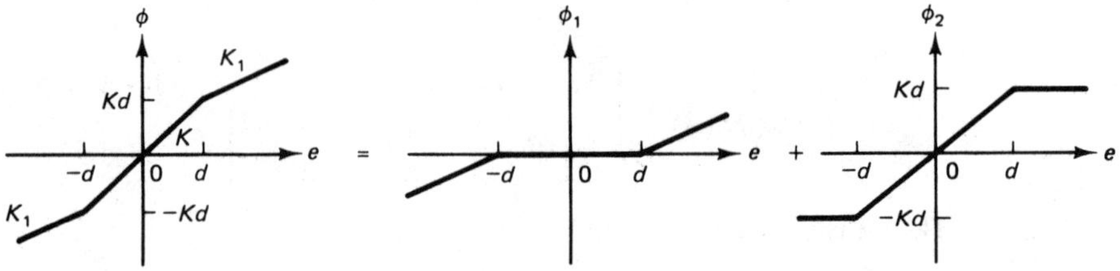

Figure 7.32 Addition of nonlinear characteristics.

Next we consider multiple nonlinear elements in series with linear elements between them. This, of course, complicates the problem considerably. Example 7.7 showed the limit-cycle determination for a single nonlinearity. Of course, a system may contain more than one nonlinearity. For these cases, the simple graphical ap-

TABLE 7.2 COMMON DESCRIBING FUNCTIONS

$f(x)$	Re $N(A)$*	Im $N(A)$
1. x^{2n-1} $(n = 1, 2, 3, \ldots)$	$\frac{2n-1}{2n}\frac{2n-3}{2n-2}\frac{2n-5}{2n-4}\cdots\frac{3}{4}A^{2n-2}$	0
2. $x^{2n-1}\lvert x\rvert$ $(n = 1, 2, 3, \ldots)$	$\frac{4}{\pi}\frac{2n}{2n+1}\frac{2n-2}{2n-1}\frac{2n-4}{2n-3}\cdots\frac{2}{3}A^{2n} - 1$	0
3. 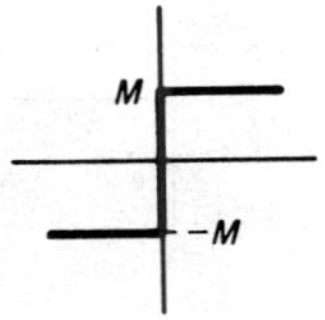	$\frac{4M}{\pi A}$	0
4. 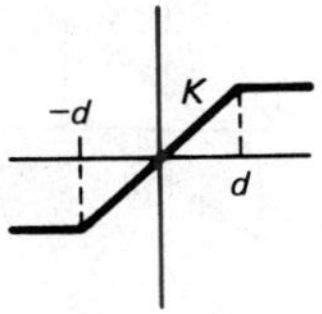	K for $\frac{A}{d} \le 1$ $KN_u\left(\frac{A}{d}\right)$ for $\frac{A}{d} \ge 1$	0
5. 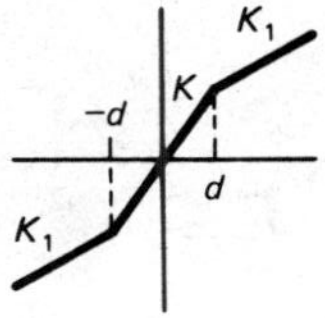	K for $\frac{A}{d} \le 1$ $K_1 + (K - K_1)N_u\left(\frac{A}{d}\right)$ for $\frac{A}{d} \ge 1$	0
6. 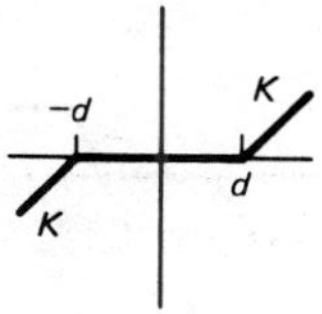	0 for $\frac{A}{d} \le 1$ $K\left[1 - N_u\left(\frac{A}{d}\right)\right]$ for $\frac{A}{d} \ge 1$	0
7. 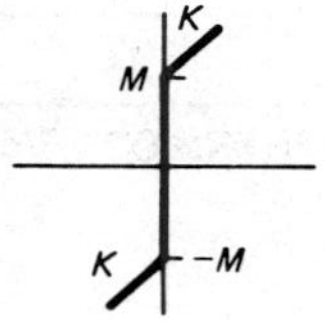	$K + \frac{4M}{\pi A}$	0

TABLE 7.2 COMMON DESCRIBING FUNCTIONS

	$f(x)$	Re $N(A)$*		Im $N(A)$	
8.	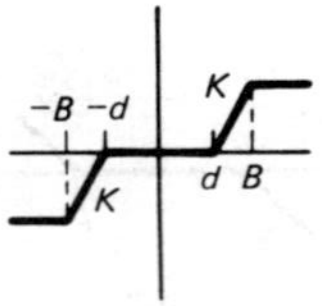	$K\left[1 - N_u\left(\frac{A}{d}\right)\right]$	for $1 \le \frac{A}{d} \le \frac{B}{d}$	0	
		$K\left[N_u\left(\frac{A}{B}\right) - N_u\left(\frac{A}{d}\right)\right]$	for $\frac{A}{d} \ge \frac{B}{d}$		
9.	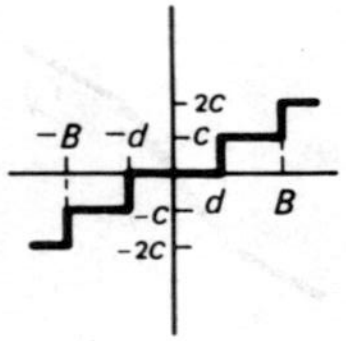	0	for $A \le d$	0	
		$\frac{4C}{\pi A}\sqrt{1 - \left(\frac{d}{A}\right)^2}$	for $d \le A \le B$		
		$\frac{4C}{\pi A}\left[\sqrt{1 - \left(\frac{d}{A}\right)^2} + \sqrt{1 - \left(\frac{d}{A}\right)^2}\right]$	for $A \ge B$		
10.	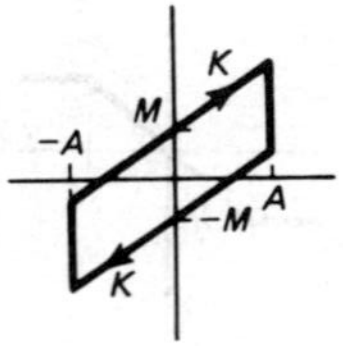	K		$\frac{4M}{\pi A}$	
11.	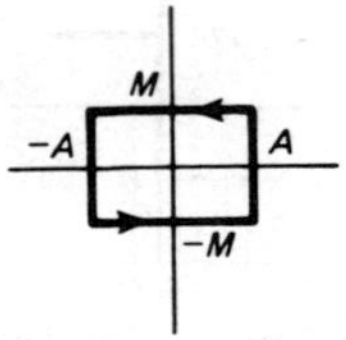	0		$-\frac{4M}{\pi A}$	
12.	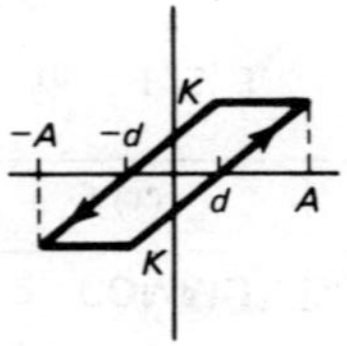	0	for $\frac{A}{d} \le 1$	0	for $A \le d$
		$\frac{K}{2}\left[1 - N_u\left(\frac{A/d}{2 - A/d}\right)\right]$	for $\frac{A}{d} \ge 1$	$\frac{4Kd(d - A)}{\pi A^2}$	for $A \ge d$

13.	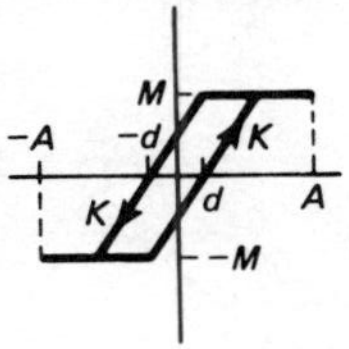	$\begin{cases} 0 & \text{for } \dfrac{A}{d} \le 1 \\ \dfrac{K}{2}\left[1 - N_u\left(\dfrac{A/d}{2 - A/d}\right)\right] & \text{for } 1 \le \dfrac{A}{d} \le \dfrac{M}{d} \\ \dfrac{K}{2}\left[N_u\left(\dfrac{A}{M}\right) - N_u\left(\dfrac{A/d}{2 - M/d}\right)\right] & \text{for } A \ge M \end{cases}$	$\begin{cases} 0 & \text{for } A \le d \\ \dfrac{4Kd(d - A)}{\pi A^2} & \text{for } d \le A \le M \\ \dfrac{4Kd(d - M)}{\pi A^2} & \text{for } A \ge M \end{cases}$
14.	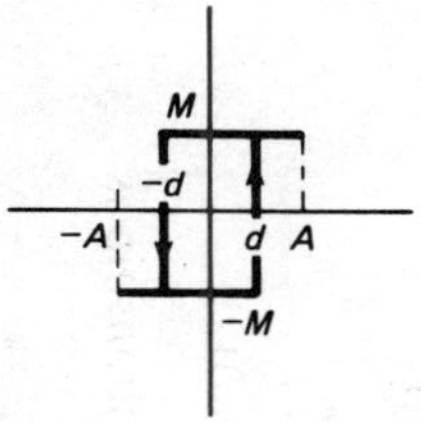	$\begin{cases} 0 & \text{for } A \le d \\ \dfrac{4M}{\pi A}\sqrt{1 - \left(\dfrac{d}{A}\right)^2} & \text{for } A \ge d \end{cases}$	$\begin{cases} 0 & \text{for } A < d \\ -\dfrac{4dM}{\pi A^2} & \text{for } A > d \end{cases}$
15.	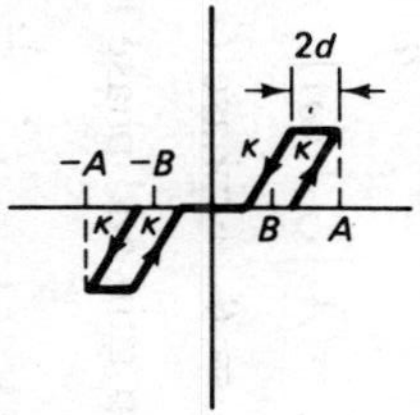	$\begin{cases} 0 & \text{for } A \le d + B \\ \dfrac{K}{2}\left[1 - N_u\left(\dfrac{A}{d + B}\right) - N_u\left(\dfrac{A}{A - 2d}\right) - N_u\left(\dfrac{A}{B - d}\right)\right] & \text{for } A \ge d + B \end{cases}$	$\begin{cases} 0 & \text{for } A \le d + B \\ \dfrac{4Kd(d + B - A)}{\pi A^2} & \text{for } A \ge d + B \end{cases}$
16.	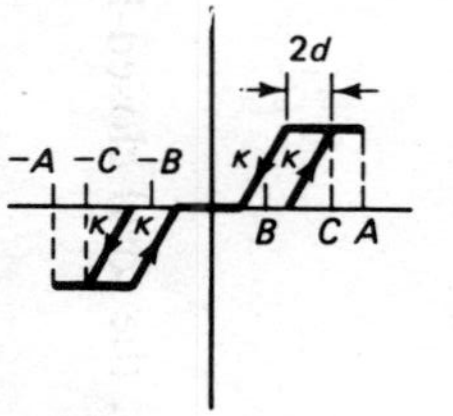	$\begin{cases} 0 & \text{for } A \le d + B \\ \dfrac{K}{2}\left[1 - N_u\left(\dfrac{A}{d + B}\right) + N_u\left(\dfrac{A}{A - 2d}\right) - N_u\left(\dfrac{A}{B - d}\right)\right] & \text{for } d + B \le A \le C \\ \dfrac{K}{2}\left[N_u\left(\dfrac{A}{C}\right) - N_u\left(\dfrac{A}{d + B}\right) + N_u\left(\dfrac{A}{C - 2d}\right) - N_u\left(\dfrac{A}{B - d}\right)\right] & \text{for } A \ge C \end{cases}$	$\begin{cases} 0 & \text{for } A \le d + B \\ \dfrac{4Kd(d + B - A)}{\pi A^2} & \text{for } d + B \le A \le C \\ \dfrac{4Kd(d + B - C)}{\pi A^2} & \text{for } A \ge C \end{cases}$
17.	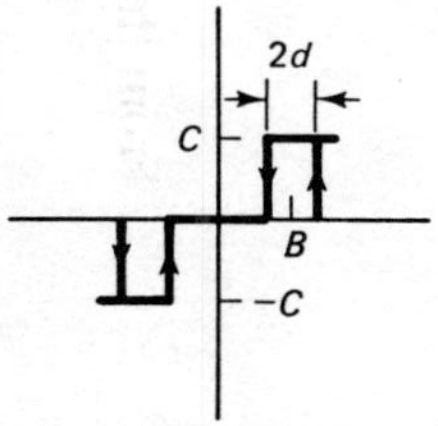	$\begin{cases} 0 & \text{for } A < d + B \\ \dfrac{2C}{\pi A}\left[\sqrt{1 - \left(\dfrac{B - d}{A}\right)^2} + \sqrt{1 + \left(\dfrac{B + d}{A}\right)^2}\right] & \text{for } A > d + B \end{cases}$	$\begin{cases} 0 & \text{for } A \le d + B \\ -\dfrac{4dC}{\pi A^2} & \text{for } A \ge d + B \end{cases}$

*$N_u(A)$ is the unit-limiter describing function, which is given by (7.71).

proach shown in Section 7.3.2 must be modified. Consider the system shown in Fig. 7.33. One approach to analyzing this system is to form a "composite nonlinearity" consisting of, for example, $N_1(A_1)$, $G_1(j\omega)$, and $N_2(A_2)$ and finding the corresponding describing function. (Note that it is assumed that the nonlinearities are functions of amplitude only. This is not as restrictive as it may seem. Many practical nonlinearities fall into this category.) The system would then consist of a single "composite nonlinearity" and a single linearity, $G_2(j\omega)$, and the analysis could proceed much as in Section 7.3.2. The obvious problem is, of course, that computing the describing function (DF) of the lumped system would generally be very difficult. Since $G_1(j\omega)$ is a function of frequency, it also follows that the "composite nonlinearity" would be a function of frequency even though the original nonlinearities are not. For these reasons, this approach is not very practical for easy solutions of typical problems.

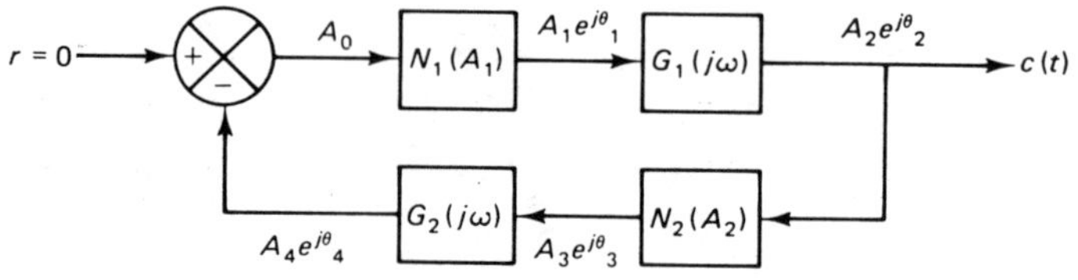

Figure 7.33 Multiple-nonlinearity loop.

Another approach is to treat each block in Fig. 7.33 individually. This can be done by assuming a sinusoidal input with amplitude A_0 and finding the resulting outputs of each element shown (e.g., A_1, A_2, etc.). Again this is valid only if $G_1(j\omega)$ and $G_2(j\omega)$ are stable and low pass. It must be remembered that the DF assumes a sinusoidal input to the nonlinearity, so this analysis is again approximate.

The calculations proceed in a straightforward manner from Fig. 7.33:

$$A_1 = A_0\gamma_1(A_0),$$

$$\theta_1 = \tan^{-1}\frac{b_{11}(A_0)}{a_{11}(A_0)},$$

$$A_2 = A_1|G_1(j\omega)|,$$

$$\theta_2 = \theta_1 + \text{phase}\{G_1(j\omega)\},$$

$$A_3 = A_2\gamma_2(A_2),$$

$$\theta_3 = \theta_2 + \tan^{-1}\frac{b_{12}(A_2)}{a_{12}(A_2)},$$

$$A_4 = A_3|G_2(j\omega)|,$$

$$\theta_4 = \theta_3 + \text{phase}\{G_2(j\omega)\}.$$

From this, the total closed-loop gain and phase relationship is

$$N_{\text{tot}}(A_0, \omega) = \frac{A_4}{A_0}e^{j\theta_4}. \tag{7.73}$$

Extending (7.62) yields

$$1 + N_{\text{tot}}(A_0, \omega) = 0. \tag{7.74}$$

Solutions to (7.74) predict limit cycles (under the DF assumptions). To solve (7.74), rewrite it as

$$1 + N_{\text{tot}}(A_0, \omega) = 1e^{-j\pi} \tag{7.75}$$

and plot $N_{\text{tot}}(A_0, \omega)$ on the gain–phase plane for set values of frequency while varying A_0 (Fig. 7.34).

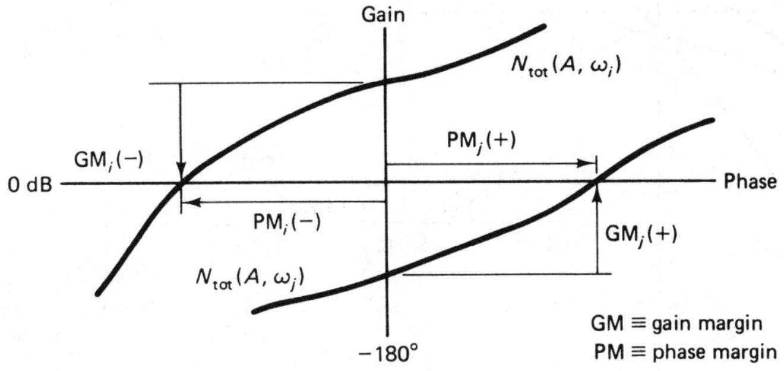

Figure 7.34 Composite gain–phase plot for Fig. 7.33.

Note that for each frequency, there exists a particular gain and phase (margin) which is the amount by which the loop gain or phase shift must be increased to produce a limit-cycle solution at that frequency. Thus a gain and phase margin of zero indicates that a limit cycle exists at that frequency. Additionally, gain and phase margins can be used in the general analysis of the nonlinear system much as they are used in linear system analysis with appropriate precautions with respect to the assumptions.

Example 7.10

Consider the system shown in Fig. 7.35. From the computer generated gain–phase plots in Fig. 7.36, it can be seen that for angular frequencies less than about 0.3 rad/sec and greater than about 0.9 rad/sec, the gain and phase margins are both positive. For angular

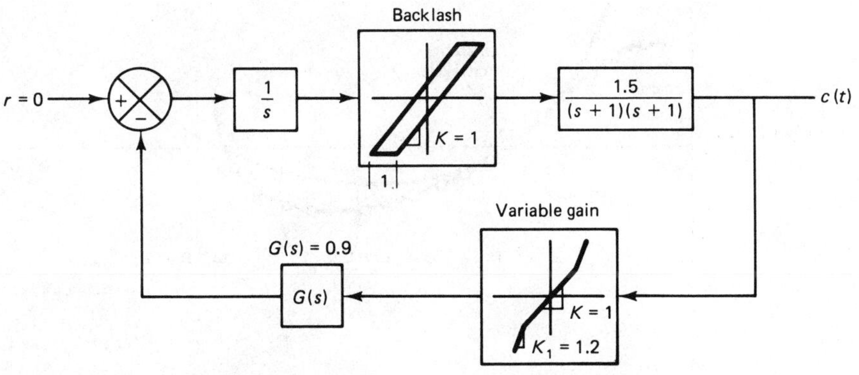

Figure 7.35 Nonlinear regulator for Example 7.10.

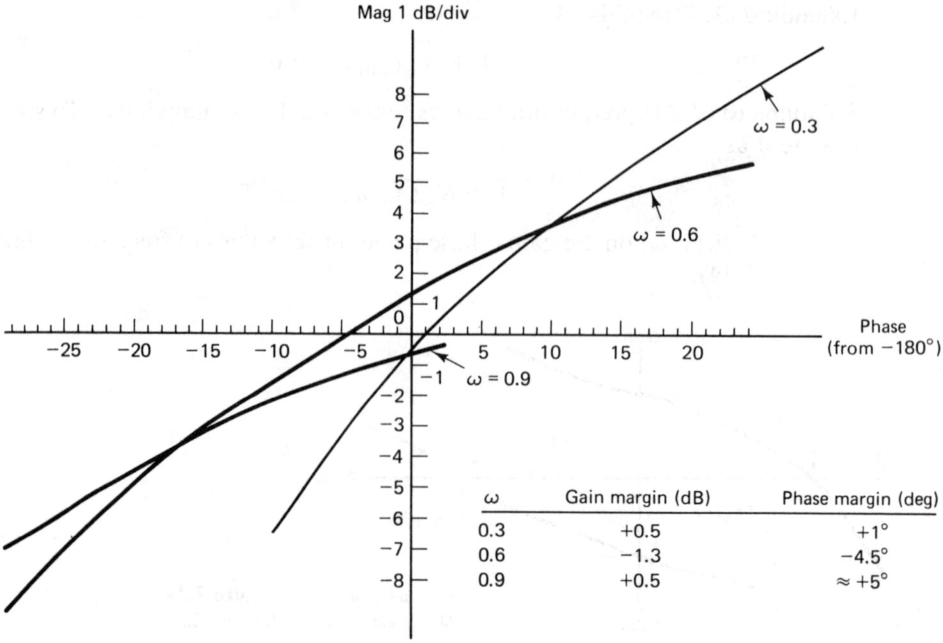

Figure 7.36 Magnitude–phase plot for Example 7.10.

frequencies greater than about 0.3 $\sec^{-1}$ and less than about 0.9 $\sec^{-1}$ (in particular, 0.6 $\sec^{-1}$), the gain and phase margins are both negative. The gain and phase margins are plotted in Fig. 7.37 and show an unstable limit cycle at 0.34 $\sec^{-1}$ with an amplitude of 0.49 at the output. Also, the analysis predicts that there is a stable limit cycle at 0.84 $\sec^{-1}$ with amplitude of 5.77.

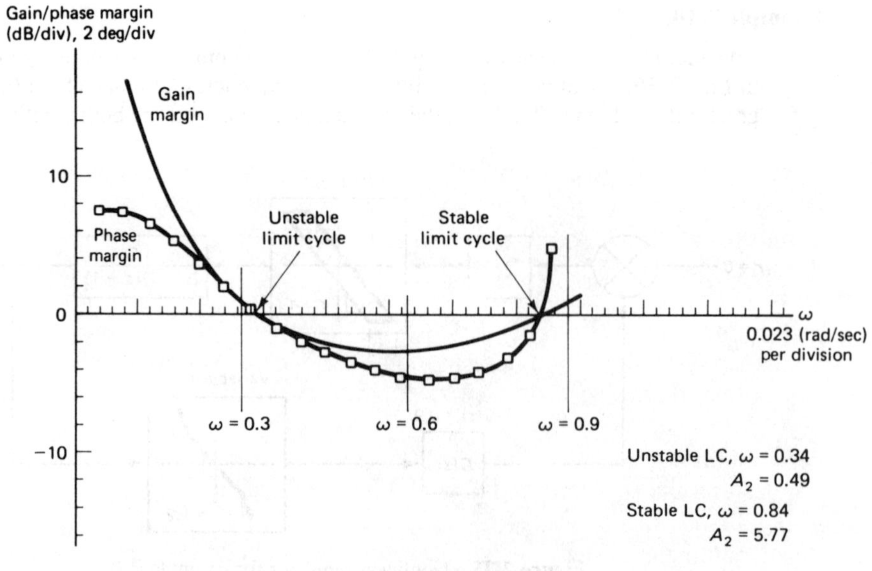

Figure 7.37 Gain and phase margin for Example 7.10.

Example 7.11

Consider the traditional first- and third-quadrant nonlinear characteristic in Fig. 7.1 so that $\phi(\cdot) \in [k_1, k_2]$. Further, assume that $\phi(\cdot)$ is odd. Is the corresponding DF similarly limited with $N(\cdot) \in [k_1, k_2]$?

Solution

$$N(A) = \frac{\omega}{\pi A} \int_0^{2\pi/\omega} \phi(A \sin \omega t) \sin \omega t \, dt$$

$$= \frac{1}{\pi A} \int_0^{2\pi} \phi(A \sin \theta) \sin \theta \, d\theta.$$

Then, since $\phi(\cdot) \geq k_1$, $\phi(A \sin \theta) \geq k_1(A \sin \theta)^2/(A \sin \theta)$, and similarly, $\phi(A \sin \theta) \leq k_2(A \sin \theta)^2/A \sin \theta$. Therefore,

$$k_1 \leq N(A) \leq k_2.$$

In this case, the describing function presumes the Aizerman conjecture. In addition, if the Aizerman conjecture indicates stability, the describing function will not locate any limit cycles. Excellent discussions of describing functions and their numerous applications are given in [A7] and [T1].

7.4 CONCLUDING REMARKS

Stability analysis of a class of nonlinear feedback control systems culminates in two-convenient frequency-domain methods in this chapter. The application of both techniques involves logical extensions of the Nyquist stability criterion, which is a classical approach to linear controller design. The describing-function method typically yields an optimistic range of acceptable design parameter values. On the other hand, the Popov approach is based on Liapunov's direct method, and thereby being only a sufficient condition, may yield somewhat overconstrained admissible parameters.

Stability analysis of nonlinear systems in the time domain is analyzed by Liapunov's direct method in Chapter 6. The utility of this method for control system design is introduced there. Further application of this powerful method for control system design is discussed in Vol. 2, which provides a focus to nonlinear system design along with new methods of analysis which are particularly convenient for bilinear systems and certain combinations of such systems that seem prevalent in practice. Many of the nonlinear systems studied in this volume may be synthesized as bilinear systems with appropriate feedback control. The Van der Pol equation is one such example.

Although the real world is composed primarily of nonlinear dynamical processes, and nonlinear system designs can yield superior performance, most engineering designs and even research continue to be based primarily on linear system theory. The demands of modern society and the availability of economical microprocessors make nonlinear system design and research on related new methodologies timely indeed.

EXERCISES

7.1. Determine a range of nonlinear gains (k_1, k_2) for which the nonlinear regulator (Fig. 7.1) is stable if

$$G(s) = \frac{1}{s(s^2 + s + 1)}.$$

Is Aizerman's conjecture valid here?

7.2. Repeat Exercise 7.1 for

$$G(s) = \frac{s + 2}{(s - 1)(s + 1)}.$$

7.3. By the Popov method, determine a range of gains (k_1, k_2) for stable nonlinear regulation (Fig. 7.1) of the linearized neutron kinetics as approximated by (3.90), or

$$G(s) = \frac{s + 0.1}{s(s + 50)} = \frac{\delta q/q_e}{\delta u}(s),$$

where $\delta u = u$ in Fig. 7.1.

7.4. Repeat Exercise 7.3 if heat-exchange dynamics is included with

$$\frac{\delta T_u}{\delta k}(s) = \frac{k_1 q_e(s + 0.1)}{s(s + 50)(s + 0.2)}$$

and

$$\delta k(s) = \delta u(s) + \alpha \delta T_u(s),$$

so that

$$G(s) = \frac{\delta T_u(s)}{\delta u(s)}.$$

Consider the effect of values of the temperature coefficient of reactivity α on the regulator system stability for conditions that establish $k_1 q_e$.

7.5 Repeat Exercise 7.3 if the feedback controller includes PI control so that

$$G(s) = \frac{(k_p s + k_i)(s + 0.1)}{s^2(s + 50)},$$

where you select k_p and k_i values for good response.

7.6. Repeat Exercise 7.5 by the describing-function method if the control is constrained by $|u(t)| \leq 1$.

7.7. Compute the describing function for a dead-zone hysteresis combination as shown in Table 7.2.

7.8. Compute the describing function for 8. in Table 7.2.

7.9. Show that the describing function for an odd nonlinearity does satisfy Aizerman's conjecture. That is, if the graph of the nonlinearity lies in the sector $[k_1, k_2]$, the corresponding describing function has values on $[k_1, k_2]$. It is assumed that e is appropriately bounded for the describing function to exist.

7.10. Consider Fig. 7.1 with $u = e^3$ and

$$G(s) = \frac{1}{(s + 1)(s + 2)}.$$

(a) Compute the describing function for e^3.

(b) Assume that $e(t) = A \sin \omega t$ and compute the frequency and amplitude of sustained closed-loop oscillation.

7.11. Consider the closed loop regulator of Figure 7.25 with the hysteresis (17) of Table 7.2 (where $c = 2$, $B = 0.75$, $d = 0.5$). The linear system is $G(s) = k/(s+1)^2$ where $k = 10$. Study the possibility of limit cycles for $k = 10$ and as a function of the gain k.

7.12. Study the regulator of Figure 7.2 with the bang-bang control (3) of Table 7.2 (with $M = 1$) and $G(s) = 1/s(s+1)^2$.

7.13. Study the regulator of Figure 7.2 with the hysteresis (10) in Table 7.2 ($k = 1, M = \pi/4$) and $G(s) = -k/(s+1)$. Analyze stability as a function of k.

7.14. Consider the architecture of Figure 7.33 where $N_1(A_1)$ is the limiter (4) in Table 7.2, $N_2(A_2)$ is the bang-bang selector (3) in Table 7.2 and $G_1(s) = 1/s(s+1)$ with $G_2(s) = 1$. Show that the block approach to the computation of the total nonlinearity implies that the limiter N_1 is irrelevant to the study of limit cycles.

7.15. Consider 7.14 where $N_2(A_2)$ is the nonlinearity (7) in Table 7.2. Compute the total nonlinearity N_{tot}. Are limit cycles possible? Assume $K = 1, M = \pi/4$.

REFERENCES

[A1] Ash, M., *Nuclear Reactor Kinetics,* McGraw-Hill, New York, 1965.

[A2] Aizerman, M. A., and Gantmacher, R. R., *Absolute Stability of Regulator Systems,* Holden-Day, San Francisco, 1964.

[A3] Aggarwal, J. R., *Notes on Nonlinear Systems,* D. Van Nostrand, Princeton, NJ, 1972.

[A4] Andronow, A. A., and Chaikin, C. E., *Theory of Oscillations,* Princeton University Press, Princeton, NJ, 1949.

[A5] Amelyanov, S. V., *Theory of Variable Structure Systems,* Science Press, Moscow, 1970.

[A6] Åstrom, K. J., Anton, J. J., and Arzen, K. E., "Expert Control," *Automatica 22,* 277–286, 1986.

[A7] Atherton, D., *Nonlinear Control Engineering* (2nd ed.), Van Nostrand Reinhold, New York, 1982.

[B1] Brockett, R. W., "Nonlinear Systems and Differential Geometry," *Proc. IEEE 64,* 61–72, 1978.

[B2] Bell, G. I., Perelson, A. S., and Pimbley, G. H., Jr. (eds.), *Theoretical Immunology,* Marcel Dekker, New York, 1978.

[B3] Bendixson, I., "Sur les courbes, définies par les équations différentielles," *Acta Math. 24,* 1901.

[B4] Bellman, R., *Introduction to Matrix Analysis,* McGraw-Hill, New York, 1960.

[B5] Berger, J. S., and Perlmutter, D. D., "Chemical Reactor Stability by Lyapunov's Direct Method," *AIChE J. 11,* 749–762, 1964.

[B6] Brockett, R. W., and Fory, L. J., "On the Stability of Systems Containing Time-Varying Gain," *Proc. Allerton Conf. on Circuits and Systems,* University of Illinois, Urbana, IL, 1964.

[B7] Brocket, R. W., and Willems, J. L., "Frequency Domain Stability Criteria, I and II," *IEEE Trans. Autom. Control AC-10,* 255–261, 407–412, 1965.

[B8] Bergen, A. R., and Baker, R. A., "On the Third-Order Systems and Aizerman's Conjecture," *IEEE Trans. Autom. Control AC-17,* 220–222, 1972.

[B9] Bergen, A. R., and Franks, R. L., "Justification of the Describing Function Method," *SIAM J. Control Optim. 9,* 568–589, 1979.

[B10] Bellman, R., *Stability Theory of Differential Equations,* McGraw-Hill, New York, 1953.

[C1] Coddington, E. A., and Levinson, N., *Theory of Ordinary Differential Equations,* McGraw-Hill, New York, 1955.

[C2] Crosbie, E. J., Hardy, J. D., and Fessenden, E., "Electrical Analog Simulation of Temperature Regulation in Man," *IRE Trans. Bio-Med. Electron. BME-8,* 4, 1961.

[C3] Cooper, L. N., "Theory of an Immune System Retrovirus," *Proc. Natl. Acad. Sci. USA 83,* 9159–9163, 1983.

[C4] Cunningham, W. J., *Introduction to Nonlinear Analysis,* McGraw-Hill, New York, 1958.

[C5] Chen, C. T., *Introduction to Linear System Theory,* Holt, Rinehart, and Winston, New York, 1971.

[D1] DeBoer, R. J., and Hogeweg, P., "Implications of Macrophage T-Lymphocyte Interactions for Tumor Rejectability," in [H5], 120–140.

[D2] Dewey, A. G., and Jury, E. I., "A Stability Inequality for a Class of Nonlinear Feedback Systems," *IEEE Trans. Autom. Control AC-11,* 54–62, 1966.

[E1] Eisen, H., *Immunology,* Harper & Row, New York, 1974.

[F1] Friedland, B., *Control System Design,* McGraw-Hill, New York, 1986.

[F2] Fu, K. S., Gonzalez, R. C., and Lee, C. S. G., *Robotics,* McGraw-Hill, New York, 1987.

[F3] Fitts, R., "Two Counter-Examples to Aizerman's Conjecture," *IEEE Trans. Autom. Control AC-11,* 553–556, 1966.

[F4] Fuji, K., and Shoji, K., "Verification of the Aizerman and/or Kalman Conjecture," *IEEE Trans. Autom. Control AC-17,* 406–408, 1972.

[G1] Goldstein, H., *Classical Mechanics,* Addison-Wesley, Reading, MA, 1964.

[G2] Guillemin, E. A., *The Mathematics of Circuit Analysis,* Wiley, New York, 1949.

[G3] Grodins, F. S., *Control Theory and Biological Systems,* Columbia University Press, New York, 1963.

[G4] Grodins, F. S., and James, G., "Mathematical Models in Respiratory Regulation," *Ann. N.Y. Acad. Sci. 109,* 852–863, 1963.

[G5] Gantmacher, F. R., *The Theory of Matrices,* Vols. 1 and 2, Chelsea, New York, 1959.

[G6] Grayson, L. P., "The Status of Synthesis Using Liapunov's Method," *Automatica 3,* 91–92, 1965.

[H1] Hsu, J. C., and Meyer, A. V., *Modern Control Principles and Applications,* McGraw-Hill, New York, 1968.

[H2] Hardy, J. D., "Physiology of Temperature Regulation," *Physiol. Rev. 41,* 541–606, 1961.

[H3] Hill, T. L., and Kedem, O., "Studies in Irreversible Thermodynamics. III. Models for Steady State and Active Transport Across Membranes," *J. Theor. Biol. 10,* 399–441, 1966.

[H4] Hobart, M. J., and McConnell, I., *The Immune System: A Course on the Molecular and Cellular Basis of Immunity,* Blackwell Scientific, Oxford, 1975.

[H5] Hoffmann, G. W., and Hraba, T. (eds.), *Immunology and Epidemiology,* Springer-Verlag, New York, 1986.

[H6] Hogg, B. W., "Representation and Control of Turbo Generators in Electric Power Systems," 112–148, Publication of the Institution of Electrical Engineers, (London), Peter Peregrinus, Stevenage, Hertfordshire, England, 1981.

[H7] Halanay, A., *Differential Equations, Stability, Oscillations, Time Lags,* Academic Press, New York, 1966.

[H8] Hagendorn, P., *Non-Linear Oscillations,* Clarendon Press, Oxford, 1981.

[H9] Hahn, W., *Theory and Application of Liapunov's Direct Method,* Prentice-Hall, Englewood Cliffs, NJ, 1963.

[H10] Halmos, P. R., *Finite Dimensional Vector Spaces,* Princeton University Press, Princeton, NJ, 1948.

[K1] Kolodziej, W. J., and Mohler, R. R., "State Estimation and Control of Conditionally Linear Systems," *SIAM J. Control Optim. 24,* 497–508, 1986.

[K2] Keepin, G. R., *Physics of Nuclear Kinetics,* Addison-Wesley, Reading, MA, 1965.

[K3] Krasovskii, N. N., "On the Stability in the Large of a System of Nonlinear Differential Equations," *Prikl. Mat. Meh. 18,* 735–737, 1954.

[K4] Kalman, R. E., and Bertram, J. E., "Control System Analysis and Design via the Second Method of Lyapunov," *J. Basic Eng. 82,* 371–399, 1960.

[K5] Kalman, R. E., "Liapunov Functions for the Problem of Luré in Automatic Control," *Proc. Natl. Acad. Sci. USA 49,* 201–205, 1963.

[L1] Ljung, L. (ed.), *Control Theory 1984–1986: A Progress Report from IFAC Technical Committee on Theory,* International Federation of Automatic Control, Laxenberg, Austria, 1987.

[L2] Langer, G. A., and Brady, A. J., "The Effects of Temperature upon Contraction and Ionic Exchange in Rabbit Myocardium," *J. Gen. Physiol. 52,* 682–713, 1958.

[L3] Lefschetz, S., *Differential Equations: Geometric Theory,* Interscience, New York, 1957.

[L4] Levinson, N., and Smith, D. K., "A General Equation for Relaxation Oscillations," *Duke Math. J. 9,* 382–403, 1942.

[L5] Lewis, J. B., "The Use of Nonlinear Feedback to Improve the Transient Response of a Servomechanism," *Trans. Am. Inst. Electr. Eng. 71,* 449–453, 1953.

[L6] LaSalle, J., and Lefschetz, S., *Stability by Liapunov's Direct Method with Applications,* Academic Press, New York, 1961.

[L7] Liapunov, M. A., "Problème générale de la stabilité du mouvement (1893)," reprinted in *Ann. Math. Stud. 17,* Princeton University Press, Princeton, NJ, 1893.

[L8] Leigh, J. R., *Essentials of Nonlinear Control Theory,* IEE Control Series, No. 2, Institution of Electrical Engineers, London, 1983.

[L9] Luré, A. I., *Some Non-linear Problems in the Theory of Automatic Control,* (English translation of 1951 Russian publication), Her Majesty's Stationery Office, London, 1957.

[L10] Luré, A. I., and Postnikov, V. N., "On the Theory of Stability of Control Systems," *Prikl. Mat. Meh. 8,* 246–248, 1944.

[L11] Lefschetz, S., *Stability of Nonlinear Control Systems,* Academic Press, New York, 1965.

[L12] LaSalle, J. P., "Complete Stability of a Nonlinear Control System," *Proc. Natl. Acad. Sci. USA 48,* 601–603, 1963.

[M1] Mohler, R. R., *Bilinear Control Processes,* Academic Press, New York, 1973.

[M2] Mohler, R. R., "Bilinear Systems and Control," in *Encyclopedia of Physical Science and Technology,* Academic Press, New York, 1987.

[M3] Mohler, R. R., and Kolodziej, W. J., "Optimal Control of a Class of Nonlinear Stochastic Systems," *IEEE Trans. Autom. Control AC-26,* 1048–1053, 1981.

[M4] Mohler, R. R., and Shen, C. N., *Optimal Control of Nuclear Reactors,* Academic Press, New York, 1972.

[M5] Milsum, J. M., *Biological Control Systems Analysis,* McGraw-Hill, New York, 1966.

[M6] Mohler, R. R., Bruni, C., and Gandolfi, A., "A Systems Approach to Immunology," *Proc. IEEE 68,* 964–990, 1980.

[M7] Mohler, R. R., and Farooqi, Z. H., "On Migratory Lymphocyte Models," in [H5], 208–223.

[M8] Mohler, R. R., Barton, C. F., and Hsu, C. S., "T and B Cells in the Immune System," in [B2], 415–436.

[M9] Marchuk, G. I., *Mathematical Models in Immunology,* Optimization Software Publication, distributed by Springer-Verlag, New York, 1983.

[M10] Merrill, S. J., "AIDS: Background and the Dynamics of the Decline of Immunocompetence," in [P2].

[M11] Massera, J. L., "On Liapunov's Conditions of Stability," *Ann. Math. 50,* 705–721, 1949.

[M12] Mohler, R. R., and Ruberti, A. (eds.), *Theory and Applications of Variable Structure Systems,* Academic Press, New York, 1972.

[M13] Mahmoud, M. S., and Singh, M. G., *Large Scale Control Systems,* Pergamon Press, Elmsford, NY, 1981.

[M14] Mittelstadt, W., et al., "Bonneville Power Administration 1400 MW Braking Resistor," *IEEE Trans. Power Appar. Syst. PAS-94,* 602–611, 1975.

[P1] Padulo, L., and Arbib, M. A., *System Theory,* W.B. Saunders, Philadelphia, 1974.

[P2] Perelson, A. S. (ed.), *Theoretical Immunology,* Vols. 1 and 2, Addison-Wesley, Reading, MA 1988.

[P3] Popov, V. M., "Stability Criteria for Nonlinear Systems of Automatic Control Based on the Use of Laplace Transform" (Rumanian), *Stud. Cercet. Energ. 9,* 119–135, 1959.

[P4] Pliss, V. A., *Certain Problems in the Theory of Stability in the Whole,* Leningrad University Press, Leningrad, 1958.

[R1] Rink, R. E., and Le, H. Q., "Multivariable Control of Bilinear Processes in HVAC Systems," 777–781, in *Proc. 24th IEEE Midwest Symposium on Circuits and Systems,* Albuquerque, NM, 1981.

[R2] Roitt, I., *Essential Immunology,* Blackwell Scientific, Oxford, 1974.

[R3] Rink, R. E., and Mohler, R. R., "Completely Controllable Bilinear Systems," *SIAM J. Control Optim. 6,* 477–486, 1968.

[R4] Russell, D. L., *Mathematics of Finite-Dimensional Control Systems,* Marcel Dekker, New York, 1979.

[S1] Stoker, J. J., *Nonlinear Vibrations,* Interscience, New York, 1950.

[S2] Stern, T. E., *Theory of Nonlinear Systems and Circuits,* Addison-Wesley, Reading, MA, 1965.

[S3] Sandor, J., *The Analysis and Design of Polynomic Control Systems,* Ph.D. dissertation (Prof. D. Williamson), Univ. New South Wales, Sydney, Australia, 1977.

[S4] Schultz, D. G., "The Generation of Liapunov Functions," in *Advances in Control Systems,* Vol. 2 (C. Leonides, ed.), 1–64, Academic Press, New York, 1965.

[S5] Shensa, M. J., "Parasites and the Stability of Equilibrium Points of Nonlinear Networks," *IEEE Trans. Circuit Theory CT-18,* 481–484, 1971.

[S6] Sandor, J., and Williamson, D., "Design of Nonlinear Regulators for Linear Plants," *IEEE Trans. Autom. Control AC-22,* 47–50, 1977.

[T1] Thaler, G., and Pastel, M. P., *Analysis and Design of Nonlinear Feedback Control Systems,* McGraw-Hill, New York, 1962.

[V1] Vidyasagar, M., *Nonlinear Systems Analysis,* Prentice-Hall, Englewood Cliffs, NJ, 1978.

[V2] Vidyasagar, M., "New Directions of Research in Nonlinear System Theory," *Proc. IEEE 74,* 1060–1091, 1987.

[V3] Van der Pol, B., "Forced Oscillations in a Circuit with Nonlinear Resistance," *Philos. Mag. 7,* 65–80, 1927.

[W1] Willsky, A., and Lo, J. T., "Estimation for Rotational Processes with One Degree of Freedom: I, II, III," *IEEE Trans. Autom. Control AC-20,* 10–33, 1975.

[W2] Willems, J. L., *Stability Theory of Dynamical Systems,* Thomas Nelson, London, 1970.

[W3] Willems, J. L., "Perturbation Theory for the Analysis of Instability in Nonlinear Feedback Systems," *Proc. Allerton Conference on Circuits and Systems,* 836–848, University of Illinois, Urbana, IL, 1966.

[Y1] Yakubovich, V. A., "Matrix Inequalities in the Theory of Nonlinear Control Systems, III," *Avtom Telemekh. 26 (5),* 1965.

[Z1] Zadeh, L. A., and Desoer, C. A., *Linear System Theory,* McGraw-Hill, New York, 1963.

[Z2] Zames, G., and Falb, P. L., "Stability Conditions with Monotone and Slope Restricted Nonlinearities," *SIAM J. Control Optim. 6,* 89–108, 1968.

APPENDIX

A

NORMS and SPACES

A.1 LINEAR VECTOR SPACES

The concept of real linear vector spaces has been utilized throughout the book. A brief summary of vector spaces and normed vector spaces is presented here. The reader is referred to any of the excellent texts on linear algebra and vector spaces (see, e.g., [B4], [G5], and [H10].

A linear vector space is defined by a set V along with operations of addition and scalar multiplication. While such spaces may be defined on any number of fields (e.g., complex numbers, rational functions, binary numbers, etc.), our primary interest is in *real linear vector spaces*—frequently called *real linear spaces* or *real vector spaces* for short. V is a real linear vector space if, and only if, for all $\mathbf{x}, \mathbf{y} \in V$ and all $a, b \in \mathbb{R}$ (i.e., a, b are real scalar numbers).

1. $\mathbf{x} + \mathbf{y} = \mathbf{y} + \mathbf{x}$.
2. $\mathbf{x} + (\mathbf{y} + \mathbf{z}) = (\mathbf{x} + \mathbf{y}) + \mathbf{z}$.
3. There exists a $\mathbf{0} \in V$ such that $\mathbf{x} + \mathbf{0} = \mathbf{0} + \mathbf{x} = \mathbf{x}$.
4. There exists a $-\mathbf{x}$ such that $\mathbf{x} + (-\mathbf{x}) = \mathbf{0}$.
5. $a \cdot (b \cdot \mathbf{x}) = (ab)\mathbf{x}$.
6. $a \cdot (\mathbf{x} + \mathbf{y}) = a \cdot \mathbf{x} + b \cdot \mathbf{y}$.
7. $(a + b) \cdot \mathbf{x} = a \cdot \mathbf{x} + b \cdot \mathbf{x}$.
8. $1 \cdot \mathbf{x} = \mathbf{x}$.

For example, consider $\mathbf{x}, \mathbf{y} \in \mathbb{R}^n$. That is, $\mathbb{R}^n$ is the space of all ordered n-tuples of real numbers such that

$$\mathbf{x} = \begin{bmatrix} x_1 \\ x_2 \\ \vdots \\ x_n \end{bmatrix}, \qquad \mathbf{y} = \begin{bmatrix} y_1 \\ y_2 \\ \vdots \\ y_n \end{bmatrix},$$

or

$$\mathbf{x}^{\mathrm{T}} = [x_1 \quad x_2 \quad \cdots \quad x_n], \qquad \mathbf{y}^{\mathrm{T}} = [y_1 \quad y_2 \quad \cdots \quad y_n].$$

Here **x, y** are called *vectors* or *column vectors,* and their transposes are called *row vectors*. Then for all $\mathbf{x}, \mathbf{y} \in \mathbb{R}^n$ and all real numbers a,

(a) $$\mathbf{x} + \mathbf{y} = \begin{bmatrix} x_1 & + & y_1 \\ & \vdots & \\ x_n & + & y_n \end{bmatrix}$$

and

(b) $$a\mathbf{x} = \begin{bmatrix} ax_1 \\ \vdots \\ ax_n \end{bmatrix}.$$

Similarly, the set of all real-valued functions $f(\cdot)$ that map an interval $[t_0, t_1] \in \mathbb{R}$ into $\mathbb{R}^n$ form a real linear vector space if for all elements $\mathbf{x}(\cdot)$, $\mathbf{y}(\cdot)$, of $\mathbf{f}(\cdot)$ and any real number a,

(a) $(\mathbf{x} + \mathbf{y})(t) = \mathbf{x}(t) + \mathbf{y}(t)$

and

(b) $(a\mathbf{x})(t) = a\mathbf{x}(t)$

for all $t \in [t_0, t_1]$. Any set of n linearly independent vectors in an n-dimensional space forms a basis for the n-dimensional space. That is, the n linear-independent vectors span the n-space.

A.2 NORMED VECTOR SPACES

Next it is desired to add the concept of distance or length to the linear vector space. Here a norm $\|\cdot\|$ is a real-valued function associated with all $\mathbf{x}, \mathbf{y} \in X$, a real linear vector space, such that for all scalars α,

1. $\|\mathbf{x}\| \geq 0$.
2. $\|\mathbf{x}\| = 0$ if and only if $\mathbf{x} = 0$.

3. $\|\alpha \mathbf{x}\| = |\alpha| \cdot \|\mathbf{x}\|$.
4. $\|\mathbf{x} + \mathbf{y}\| \le \|\mathbf{x}\| + \|\mathbf{y}\|$.

Apparently, this definition of length or distance (e.g., $\|\mathbf{x} - \mathbf{y}\|$ measures the distance between $\mathbf{x}$ and $\mathbf{y}$) satisfies our intuitive notions. For example, the traditional Euclidean norm (or length) is given by

$$\|\mathbf{x}\| = \left[\sum_{i=1}^{n} x_i^2\right]^{1/2}. \tag{A.1}$$

Other convenient norms are

$$\|\mathbf{x}\| = \max |x_i|$$

and

$$\|\mathbf{x}\| = \left[\sum_{i=1}^{n} |x_i|^p\right]^{1/p}.$$

Here x_i is the ith component of $\mathbf{x}$, and $1 \le p \le \infty$. Note that $p = 2$ corresponds to the Euclidean norm.

Consequently, the normed vector space allows the definition of such concepts as convergence, continuity, and stability. By definition, a sequence $\{\mathbf{x}_i\}$, $i = 1, \ldots, n$, which is defined on a normed vector space, *converges* to an element $\mathbf{x}_0$ if for every $\epsilon > 0$, there exists an integer $N(\epsilon)$ such that $\|\mathbf{x}_i - \mathbf{x}_0\| < \epsilon$ if $i \ge N(\epsilon)$, that is, $\|\mathbf{x}_i - \mathbf{x}_0\| \to 0$ as $i \to \infty$.

We say that $\mathbf{f}(\cdot)$ is *continuous* at $\mathbf{x}_0$ if for every $\epsilon > 0$ there exists a $\delta(\epsilon, \mathbf{x}_0)$ such that

$$\|\mathbf{f}(\mathbf{x}) - \mathbf{f}(\mathbf{x}_0)\|_y < \epsilon$$

if

$$\|\mathbf{x} - \mathbf{x}_0\|_x < \delta(\epsilon),$$

where $\|\cdot\|_x$, $\|\cdot\|_y$ refer to the appropriate norms. $\mathbf{f}(\cdot)$ is *continuous* if it is continuous at every $\mathbf{x}_0 \in X$. $\mathbf{f}(\cdot)$ is *uniformly continuous* if it is continuous such that $\delta(\epsilon, \mathbf{x}_0) = \delta(\epsilon)$.

A.3 MATRIX NORMS

The norm $\|\mathbf{A}\|$ of an $n \times m$ matrix $\mathbf{A}$ can be defined similarly in an $(n \times m)$-dimensional vector space. $\|\mathbf{A}\|$ is a matrix norm if and only if

1. $\|\mathbf{A}\| \ge 0$.
2. $\|\mathbf{A}\| = 0$ if and only if all elements $a_{ij} = 0$.
3. $\|a\mathbf{A}\| = |a|\,\|\mathbf{A}\|$ for any constant a.
4. $\|\mathbf{A} + \mathbf{B}\| \le \|\mathbf{A}\| + \|\mathbf{B}\|$, where $\mathbf{A}$ and $\mathbf{B}$ are of identical dimensions.

Sometimes, it is useful to add the additional postulate that

$$\|\mathbf{AB}\| \leq \|\mathbf{A}\| \cdot \|\mathbf{B}\|.$$

Examples of useful matrix norms are given by

$$\max_{i,j} |a_{ij}|, \qquad \max_{i} \left(\sum_{j} |a_{ij}|\right), \qquad \sum_{i,j} |a_{ij}|, \qquad \left(\sum_{i,j} a_{ij}^2\right)^{1/2}.$$

Generally, it is desired that a matrix norm be defined so that there is a consistent measurement of distance from the original vector space X to the new space Y, where $\mathbf{x} \in X$, $\mathbf{y} \in Y$, and $\mathbf{y} = \mathbf{Ax}$. This is accomplished by means of the so-called induced norm. For a given vector norm $\|\mathbf{x}\|$ in $\mathbb{R}^n$, the induced matrix norm for $\mathbf{A} \in \mathbb{R}^{n\times n}$ is given by

$$\begin{aligned} \|\mathbf{A}\| &= \sup_{\mathbf{x}\neq 0} \frac{\|\mathbf{Ax}\|}{\|\mathbf{x}\|} = \sup_{\|\mathbf{x}\|=1} \|\mathbf{Ax}\| \\ &= \sup_{\|\mathbf{x}\| \leq 1} \|\mathbf{Ax}\|. \end{aligned} \tag{A.2}$$

Consequently, $\|\mathbf{A}\|$, the induced matrix norm, is the least upper bound (supremum) of $\|\mathbf{Ax}\|/\|\mathbf{x}\|$ as $\mathbf{x}$ varies over $\mathbb{R}^n$. Equation (A.2) can be interpreted as the radius of the smallest ball in $\mathbb{R}^n$ that covers the image of the unit ball ($\|\mathbf{x}\| \leq 1$) under the mapping $\mathbf{A}$. It can shown that for each $\|\mathbf{x}\|$ on $\mathbb{R}^n$, the induced $\|\mathbf{A}\|$ maps $\mathbb{R}^{n\times n}$ into $[0, \infty]$ and satisfies the foregoing conditions to be a matrix norm.

For the common Euclidean norm (A.1), the induced Euclidean norm becomes

$$\|\mathbf{A}\| = \max_{\|\mathbf{x}\|=1} \|\mathbf{Ax}\| = \left[\max_{\|\mathbf{x}\|=1} (\mathbf{x}^{\mathrm{T}}\mathbf{A}^{\mathrm{T}}\mathbf{Ax})\right]^{1/2}. \tag{A.3}$$

But since $\mathbf{A}^{\mathrm{T}}\mathbf{A}$ is a symmetric matrix, it can be diagonalized by an orthogonal matrix $\mathbf{T}$ so that

$$\mathbf{x}^{\mathrm{T}}\mathbf{A}^{\mathrm{T}}\mathbf{Ax} = \mathbf{y}^{\mathrm{T}}\mathbf{\Lambda y},$$

where $\mathbf{\Lambda}$ is a diagonal matrix with

$$\mathbf{\Lambda} = [\lambda_i(\mathbf{A}^{\mathrm{T}}\mathbf{A})].$$

Then with

$$\|\mathbf{A}\| = \max_{\mathbf{y}} \left(\frac{\mathbf{y}^{\mathrm{T}}\mathbf{\Lambda y}}{\mathbf{y}^{\mathrm{T}}\mathbf{y}}\right) = \max_{\mathbf{y}} \left(\frac{\lambda_1 y_1^2 + \cdots + \lambda_n y_n^2}{y_1^2 + \cdots + y_n^2}\right)^{1/2}$$

or

$$\|\mathbf{A}\| = \left[\max_{i} \lambda_i(\mathbf{A}^{\mathrm{T}}\mathbf{A})\right]^{1/2}. \tag{A.4}$$

The square roots of the eigenvalues of $\mathbf{A}^{\mathrm{T}}\mathbf{A}$ are called singular values of $\mathbf{A}$. If $\|\mathbf{x}\| = \max_{i} |\mathbf{x}_i|$, the induced norm becomes the row sum

$$\|\mathbf{A}\| = \max_i \sum_j |a_{ij}|, \tag{A.5}$$

and for $\|\mathbf{x}\| = \sum_{i=1} |\mathbf{x}_i|$, the induced norm is given by the column sum

$$\|\mathbf{A}\| = \max_j \sum_i |a_{ij}|. \tag{A.6}$$

APPENDIX

B

EXISTENCE and UNIQUENESS of SOLUTIONS

We summarize here conditions sufficient for the existence and uniqueness of solutions to nonlinear ordinary differential equations. The reader is referred to [C1] and [B10] for an excellent analysis of this topic. Here we consider a few essential basic definitions of relative function smoothness.

The system of interest is given by

$$\frac{d\mathbf{x}}{dt} = \mathbf{f}(\mathbf{x}, t), \tag{B.1}$$

where $\mathbf{x} \in \mathbb{R}^n$, t is real time, and $\mathbf{x}(t_0) = \mathbf{x}_0$. Continuity of $\mathbf{f}(\cdot)$ is defined in Appendix A. Roughly, $\mathbf{f}(\cdot)$ is continuous at a point $\mathbf{x}_1$, t_1 if

$$\lim_{\substack{x \to 1 \\ t \to 1}} \mathbf{f}(\mathbf{x}, t) = \mathbf{f}(\mathbf{x}_1, t_1).$$

If $\mathbf{f}(\cdot)$ is continuous except at a finite number of values of its arguments, where there are jump discontinuities, we say that $\mathbf{f}(\cdot)$ is piecewise continuous. It is natural to define degrees of continuity over some range of values of its arguments, which for many cases of practical interest is global.

Now define an important stronger condition of the vector function $\mathbf{f}(\mathbf{x}, t)$. If there exists a constant M such that

$$\|\mathbf{f}(\mathbf{x}, t) - \mathbf{f}(\mathbf{x}_1, t)\| \leq M\|\mathbf{x} - \mathbf{x}_1\| \tag{B.2}$$

for all $t \in [t_0, T]$ and $\mathbf{x}$, $\mathbf{x}_1 \in X$, written as $(\mathbf{x}, t)$, $(\mathbf{x}_1, t) \in D$, then $\mathbf{f}(\cdot)$ is said to satisfy a Lipschitz condition on $\mathbb{R}$. The constant M is known as the Lipschitz constant. If $M < 1$, the $\mathbf{f}(\cdot)$ mapping is a contraction. It is seen that the Lipschitz condition merely places a constraint on the gradient of $\mathbf{f}(\cdot)$ relative to $\mathbf{x}$. If $\mathbf{f}(\cdot)$ is continuously differentiable in $\mathbf{x}$, the Lipschitz condition is

$$\left\| \frac{\partial \mathbf{f}}{\partial \mathbf{x}} \right\| \leq M \tag{B.3}$$

for all $(\mathbf{x}, t)$, $(\mathbf{x}_1, t) \in D$. If $\mathbf{f}(\mathbf{x}, t)$ has continuous derivatives with respect to $\mathbf{x}$ on D, it satisfies a Lipschitz condition on D which in turn implies that $\mathbf{f}(\cdot)$ is continuous on D.

B.1 EXISTENCE AND UNIQUENESS CONDITIONS

For nonlinear systems, solutions may exist only over limited regions where the equations are well behaved. Suppose that $\mathbf{f}(\mathbf{x}, t)$ in (B.1) is continuous over some region D such that

$$|t - \tau| \leq a$$

and (B.4)

$$\|\mathbf{x} - \boldsymbol{\xi}\| \leq b,$$

where a, b are real positive constants and $\mathbf{x}(\tau) = \boldsymbol{\xi}$. It follows that $\mathbf{f}(\cdot)$ is bounded on D.

For the solution to (B.1) to remain in D, however, the allowable interval of time I is constrained. In this regard, let

$$K = \max_D \|\mathbf{f}(\mathbf{x}, t)\|$$

and (B.5)

$$\alpha = \min\left(a, \frac{b}{K}\right).$$

The geometry of the region D and the admissible interval I is constructed in Fig. B.1 for scalar $x(t)$ such that

$$|t - \tau| \leq \alpha. \tag{B.6}$$

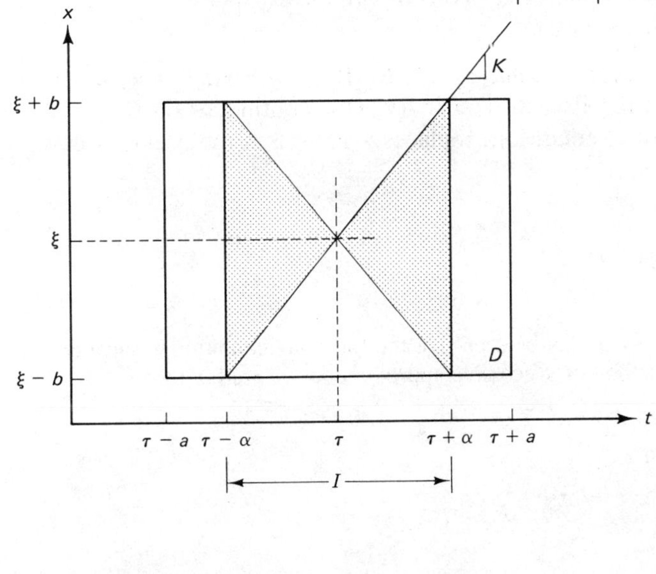

Figure B.1 Geometrical interpretation of ***I*** and ***D***.

If $t \in I$, $x(t)$ remains in the shaded region within D such that $x(\tau) = \xi$ as in Fig. B.1.

Existence and uniqueness of solutions to (B.1) generally are based on conditions of convergence of a sequence of approximating solutions to the integral form of (B.1) which are constrained to the shaded region shown in Fig. B.1 for the scalar case. Typically, I is divided into N segments of equal length h, so that

$$t_k = \tau + kh,$$

$k = \ldots, -1, 0, 1, \ldots$. Then the simple Euler method of approximations might be used with

$$x_{k+1} = x_k + hf(x_k, t_k), \tag{B.7}$$

where $x_0 = \xi$. Of course, there are many better approximations available which have been developed for numerical computations with modern high-speed digital computers.

Since $f(\cdot)$ is continuous on a closed, bounded region, it is also uniformly continuous. Consequently, it can be shown from the definition of uniform continuity in Section A.2 that the discretization satisfies

$$h \leq \min\left(\delta, \frac{\delta}{M}\right). \tag{B.8}$$

Here the appropriate $\delta(\epsilon)$ is established from the definition of uniform continuity. That is, for any $\epsilon > 0$ it is possible to find a $\delta > 0$ such that $|f(x, t) - f(x_k, t_k)| < \epsilon$ for (x, t), $(x_k, t_k) \in D$ with $|t - t_k| < \alpha$, and $|x - x_k| < \delta$. Then the resolution is determined by ϵ so that for a given gradient of $f(\cdot)$ with respect to its arguments, h is set accordingly.

Proof of convergence of an approximating sequence to $x(t)$ under the required conditions of continuity can be established by Ascoli's lemma [C1]. According to this lemma, an infinite set of uniformly bounded, equicontinuous functions on a bounded interval I does contain a uniformly convergent sequence on I. Then the following existence theorem can be proven.

Theorem B.1. There exists a solution $\mathbf{x}(t)$ to (B.1) with $\mathbf{x}(t_0) = \mathbf{x}_0$ which is continuous and differentiable for all t_0, $t \in I$ if $\mathbf{f}(\mathbf{x}, t)$ is continuous on D.

Here $\mathbf{x}(t_0) = \mathbf{x}_0$, the initial condition, replaces $\mathbf{x}(\tau) = \boldsymbol{\xi}$ in the previous development.

Example B.1

Consider the nonlinear equation

$$\dot{x} = x^{1/2}, \tag{B.9}$$

where $x(0) = 0$, $t \in [0, \infty)$. It can be seen that the Lipschitz condition is satisfied for $t \in [0, \infty)$. If Picard's method of successive approximation is used with

$$x = 0,$$

and a nonsingular solution

$$x = \frac{t^2}{4},$$

where $t \geq 0$. For this problem $f(x) = x^{1/2}$ is continuous in x, and a solution does exist. In fact, there are two solutions. Obviously, a stronger condition is required for the solution to be unique.

For this problem it is seen that

$$\frac{\partial f}{\partial x} = \frac{1}{2}\left(x\right)^{-1/2}$$

which becomes infinite at $x = 0$. Thus we might suspect that the existence of two solutions at $x = 0$ may be caused by this lack of differentiability. Indeed, it can be shown that the existence of such derivatives is sufficient for a unique solution. However, it can be shown that the aproximating sequence used for proving existence of a solution under conditions of continuity will converge to a unique solution if $\mathbf{f}(\mathbf{x}, t)$ in (B.1) satisifes a Lipschitz condition (B.2). As noted with (B.3), this is a somewhat weaker condition than differentiability. For example (B.1), it is seen from (B.2) with $x_1 = 0$ that in the limit as $x \to 0$, no Lipschitz constant M exists to satisfy the sufficient condition.

In general, the following theorem may be derived [C1].

Theorem B.2. A unique solution $\mathbf{x}(t)$ to (B.1) exists with $\mathbf{x}(t_0) = \mathbf{x}_0$ and $t_0, t \in I$ if (i) Theorem B.1 is satisfied, and (ii) $\mathbf{f}(\mathbf{x}, t)$ satisfies a Lipschitz condition on D with appropriate constant M.

Example B.2

Consider

$$\dot{x} = tx + 2t - t^3, \tag{B.10}$$

where $x(0) = 0$, $t \in [0, \infty)$. It can be seen that the Lipschitz condition is satisfied for $t \in [0, \infty)$. If Picard's method of successive approximation is used with

$$x_k(t) = \int_0^t f(x_{k-1}, \sigma)\, d\sigma + x_0, \tag{B.11}$$

then

$$x_1(t) = t^2 - \frac{t^4}{4},$$

$$\vdots$$

$$x_N(t) = t^2 - \frac{t^{2N+2}}{4 \cdot 6 \cdots (2N + 2)}$$

and

$$\lim_{N\to\infty} x_N(t) = x(t) = t^2$$

for all $t \in [0, \infty)$.

B.2 REMARKS

Nonlinear system (B.1) with $\mathbf{x}(t_0) = \mathbf{x}_0$ has a unique solution if Theorem B.2 is satisfied. It should be noted, however, that while the theorem presents sufficient conditions, they are not necessary. On the other hand, continuity of $\mathbf{f}(\cdot)$ over its arguments as stipulated by Theorem B.1 is sufficient for the existence of any solution. For linear systems $\mathbf{f}(\mathbf{x}, t) = \mathbf{A}(t)\mathbf{x}$, and continuity of $\mathbf{A}(t)$ is sufficient for the presence of a unique solution.

Several remarks follow for the autonomous system

$$\frac{d\mathbf{x}}{dt} = \mathbf{f}(\mathbf{x}). \tag{B.12}$$

These are given here without proof but are mostly obvious.

1. An equilibrium state for (B.12) can be reached or left in finite time if a Lipschitz condition is not satisfied.
2. Jump solutions may occur when the coefficient of the highest derivative term vanishes (e.g., inertialess system).
3. First-order systems cannot have continuous periodic solutions but can have periodic jump (piecewise continuous) solutions.
4. Second-order systems cannot have trajectories that spiral inward and then back out, but third-order and higher-order systems can exhibit such phenomena.
5. The general presence of unique solutions in practice may dictate insufficiency of a model order by the presence of overlapping or jump trajectories.

INDEX

S

T

U

V

W

Z